FUNDAMENTALS OF POLYMER SCIENCE

SECOND EDITION

FUNDAMENTALS OF POLYMER SCIENCE

AN INTRODUCTORY TEXT

PAUL C. PAINTER
MICHAEL M. COLEMAN
The Pennsylvania State University

TECHNOMIC
PUBLISHING CO., INC.
LANCASTER · BASEL

Fundamentals of Polymer Science

a **TECHNOMIC®** publication

Published in the Western Hemisphere by
Technomic Publishing Company, Inc.
851 New Holland Avenue, Box 3535
Lancaster, Pennsylvania 17604 U.S.A.

Distributed in the Rest of the World by
Technomic Publishing AG
Missionsstrasse 44
CH-4055 Basel, Switzerland

Printed in the United States of America
10 9 8 7 6 5 4 3 2

Main entry under title:
  Fundamentals of Polymer Science: An Introductory Text—Second Edition

A Technomic Publishing Company book
Bibliography: p.
Includes index p. 471

Library of Congress Catalog Card No. 97-60515
ISBN No. 1-56676-559-5

DEDICATION AND ACKNOWLEDGEMENT

This is the book that we swore we would never write after we finished our last book. But now that we have written it and the frustrations have passed, we gratefully dedicate the book to our fathers, Charles Painter and Ronald Coleman, whose lives were and much more difficult than ours and who sacrificed many things to give us a start.

We would also like to take this opportunity to acknowledge our friends at Technomic, Mike Margotta and Tony Deraco, who shrewdly managed to hide their superior golfing abilities in order not to embarrass the authors and also persuade them to complete this second edition.

TABLE OF CONTENTS

PREFACE TO THE SECOND EDITION

Writing a book, particularly a textbook, can take a lot out of you:

Writing a book is an adventure: it begins as an amusement, then it becomes a mistress, then a master and finally a tyrant.

Winston S. Churchill

Things get particularly hard near the end, when you are confronted by the tedious tasks of preparing an index, editing and proof reading; especially when you look out the window, the sun is shining and you would rather be playing golf.

All morning I worked on the proof of one of my poems, and I took out a comma; in the afternoon I put it back.

Oscar Wilde

So it is perhaps inevitable that your attention slips and errors creep in. You don't realize this until later, of course, when you start to find some minor typo's or garbled sentences. This is not too bad and you can always blame your publishers:

It is with publishers as with wives: one always wants somebody else's.

Norman Douglass

But if your book is used as a text, each and every mistake will be found by your students and the number of revealed errors becomes daunting. Even worse, these errors are often brought to your attention in a manner that you believe reflects a large measure of glee:

There are those who notice mistakes in passing and those who buy books for the sole purpose of finding them.

A. K. Dewdney

It is particularly galling when your so-called friends participate in the criticism:

Of all plagues, good heaven, thy wrath can send,
Save me, oh save me from the candid friend.

George Canning

Naturally, your first instinct is to attribute their comments to mere envy:

Nothing is more humiliating than to see idiots succeed in enterprises we have failed in.

Gustave Flaubert

This base emotional response is quickly replaced by denial:

If at first you don't succeed, destroy all evidence that you have tried.

Anonymous

But in the end a measure of objectivity returns and you start to make a list of corrections. You then also notice that there are certain topics that you really made a mess of:

I know you believe you understand what you think I said, but I am not sure you realize that what you heard is not what I meant.

Anonymous

So you end up writing a second edition. You hopefully fix all the errors in the first edition, modify a little here and there to improve clarity, add some study questions (for those idle teachers who cannot be bothered to make them up!) and end up totally rewriting and reorganizing two chapters. But you no longer have the illusion that your efforts will be remotely error-free; you just hope to get some minor satisfaction from your work:

Some day I hope to write a book where the royalties will pay for the copies I give away.

Clarence Darrow

February 1997

PAUL PAINTER
MIKE COLEMAN

PREFACE TO THE FIRST EDITION

If you, dear reader, ever write a book, you will eventually find that having labored long and hard to write down and give shape to your view of a subject, an effort that can result in hundreds of pages of text and figures that perhaps (if you are lucky) only a few hundred people might ever read, you are then faced with the oppressive but suddenly daunting task of writing a preface. This is where you are supposed to explain what the book is about and why you wrote it in the first place. The problem is that you no longer remember.

But a preface has its uses. It can give readers a whiff of the author's prose style and an indication of his potential as an inducer of tedium, thus enabling them to moderate their enthusiasm, lower their sights, and so prepare themselves for the main body of the work.

Frank Muir (An Irreverent and Thoroughly Incomplete
Social History of Almost Everything)

In addition, you are by now tired and fed up with the whole enterprise;

How unfit, and how unworthy a choice I have made of myself, to undertake a work of this mixture.

Sir Walter Raleigh

and you are starting to suspect that your explanation of what you know reveals even more about what you don't know;

A person who publishes a book willfully appears before the populace with his pants down.

Edna St. Vincent Millay

Even worse, you find that if your scribblings have been a collaborative effort you might no longer be on speaking terms with your co-author and one-time friend;

I have known two professors of Greek who ceased speaking to one another because of divergent views on the pluperfect subjunctive.

Stephen Leacock

Fortunately, you realize that this too will pass and you must save your disagreements for more vital matters;

I've seen lifelong friends drift apart over golf just because one could play better, but the other counted better.

Stephen Leacock

What really happens is that if you teach, then at some point you become so frustrated with what you perceive as the shortcomings of the available texts that you decide to write your own. And that is what happened to us. Of course, you then find that the books you were using were not so bad after all, its just that you wanted to emphasize or include subjects that perhaps others found uninteresting or irrelevant, but which to you are of consuming interest and central to the whole field.

So, here is our effort; the essential parts of what we teach in two courses to juniors, seniors and some first-year graduate students that have drifted into polymers from other fields. It is an introduction and survey of the subject based on a refinement of what we have taught over the last ten years. As with all books of this type, it is difficult to know what to assume as prior knowledge. We naturally suppose that the student has digested some fundamentals of organic chemistry,

Organic chemistry is the study of organs; inorganic chemistry is the study of the insides of organs.

Max Shulman

and has enough physics and physical chemistry to be comfortable with basic molecular concepts;

MOLECULE,*n*. The ultimate, indivisible unit of matter. It is distinguished from the corpuscle, also the ultimate, indivisible unit of matter, by a closer resemblance to the atom, also the ultimate, indivisible unit of matter. Three great scientific theories of the structure of the universe are the molecular, the corpuscular and the atomic. A fourth affirms, with Haeckel, the condensation or precipitation of matter from ether—whose existence is proved by the condensation or precipitation. The present trend of scientific thought is toward the theory of ions. The ion differs from the molecule, the corpuscle and the atom in that it is an ion. A fifth theory is held by idiots, but it is doubtful if they know any more about the matter than the others.

Ambrose Bierce
(The Devil's Dictionary)

However, it has been our experience that the average student has only imperfectly digested certain fundamentals before starting their studies of polymers, so in our lectures and in this book we attempt to start at a simple level with brief reviews of what should be (but usually isn't) well-known material. Of course, we have to guard against carrying this to extremes and we have tried to overcome a tendency to over-explain;

I am standing on the threshold about to enter a room. It is a complicated business. In the first place I must shove against an atmosphere pressing with a force of fourteen pounds on every square inch of my body. I must make sure of landing on a plank travelling at twenty miles a second round the sun–a fraction of a second too early or too late, the plank would be miles away. I must do this whilst hanging from a round planet head outward into space, and with a wind of aether blowing at no one knows how many miles a second through every interstice of my body. The plank has no solidity of substance. To step on it is like stepping on a swarm of flies. Shall I not slip through? No, if I make the venture one of the flies hits me and gives a boost up again; I fall again and am knocked upwards by another fly, and so on. I may hope that the net result will be that I remain about steady; but if unfortunately I should slip through the floor or be boosted too violently up to the ceiling, the occurrence would be, not a violation of the laws of Nature, but a rare coincidence. These are some of the minor difficulties. I ought really to look at the problem four-dimensionally as concerning the intersection of my world-line with that of the plank. Then again it is necessary to determine in which direction the entropy of the world is increasing in order to make sure that my passage over the threshold is an entrance not an exit.
Verily, it is easier for a camel to pass through the eye of a needle that for a scientific man to pass through a door.

Sir Arthur Eddington
(The Nature of the Physical World)

Keeping this warning in mind, we have tried to say what we have to say simply;

Don't quote Latin; say what you have to say, and then sit down.

Arthur Wellesley
Duke of Wellington

and in so doing we have deliberately adopted a conversational tone to our writing. Some might find this irritating and unrigorous;

This is the sort of English up with which I will not put.

Winston S. Churchill

but in the final analysis this is the way we like to write and if you don't like it, write your own damned book!

February 1994

PAUL PAINTER
MIKE COLEMAN

CHAPTER 1

The Nature of Polymeric Materials

"He fixed thee mid this dance
of plastic circumstance"
—Robert Browning

A. WHAT ARE POLYMERS—WHAT IS POLYMER SCIENCE?

Simply stated, polymers are very large molecules (macromolecules) that are comprised or built up of smaller units or monomers. The arrangements of these units, the various types of chains that can be synthesized and the shapes that these chains can bend themselves into, result in a class of materials that are characterized by an enormous and intriguing range of properties. Some of these are unique to polymers (e.g., rubber elasticity) and, as we shall see, are simply a consequence of their size and chain-like structure.

Polymer science is also a relatively new discipline and one that is characterized by extraordinary breadth. It involves aspects of organic chemistry, physical chemistry, analytical chemistry, physics (particularly theories of the solid state and solutions), chemical and mechanical engineering and, for some special types of polymers, electrical engineering. Clearly, no one person has an in-depth knowledge of all these fields. Most polymer scientists seek a broad overview of the subject that is then usually supplemented by a more detailed knowledge of a particular area. This book is a first step towards the former and to give a flavor for the diversity of this subject matter we will commence with an outline of some of the areas we will cover.

Polymer Synthesis

Many polymer scientists think that it is unlikely that we will ever again see any new thermoplastic take the world by storm (i.e., achieve levels of production comparable to polyethylene or polystyrene), but it should be kept in mind that similar things were being said round about 1950, just before high density polyethylene and isotactic polypropylene made their debut (some of this terminology will be defined shortly). Today, there are two good reasons to think they may be right, however. First, all the monomers that can be readily polymerized already have been; second, commercializing a new *commodity* plastic would probably cost more than $1 billion (*The Economist*, May 1980). This is not promising in an industry that has become infected by MBA's with a six month time-horizon. Fortunately, this does not mean that polymer synthetic chemists are out of business. There is considerable interest in using new catalysts (for example) to produce commodity plastics more cheaply, or in producing better defined chain structures to give controlled properties; synthesizing "specialty" polymers such as those with stiff chains and strong intermolecular attractions to give thermal resistance and high strength; or chains with the types of delocalized electronic structures that result in unusual electrical and optical properties. These materials would be produced in smaller quantities than "bulk" plastics, but can be sold at a much higher price. Our intention is only to give you a start in this area, however, so you won't find any discussion of advanced synthetic methods in this book. In chapter 2 we simply cover the basics of polymer synthesis, but this is enough for you to grasp the essentials of how the majority of commercial polymers are produced.

Polymer Characterization

What a chemist *thinks* he or she has made is not always the stuff that is lying around the bottom of his or her test tube. Accordingly, there is an enormous field based on characterization. This is now a particularly exciting area because of recent advances in instrumentation, particularly those interfaced with high-powered yet small and relatively cheap dedicated computers. These novel analytical techniques are not only useful in studying new materials, but answering questions that have intrigued polymer scientists for decades. For example, spectroscopic techniques are used to examine "local" chemical structure and interactions in polymer systems. Electron and other microscopies and the scattering of electromagnetic radiation are used to characterize overall structure or morphology; i.e., how components of a system phase separate into various types of structures; how chains fold into crystals; and the shape of an individual chain in a particular environment. Some techniques are so expensive that national facilities are required, e.g., synchotron radiation and neutron scattering. Our focus in this book will be on the basics, and in particular we will discuss the measurement of molecular weight and the use of molecular spectroscopy to characterize chain structure.

Polymer Physical Chemistry

Paul Flory was awarded the Nobel Prize in Chemistry for his work in this area and we will mention his name often in this book. Polymer physical chemistry is a subject that demands a knowledge of theory and the ability to perform carefully controlled experiments, often using the types of instruments mentioned above. (The areas of characterization and polymer physical chemistry overlap considerably and they are artificially separated here merely to illustrate the different types of things polymer scientists do.) The simplest way to get a "feel" for this subject is get a copy of Flory's book *Principles of Polymer Chemistry*, still a classic after forty years, and scan the chapters on, for example, the theories of rubber elasticity, solution thermodynamics, phase behavior, etc.

This subject is still attracting enormous interest, and there has been much recent emphasis on polymer blends or alloys and polymer liquid crystals.

Polymer Physics

Polymer physics and polymer physical chemistry are overlapping disciplines that are not, in many cases, easily delineated. Historically, however, it is possible to point to an enormous impact by theoretical physicists starting in the late 1960's and early 1970's. Until then most theory was based on almost classical physical chemistry, but a number of leading physicists (notably de Gennes and Edwards) started to apply modern theories of statistical physics to the description of long chain molecules. The result has been a revolution in polymer theory, one that is not easily assimilated by "traditional" polymer scientists and is still ongoing.

Polymer physics is not confined to theory, however. Experimental polymer physics continues to focus on areas such as chain conformation, viscoelastic and relaxation properties, phenomena at interfaces, kinetics of phase changes and electrical and piezoelectric properties. In our discussions of polymer physics and physical chemistry we will focus on chain conformation and morphology (structure), thermal properties (crystallization, melting and the glass transition), solution properties and the determination of molecular weight.

Polymer Engineering

Last, but by no means least, there is a vast area that is involved with chemical engineering (the processing of polymers) and mechanical engineering (studies of strength, fatigue resistance, etc.) as applied to polymeric materials. For example, there is enormous interest in producing ultra-high strength polymer fibers. It turns out that even "common or garden" polymers like polyethylene or polypropylene can be processed to give "high-tech" properties. The trick is to align the chains as perfectly as possible; not an easy task. Once made of course, the mechanical properties of these materials have to be determined. This involves more than stretching a fiber until it breaks. The mechanical properties of polymers are complicated by all sorts of factors (defects, relaxation processes, etc.) and the field is an intriguing combination of mechanical measurements, structural characterization and theory. The last chapter in this book will deal with mechanical properties of polymers.

B. SOME BASIC DEFINITIONS - THE ELEMENTS OF POLYMER MICROSTRUCTURE

It may seem a bit like putting the cart before the horse, but we will first describe some basic molecular characteristics of the chain structure of polymers before we consider the chemistry of how these materials are produced. By arranging the subject matter in this manner it immediately becomes clear why certain synthetic methods have become important and how polymer chemistry can now be used (in principle) to obtain designed materials, with particular types of molecular groups arranged in a chosen pattern.

In many texts it is also customary to introduce various classification schemes at this point. Many of these are quite arbitrary and the subject has outgrown them. There is one that is both descriptive and remains widely used, however, and this divides polymers into *thermoplastics* and *thermosets*. As the names suggest, the former flow or, more precisely, flow more easily, when squeezed, pushed or stretched by a load, usually at elevated temperatures. Most of them hold their shape at room temperature, but can be reheated and formed into different shapes. Thermosets are like concrete, they flow and can be molded when initially constituted, but then become set in their shape, usually through the action of heat and (often) pressure, this process is often called "curing". Reheating a "cured" sample serves only to degrade and generally mess-up the article that has been made. This difference in behavior is, of course, a direct consequence of the various types of arrangements of units in polymer chains (polymer *microstructure*) and we will now take a look at these structures.

Linear, Branched and Cross-Linked Polymers

The simplest type of polymer is a *linear homopolymer*, by which we mean a chain that is made up of identical units (except for the end groups) arranged in a linear sequence. If we could see such a chain under a microscope of some kind it would appear, relative to small roughly spherical molecules such as H_2O, like a tangled up kite string or fishing line. Polymers where there are different types of units in the same chain are called *copolymers* and we will describe these in more detail shortly.

An example of a polymer that can be synthesized in the form of a linear chain is polyethylene (usually denoted PE). A part of a typical chain is shown in figure 1.1. Normally, commercial samples of this polymer are very long (have a high molecular weight), so that figure 1.1 only shows a small part of such a chain.

It is usual to represent such a polymer in terms of its *chemical repeat unit*, as follows:

$$-(CH_2 - CH_2)_n-$$

where n can be very large. The chemical repeat units of a few common polymers are listed at the end of this chapter.

It turns out that when polyethylene was first synthesized on an industrial scale (in the 1930's) a high pressure process was used (at low pressures ethylene does not polymerize at all) and for various reasons (that we will consider when we discuss synthesis) *branched polymers* were produced. The chemical structure of a branched polyethylene is also illustrated in figure 1.1. It wasn't until various catalytic processes were discovered (in the early 1950's) that linear polyethylene was synthesized[*].

---

[*] The history of these discoveries and the interplay of the personalities involved is both fascinating and intriguing. We will discuss some of the history in chapter 2 and the interested reader should consult the references listed in that chapter.

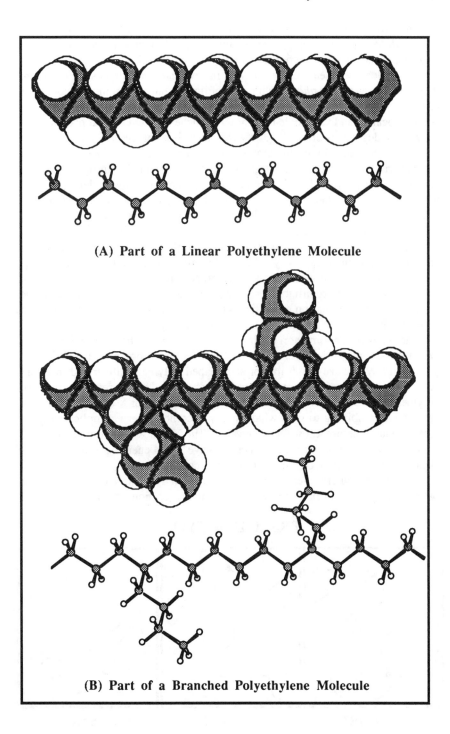

(A) Part of a Linear Polyethylene Molecule

(B) Part of a Branched Polyethylene Molecule

Figure 1.1 *"Ball and stick" and space filling models of polyethylene.*

The difference in chemical structure between linear and branched polyethylene appears to be quite small and localized to those units that form the branch points. Such differences can have a profound affect on properties, however. Because linear chains can pack in a regular three dimensional fashion, they form a *crystalline phase* (while part of it stays tangled up in an *amorphous phase*). A highly branched polymer is incapable of doing this, because the random location and varying lengths of the branches prevent regular packing.

At low degrees of branching, however, there is a degree of crystallinity, but it is lower than that found in samples of linear polyethylene. Because of the regular arrangement and hence tighter packing of the chains in the crystalline regions, linear polyethylene is often called high density polyethylene (HDPE), while slightly branched material is called low density polyethylene (LDPE). (Highly branched and hence completely amorphous samples are not (presently) of any commercial significance.)

Properties such as stiffness, strength, optical clarity, etc. are profoundly affected by the changes in crystallinity that are a consequence of branching. Obviously, crystallization in polymers is a broad and extensive subject in its own right and one we will return to later. The point that should be grasped here is the effect chain structure has on properties and hence the importance of having a firm understanding of these first principles and definitions.

The types of random short chain branches described above are not the only type that occur. Various reactions lead to long chain branches and it is also possible to synthesize comb-like or star-shaped polymers (illustrated in figure 1.2). Certain types of monomers form highly branched structures at the onset of polymerization. This is illustrated in figure 1.3, where Y or X shaped monomers are shown combining to give what usually becomes a very complicated three dimensional structure known as a *network*. The Y and X shaped molecules are *trifunctional* or *tetrafunctional* molecules (have 3 or 4 chemical groups that are capable of reacting with one another to form a covalent bond).

A network, where every unit is interconnected through various tortuous pathways, is only formed at a certain point in the reaction and always remains

BRANCHING TYPES

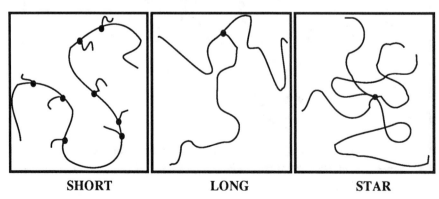

SHORT   LONG   STAR

Figure 1.2 Schematic diagram depicting short, long and star branching.

NETWORK FORMATION

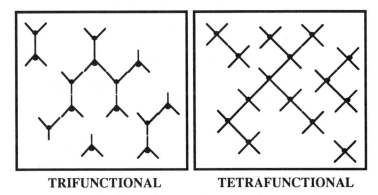

TRIFUNCTIONAL   **TETRAFUNCTIONAL**

Figure 1.3 *Schematic diagram depicting the development of branched chains and networks.*

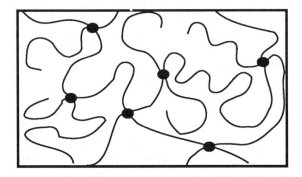

Figure 1.4 *Schematic diagram depicting the cross-linking of linear polymer chains.*

mixed with monomers and other small molecules, unless the reaction is driven to ultimate completion (i.e., all the original functional groups are reacted). This type of network is typical of various thermosets.

As the old saying has it, there is more than one way to skin a cat, and networks can also be formed by taking linear polymer chains and chemically linking them together, as illustrated in figure 1.4. This process is called cross-linking (or vulcanization) and is crucial in forming various types of elastomeric materials. If all the chains become interconnected by such reactions, then a so-called infinite network is formed.

Homopolymers and Copolymers: A Question of Semantics?

Synthetic polymers have traditionally been named for their respective monomeric precursors; for example, polyacrylonitrile $[-CH_2-CHCN-]_n$ from acrylonitrile $CH_2=CHCN$; poly(vinylidene chloride) $[-CH_2-CCl_2-]_n$ from vinylidene chloride $CH_2=CCl_2$; polystyrene $[-CH_2-CH(C_6H_5)-]_n$ from styrene $CH_2=CH(C_6H_5)$; ethylene-co-vinyl acetate copolymers from ethylene $CH_2=CH_2$

and vinyl acetate $CH_2=CH(OCOCH_3)$ etc. To reiterate, polymers synthesized from a single monomer are commonly referred to as *homopolymers* but, as we shall see, this term is not necessarily an adequate description of such polymers and may lead to serious misconceptions. The term *copolymer* is usually used to describe a polymer derived from two or more monomers, as in the case of ethylene-co-vinyl acetate mentioned above.

The problem is that many polymers synthesized from a single monomer might be better described as "pseudo-copolymers". For example, we mentioned above that polyethylene may be synthesized from ethylene via a number of techniques including high-pressure, free radical and catalytic methods (Ziegler-Natta and supported metal oxides). A range of polyethylenes, usually denoted in terms of density (i.e., low, medium and high density), are commercially available. There are large differences in the physical and mechanical properties of these polyethylenes which may be primarily attributed to variations in the polymer chain microstructure. During the polymerization of ethylene, structural units other than methylene groups are incorporated into the polymer backbone. These structural irregularities or defects, are alkyl substituted groups ($-CH_2CHR$; where R = methyl, ethyl etc.), which are the short-chain branches described above. Thus, if the term homopolymer implies a polymer chain containing identical structural units, it follows that polyethylenes can be, at best, only loosely described as homopolymers. In fact, it might be better to describe polyethylene in terms of a copolymer of ethylene and other α-olefins. It is interesting to note that "actual" copolymers of ethylene and butylene, synthesized at relatively low pressures using Ziegler-Natta catalysts, have been commercially introduced and called "linear low density polyethylenes" (LLDPE). This is somewhat confusing, but there is a degree of logic in naming the polymer as such, as it is analogous to low density polyethylene (LDPE). The main difference being that short chain branching units are incorporated either by polymerization conditions using a single monomer, or by the deliberate copolymerization of a second comonomer.

Isomerism in Polymers

Just because a polymer is synthesized from a pure monomer, where all the molecules are initially identical, does not mean that the final product consists of chains of regularly arranged units. We have already seen that branching can occur in a polymer such as polyethylene and it turns out that if we start with an asymmetric monomer we can obtain a variety of microstructures consisting of various geometric and stereo isomers. The most important types are sequence isomerism, stereoisomerism and structural isomerism.

Sequence Isomerism

When a monomer unit adds to a growing chain it usually does so in a preferred direction. Polystyrene, poly(methyl methacrylate) and poly(vinyl chloride) are only a few examples of common polymers where addition is almost exclusively *head-to-tail*. However, for several specific polymers there are significant numbers of structural units that are incorporated "backwards" into the chain leading to *head-to-head* and *tail-to-tail* placements. This is referred to as *sequence isomerism*. Classic examples include poly(vinyl fluoride), poly(vinylidine fluoride), polyisoprene and polychloroprene.

In the general case of a vinyl monomer $CH_2=CXY$, the head of the unit is arbitarily defined as the CXY end while the tail is defined as the CH_2 end. A unit may add onto the chain in two ways:

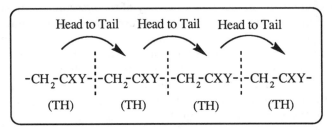

For convenience we will define "normal" addition as (TH) and "backward" addition as (HT). In poly(vinyl chloride) (X = H, Y = Cl), poly(methyl methacrylate) (X = CH_3, Y = $COOCH_3$) or polystyrene (X = H, Y = C_6H_5), for example, the units are arranged almost exclusively in a head-to-tail fashion, thus:

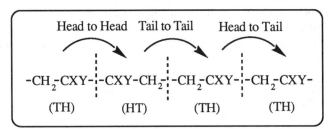

Some polymers, e.g. poly(vinyl fluoride) (X = H, Y = F) or poly(vinylidine fluoride) (X = Y = F), have a significant number of units incorporated "backwards" into the chain, giving head to head and tail to tail placements:

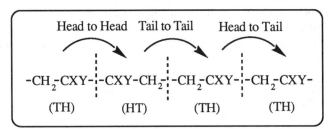

Stereoisomerism

Continuing with our discussion of polymers derived from a single monomer, let us now consider another major facet of polymer microstructure, namely *stereoisomerism*. Polymerization of a vinyl monomer, $CH_2=CHX$, where X may be a halogen, alkyl or other chemical moiety except hydrogen, leads to polymers with microstructures that are described in terms of *tacticity*. The substituent placed on every other carbon atom has two possible arrangements relative to the chain and the next X group along the chain. At one extreme, the monomer unit, which contains an asymmetric carbon atom, may be incorporated into the polymer chain in a manner such that each X group is *meso* to the preceding X group (i.e., on the same side of the extended chain - see figure 1.5). This polymer is described as *isotactic*.

Conversely, at the other extreme, addition of a monomer unit to the growing polymer chain may be directed to yield a polymer in which each successive X group is *racemic* to the preceding one (i.e., on the opposite side of the stretched out chain—again, see figure 1.5). This is called a *syndiotactic* polymer. Polypropylene ($X=CH_3$), for example, may be synthesized to yield an essentially isotactic or syndiotactic polypropylene by varying the polymerization conditions. These structures are illustrated in figure 1.6 for an *extended* polypropylene chain. (Because of steric repulsions the isotactic chain does not like this zig-zag shape, but would prefer to fold into a helix. We will consider such *conformations* later). Between these two extremes there is an enormous number of polymer chain microstructures based on different distributions of meso and racemic placements of structural repeating units. If there is no preferred direction to the addition of a monomer to the growing chain, i.e., the monomers are incorporated randomly with respect to the stereochemistry of the preceding unit, an *atactic* polymer is formed. Commercial polystryene, used for the manufacture of clear plastic beer glasses, and poly(methyl methacrylate), or Plexiglas®, are examples of a commonly encountered atactic polymers. In common with our general theme, it is also possible to conceive of vinyl homopolymers in terms of "copolymers" in which the comonomers are represented by the meso and racemic placements of the structural units of the polymer chain. This will prove a useful approach when we apply statistical methods to the description of microstructure in chapters 4 and 5.

Structural Isomerism—The Microstructure of Polydienes

Another outstanding example of the complex polymer chain microstructure that can result from the polymerization of a single monomer is seen in the synthesis of polydienes from conjugated dienes. Isoprene, for example, may be

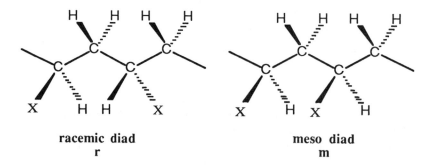

racemic diad
r

meso diad
m

Figure 1.5 *Schematic diagram depicting racemic and meso diads.*

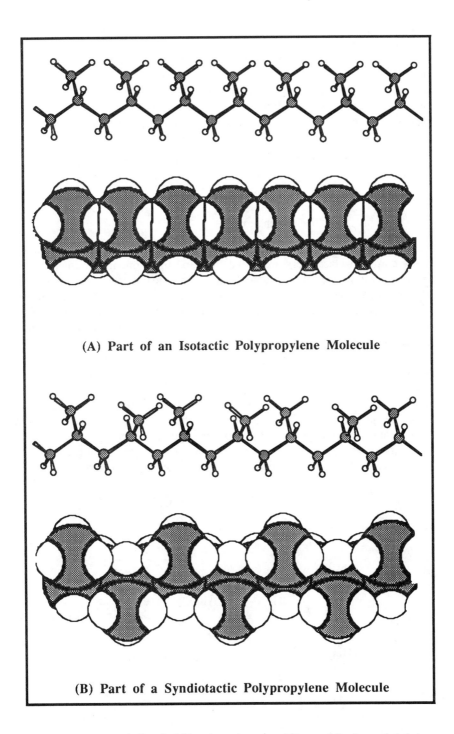

(A) Part of an Isotactic Polypropylene Molecule

(B) Part of a Syndiotactic Polypropylene Molecule

Figure 1.6 *Side view "ball and stick" and top view space filling models of extended chain isotactic and syndiotactic polypropylenes.*

polymerized to polyisoprene by a variety of techniques including free radical, ionic and Ziegler-Natta catalysis. Mother Nature also synthesizes "natural" polyisoprenes, Heavea (Natural Rubber) and Gutta Percha, but uses trees instead of round bottom flasks as polymerization vessels. The way in which isoprene is incorporated into a growing polymer chain is dependent upon the method of polymerization and the precise experimental conditions. Three common examples of diene monomers can be represented by the general formula:

$$\overset{1}{CH_2} = \overset{2}{CX} - \overset{3}{CH} = \overset{4}{CH_2}$$

where:

X = H	butadiene
X = CH$_3$	isoprene
X = Cl	chloroprene

We have numbered the carbon atoms for reasons that will soon be apparent. Many common elastomers are polydienes and therefore important polymer materials, but they're often regarded with fear and loathing by new students of polymer science because a description of their chain structure requires learning some nomenclature. The geometric and stereoisomers that are found in polydienes are a result of the configuration of the double bond and whether the polymerization has proceeded by opening both monomer double bonds or just one.

Four major *structural isomers*, shown schematically in figure 1.7, are possible. The so-called 1,4 polymers are formed by linking carbon atom 4 of one monomer to carbon atom 1 of the next, and so on. The arrangement of the substituent carbon atoms relative to the double bond in 1,4 polymers can be in either the *cis* or *trans* configuration. The 1,2 or 3,4 polymers are formed when incorporation into the polymer chain occurs through either the first or second double bond, respectively. These have the same configurational properties as vinyl polymer chains and may occur in isotactic or syndiotactic sequences.

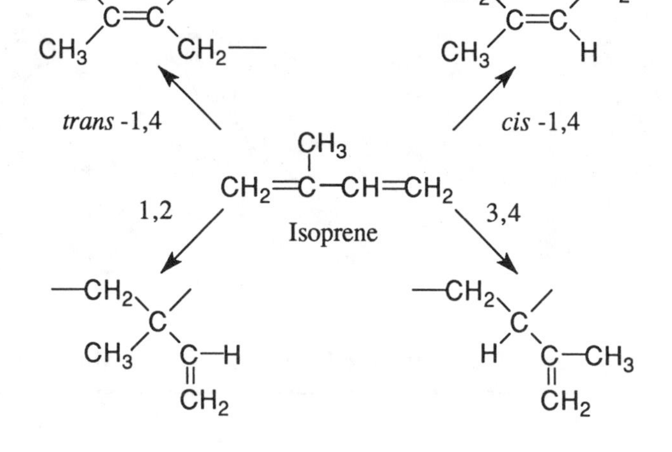

Figure 1.7 Schematic representation of the structural isomers of polyisoprene.

Table 1.1 *Microstructure of Polychloroprenes.*

Polym. Temp. (°C)	*trans* -1,4 (TH) (%)	*trans* -1,4 (HT) (%)	*cis*-1,4 (%)	1,2 (%)	*iso*-1,2 (%)	3,4 (%)
+40	76.1	13.4	5.2	2.1	1.7	1.5
+20	79.2	11.8	4.1	1.9	1.6	1.4
0	83.6	10.3	3.1	1.2	0.7	1.1
-20	86.4	9.6	1.7	1.1	0.5	0.7
-40	87.7	8.7	1.7	0.8	0.6	0.5

Natural rubber is high molecular weight, practically pure *cis*-1,4-polyiso-prene, but many synthetic elastomers have various proportions of the isomers described above. Even when the polymerization conditions dictate that a particular structural isomer is overwhelmingly favored, there are still some of the other structural isomers incorporated. The average composition of the chains varies with the temperature of polymerization, whether or not a catalyst has been used, and so on. Using free radical methods of polymerization a wide distribution of all four structural isomers is obtained, which may be varied significantly by changing the temperature of polymerization. And structural isomerism is not the only complication occurring in the microstructure of polydienes. Sequence isomerism is also common. Consider, for example, the results shown in table 1.1 which one of us (the better looking one) obtained from an NMR analysis (see Chapter 6) of polychloroprenes synthesized at various temperatures[*].

"Real" Copolymers—Additional Complexity or More of the Same?

As mentioned above, "real" copolymers are synthesized from two or more comonomers. (For completeness, the terms *terpolymer* and *tetrapolymer* etc., are also used to describe polymers derived, respectively, from three, four or more comonomers.) In the simplest case, we must consider the microstructure of a copolymer, synthesized from say the two monomers A and B, in terms of the concentration of A and B units incorporated into the polymer chain and how these units are distributed. However, we must remember that the actual microstructure of the copolymer can be far more complex and all of the micro-structural variations that we have considered for homopolymers can be present in copolymers.

Two monomers may be distributed in a polymer chain in a number of well-defined ways. For example, consider copolymers synthesized from the monomers styrene (St) and butadiene (Bd). At one extreme, we can synthesize a *block copolymer*, (using "living" anionic polymerization - see Chapter 2), which may resemble $(St)_p$-$(Bd)_q$-$(St)_r$, a *triblock* copolymer where p, q and r represent blocks containing different numbers of St or Bd structural units. Such

* M. M. Coleman, D. L. Tabb and E. G. Brame, Jr., *Rubber Chem. Technol.* **50**(1), 49 (1977).

copolymers are commercially manufactured by the Shell Chemical Company under the trade name Kraton®, and fall into the general category of *thermoplastic elastomers*.

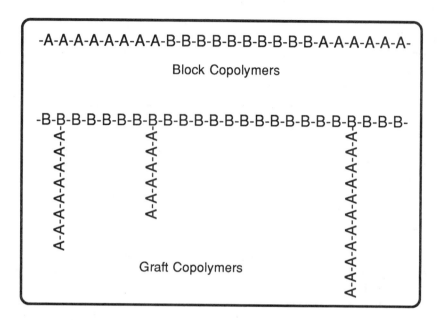

Block copolymers can also be made with two blocks of units (*diblock*), three (*triblock*), and so on. There are even exotic arrangements such as star copolymers whose arms are blocks. (The chemistry involved in synthesizing such chains is very clever and polymer chemists get up to all sorts of weird and wonderful things these days!) *Graft* copolymers can also be regarded as a type of block copolymer and they have taken on increasing significance in producing commercially useful materials. Here chains of one type of unit (say A) are "grafted" onto the backbone of a chain of a different type (B) as illustrated in the above schematic representation.

At the other extreme, it is possible to synthesize an *alternating copolymer* of the two monomers which may be represented by $-(AB)_n-$. In this case, each A unit incorporated into the polymer chain is immediately followed by a B unit and vice versa. Fine examples are the copolymers of styrene and maleic anhydride synthesized by free radical polymerization, which are essentially alternating (see later in chapter 5). (Note also that it would be perfectly reasonable to consider this as a "homopolymer" of polyAB - the polymer chain certainly doesn't know the difference!)

-A-B-A-B-A-B-A-B-A-B-A-B-A-B-A-B-A-B-A-B-A-B-

Alternating Copolymers

Between these two extremes we can also define *random copolymers*, although this is to some degree misleading and they are more accurately described as *statistical* copolymers. If the copolymer chain consisted of exactly 50% A units and 50% B units and the sequence of units followed the probabilities of coin toss (Bernoulian) statistics, then the copolymer would indeed be truly random.

-A-B-B-B-A-A-B-A-B-A-A-A-B-A-B-B-A-B-B-A-A-A-B-

Random Copolymers

However, we must recognize that random copolymers do not have to have a 50:50 % composition. As long as the monomers are incorporated into the growing chain entirely at random then the probability of adding either an A or B unit solely depends upon the relative concentration of each. For example, if the chain consisted of 75% B units and 25% A units it can still be designated random as long as the placement probabilities were adjusted to account for the difference in composition (i.e., one could build a hypothetical chain by assuming the probability that the first unit is an A is equal to 0.25 or that of a B is equal to 0.75, and so on for the second, third and up to the last unit).

Although we have defined three distinct types of copolymer microstructure, there are obviously a myriad of intermediate microstructures. Thus, we see in the literature descriptive terms describing copolymer microstructure in terms of an alternating or "blocky" tendency. This is not really an adequate description and it is necessary to describe the microstructure in terms of the sequence distribution (usually diads, triads and sequences of higher order). We will discuss this in more depth when we consider copolymerization, but we must clearly be careful in our use of the word *random*.

Polymer Blends

Copolymers are often synthesized in an attempt to obtain properties that are intermediate, superior or just different from those of the homopolymers that can be produced from the same monomers. Another way of doing this is to simply mix the homopolymers together to form *blends*. Most non-polar polymers do not mix, for reasons we will discuss in the chapter on solutions and blends. Such *immiscible* blends are not necessarily useless, however. Certain types of mixtures can have enhanced impact resistance, for example, although this depends upon the character of the components and the size of the phase separated domains. In recent years an increasing number of miscible (single phase) systems have been reported and this is presently (1996) an area of intense research activity.

The reluctance of most polymers to mix with one another has important ramifications in the plastics recycling industry. While some believe it is simply an industrial conspiracy that prevents us from recycling all plastics materials, it is actually a formidable scientific problem. Consider, for example, the ubiquitous soda bottle, illustrated schematically in figure 1.8.

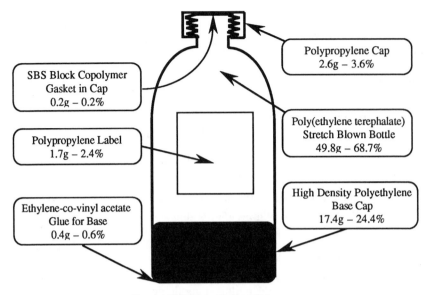

Source: Phil Blatz, DuPont Polymers

Figure 1.8 Schematic representation of a standard 2 litre soda bottle.

This is quite a complex piece of engineering, involving many different plastics materials, which are necessary because of the different chemical, physical and mechanical requirements of the various different parts of the bottle.

Soda bottles are recycled in the USA, but not before the different plastics materials are separated from one another, and then only the poly(ethylene terephthalate) (PET) blown bottle and the high density polyethylene (HDPE) base cup are actually recycled; the remaining plastics are committed to landfill. Why can't we just grind up the whole bottle and put the mixture into an extruder or injection molding machine and make something useful? It just doesn't work. PET is completely immiscible with HDPE. For that matter, HDPE is immiscible with polypropylene (PP); two polymers that are chemically quite similar. Why is it that the vast majority of polymer are immiscible with one another? We will attempt to answer that question in Chapter 9.

C. MOLECULAR WEIGHT—SOME INITIAL OBSERVATIONS

A key factor in determining polymer properties is polymer chain length, or molecular weight and its distribution, and we will be discussing the measurement of molecular weight in some detail in Chapter 10. We need to consider some basic definitions here, however, because our discussion of polymerizations will involve the relationship of mechanisms to the *degree of polymerization* (which is equal to the number of structural units in the chain and is another way of describing chain length).

We will first consider the ways in which the molecular weight of a high polymer differs from that of a low-molecular-weight substance. The most

important characteristic of the vast majority of synthetic high polymers is that the individual chain molecules are not all of the same weight, and there is a spread around some mean value. [In contrast, for a material of low molecular weight such as benzene or water, *every* molecule is identical (ignoring isotope effects)]. The polymer is said to have a molecular weight *distribution*. We shall see that this results in different molecular weight *averages*. This might at first be confusing, but the average molecular weight is not a unique thing, it can be defined in a number of different ways. Two of the most commonly used averages are the *number average* molecular weight and the *weight average* molecular weight, defined as:

Number average:

$$\overline{M}_n = \frac{\sum N_x M_x}{\sum N_x} \tag{1.1}$$

Weight average:

$$\overline{M}_w = \frac{\sum N_x M_x^2}{\sum N_x M_x} \tag{1.2}$$

where M_x is the molecular weight of a molecule corresponding to a degree of polymerization x (i.e., consisting of x monomer units of molecular weight M_o, thus $M_x = x M_o$), n_x is the number (or number of moles) of such molecules and N_x and W_x are the total number and weight of the molecules of "length" x, respectively.

It has been our experience that students are confused by these definitions the first few times they encounter them, so it is worth spending a little time discussing averages and distributions in more detail and to make a start we will consider an example worthy of Monty Python's Flying Circus. Imagine an elephant with four mosquitoes perched upon its bum*. Let's say the elephant weighs 10,000 lbs and the mosquitoes are the size of small birds weighing, say, 1 lb each (you know, the type you encounter every time you go camping!). What is the average weight of the elephant and the four mosquitoes? The average that is familar to most people is the number average (equation 1.1), which is obtained by simply adding up the weights of all the things present (elephants and mosquitoes) and dividing by the total number of these things (elephants and mosquitoes). Since there are four mosquitoes their total weight is 4 x 1 = 4 lb, which together with the 10,000 lb weight of the one elephant yields a combined total weight of 10,004 lbs. We can obtain this result using equation 1.1:

$$\text{Total weight} = \sum N_x M_x = (4 \times 1) + (1 \times 10,000)$$

$$= 10,004 \text{ lbs} \tag{1.3}$$

where x now represents a species, and is either a mosquito or an elephant**. The total number of things present is:

---

* A mildly vulgar English term for rear-end.
** If we were talking about chains, N_x would be the number of chains that had x repeating units.

$$\text{Total number} = \sum N_x = (4 + 1) = 5 \qquad (1.4)$$

The number average weight, $\overline{M}_n$, is therefore about 2,000 lbs.

$$\overline{M}_n = \frac{\sum N_x M_x}{\sum N_x} = \frac{10,004}{5} \cong 2,000 \qquad (1.5)$$

Clearly, the number average is obtained by "sharing" the total weight of all the species present equally amongst each of the species.

Now, suppose the elephant is startled by a mouse, goes on a rampage and tramples the poor little beast. The effect of the mosquitoes sitting on the elephant's bum would be negligible compared to the stomping administered by the elephant. The weight average is a much more realistic indicator of this gross mechanical property. A weight average does not "count" species just by their number, but takes into consideration their total weight. Thus we replace N_x by W_x, the total weight of each species, in equation 1.3. There are 4 lbs of mosquitoes and 10,000 lbs of elephant present, so if we "count" in terms of weight instead of numbers, we obtain:

$$\sum W_x M_x = (4 \times 1) + (10,000 \times 10,000) \cong 10^8 \qquad (1.6)$$

Now, in order to obtain an average we must divide by the total weight of everything present, so that our new average, $\overline{M}_w$, is given by:

$$\overline{M}_w = \frac{\sum W_x M_x}{\sum W_x} \cong \frac{10^8}{10,004} \cong 10,000 \qquad (1.7)$$

where we use the symbol $\cong$ to mean "approximately equal to" or "of the order of". Clearly, the weight average in this example is close to the weight of the largest species, the elephant, and is a much better measure of stomping ability than the number average. Note also that equation 1.7 does not look like equation 1.2. They are equivalent, however, as the total weight of all species x present is $\sum W_x = \sum N_x M_x$ and substitution into equation 1.7 leads to the same result as equation 1.2.

If this idea of averaging by weight instead of number is still giving you concern, try one more way of looking at it. We can determine an average weight per species by multiplying the fraction of each species present, by the weight of each species, M_x, and then sum over all species (i.e., the average weight of each species is just the sum of the fractional contributions of each of the components in the mixture). We can determine the fractional contribution by number, so that in our absurd example one-fifth of our sample is elephant and four-fifth's mosquito, so that the number average weight is:

$$\overline{M}_n = \sum X_x M_x = \left(\frac{4}{5} \times 1 \right) + \left(\frac{1}{5} \times 10,000 \right) \cong 2,000 \qquad (1.8)$$

Here we use X_x to denote the number fraction (or mole fraction) which is defined as $N_x / \sum N_x$.

If we use a weight fraction, $w_x = W_x/\Sigma\, W_x$, instead of a number fraction, however, we obtain:

$$\overline{M}_w = \Sigma\, w_x\, M_x = \left(\frac{4}{10,004} \times 1\right) + \left(\frac{10,000}{10,004} \times 10,000\right)$$

$$\cong 10,000 \tag{1.9}$$

If you still don't get it then there is no hope for you when it comes to understanding averages and you might as well give up and become a sociologist. Getting back to polymers, it will become apparent to you once you have read Chapters 2 through 4 that most methods of synthesis lead to a statistical distribution of molecular weights. Let us consider a sample where by careful fractionation we can separate the molecules by size, such that the first fraction contains molecules with a molecular weight less than 10,000 g/mole, the second contains those molecules with a molecular weight between 10,000 and 30,000 g/mole, the third between 30,000 and 50,000 g/mole, and so on. In this sample there are no molecules that have a molecular weight of greater than 200,000 and the number of moles of polymer in each fraction are given in table 1.2. (It is assumed that the fraction with a molecular weight between 10,000 and 30,000 g/mole has an average value of 20,000 g/mole etc.) The number of moles of each fraction can be plotted as a function of the molecular weight of that fraction in the form of a histogram, as shown in figure 1.9. (Note that the mole fraction of polymers with molecular weights less than about 50,000 or more than 150,000 g/mole are so small that they don't register on this plot.) Using the data in table 1.2 and equations 1.1 and 1.2, the number and weight averages can be calculated. The sums necessary to calculate $\overline{M}_n$ are given in table 1.2, but the calculation of $\overline{M}_w$ is left as an exercise for the student (your answer should be of the order of 103,000).

Table 1.2 Molecular Weight Data.

Fraction	Molecular Weight M_x	Number of Moles N_x	Weight of Fraction $(N_x M_x)$
1	0	0.00000	0
2	20000	0.00015	3
3	40000	0.01953	781
4	60000	0.62500	37500
5	80000	5.00000	400000
6	100000	10.00000	1000000
7	120000	5.00000	600000
8	140000	0.62500	87500
9	160000	0.01953	3125
10	180000	0.00015	27
11	200000	0.00000	0

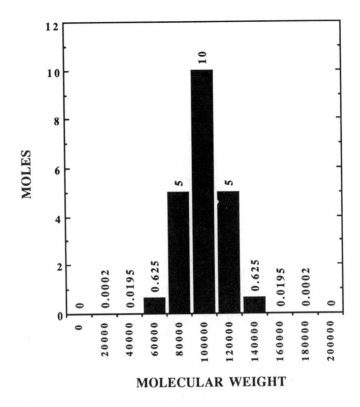

Figure 1.9 *Histogram of the distribution of molecular weights.*

Clearly, the answer you get may not be precise in that we have separated the sample rather crudely into fractions that each contain a significant range of molecular weights. It would obviously be much better if we could obtain the distribution as a continuous curve, or at least one where the separation is so fine that to all intent and purposes we can treat it as continuous. Later on in this book we will show how a technique called size exclusion (or gel permeation) chromatography (SEC/GPC) can be used to accomplish this. At this juncture we wish to make some additional points about averages, however, and to do this we will next consider the distribution of the same sample separated into much finer fractions, as illustrated in figure 1.10.

This figure plots the number of moles of a fraction with a particular molecular weight (N_x) as a function of molecular weight (M_x), and the position of the peak maximum corresponds *in this case* to the number average molecular weight, $\overline{M}_n$, and is equal to 100,000 g/mole. This is because the distribution is symmetric about this maximum position (it is actually a Gaussian distribution, for those of you interested in such things). We calculated the value of $\overline{M}_n$ as before, by summing the values of $N_x M_x$ for all the points shown in figure 1.10 and then divided by $\Sigma\ N_x$.

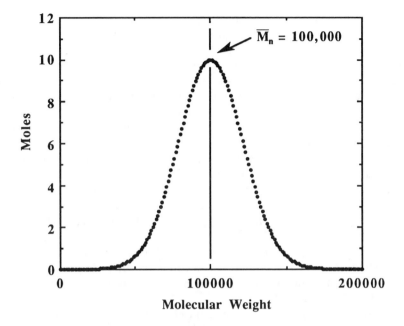

Figure 1.10 *Plot of moles of polymer fractions versus molecular weight.*

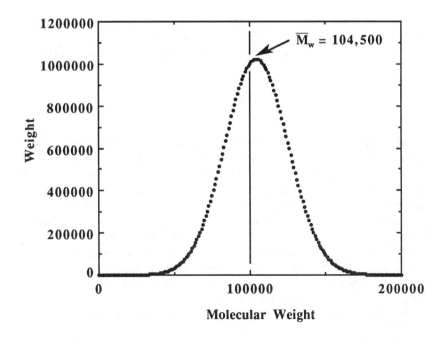

Figure 1.11 *Plot of the weight of polymer fractions versus molecular weight.*

We mentioned this to make an additional point concerning the *moments* of a distribution. If we take any one of the points on the curve shown in figure 1.10 and multiply the value N_x by M_x for this point, we are actually taking a *moment* about the origin (think back to your old high school stuff on weights placed on a balanced beam or kids sitting on a see-saw). The *first moment* of the distribution as a whole is defined as the sum of all individual moments (i.e., adding up the values of $N_x M_x$ for all the points shown in figure 1.10, $\Sigma\, N_x M_x$, and normalizing by dividing by $\Sigma\, N_x$). This is of course the number average molecular weight, $\overline{M}_n$.

If instead of plotting a *number* distribution (i.e. number of moles versus molecular weight) we plot a *weight* distribution (the weight of polymer in fraction x ($W_x = N_x M_x$) versus the molecular weight of the polymers in that fraction, M_x) we obtain the distribution shown in figure 1.11. The position of the number average molecular weight, $\overline{M}_n = 100,000$ g/mole, is shown as a vertical line which you will note does not go through the peak maximum. Actually, the peak position of the weight distribution is shifted to about 104,500 g/mole (the distribution for this example is still symmetric). This peak position is the first moment of the weight distribution about the origin and is equal to $\overline{M}_w$ (i.e. $\Sigma\, (N_x M_x) M_x / \Sigma\, N_x M_x$).

So, we have seen that the number and weight averages are nothing more than the first moments of their respective distributions. It turns out that when describing distributions, higher order moments (second moment, third moment and so on) can be useful as they describe various properties of the distribution, such as its breadth (second moment), skewness or asymmetry (third moment) etc.[*].

In polymer science it is common to use the ratio of the weight average to the number average molecular weight as a measure of the breadth of the distribution rather than the moments (although they are related) and this ratio is called the *polydispersity* of the sample.

$$\text{Polydispersity} = \frac{\overline{M}_w}{\overline{M}_n} \geq 1 \qquad (1.10)$$

For a homogeneous sample where the polymer chains are all the same length, which is often referred to as a *monodisperse* sample, $\overline{M}_w = \overline{M}_n$ and the polydispersity is equal to 1. But if there is a distribution of molecular weights, $\overline{M}_w$ is always greater than $\overline{M}_n$. You should also note that in the simple example (figure 1.11) the polydispersity is actually very small (≈ 1.05), although plots of the distribution *look* quite broad. It is actually very difficult to get distributions this narrow, but it can be achieved using some clever chemistry (as we will see later in Chapter 2). Most commercial polymers have much broader distributions (polydispersities of the order of 5-10 are not uncommon).

There are two things that remain before we conclude this initial discussion of molecular weight. First we have defined two ways of averaging molecular weights ($\overline{M}_w$ and $\overline{M}_n$); are there more? If you go back and compare equations 1.1

---

[*] If you get into statistics in more depth you will find that these properties are related to moments about the mean, rather than the origin, but we have no need to worry about that here.

and 1.2 you should notice that $\overline{M}_w$ can be obtained from $\overline{M}_n$ by multiplying each term in the summations by M_x. We can also multiply each term in the expression for $\overline{M}_w$ by M_x to obtain a so-called z-average molecular weight,

$$\overline{M}_z = \frac{\sum N_x M_x^3}{\sum N_x M_x^2} \qquad (1.11)$$

and so on to obtain higher order molecular weight averages. The importance of $\overline{M}_w$, $\overline{M}_n$ and $\overline{M}_z$ is that there are experimental methods that give these averages directly and we will describe some of these in Chapter 10.

The second point we wish to make before concluding is that so far we have only considered symmetric distributions and in the examples we gave, $\overline{M}_w$ and $\overline{M}_n$ were not very different. However, as soon as we consider asymmetric or skewed distributions, such as that shown in figure 1.12, then the difference becomes larger and the average values ($\overline{M}_n$ for the number distribution, $\overline{M}_w$ for the weight distribution) no longer correspond to the peak positions. We will see later when we discuss the kinetics and statistics of polymerizations that some quite unusual looking distributions can be obtained, such as that illustrated in figure 1.13 which shows both the number fraction and weight fraction distribution of chains found at a certain stage in a polycondensation reaction. More on this later!

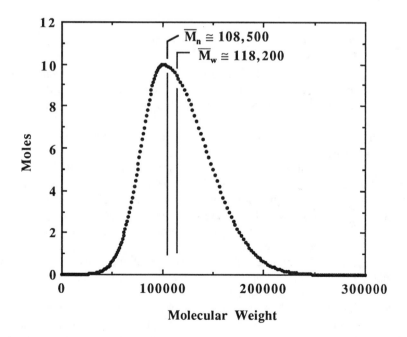

Figure 1.12 *Plot of moles of polymer fractions versus molecular weight for an asymmetric distribution of molecular weights.*

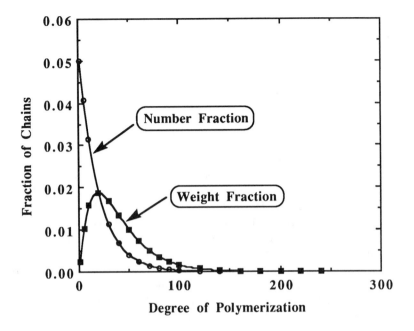

Figure 1.13 *Plot of the number and weight fraction distributions for polymer chains found in a typical polycondensation reaction.*

Polymer Analysis (or How Do We Know What We've Got?)

One of the more difficult and important problems in polymer science is the characterization of the structure of a macromolecule. If polymer chains were made up of identical monomers, the determination of the molecular weight and molecular weight distribution would be sufficient to characterize the structure of the chain. However, a polymer chain made up of a single well-defined type of unit is a rarer beast than one would first imagine. Conceptually, the simplest polymer would consist of an infinite chain of structurally and configurationally perfect repeating units. We can approximate the ideal case in, for example, an ultra-high molecular weight linear polyethylene (PE). However, even in this case the polymer by necessity has *end groups* (normally methyl or vinyl groups) and inevitably there are chemical and structural defects present in the chain resulting from oxidation, degradation and the incorporation of impurities. Also, for the majority of synthetic polymers, because of the nature of the poly-merization process, other competing reactions occur such as branching and the various types of isomerization. We have defined these and other aspects of polymer microstructure in this chapter, but we have not mentioned how we would experimentally determine the number, type and distribution of units present in the chains of a given sample. This is a difficult, but fascinating, analytical problem and one that we address later in this book. We will simply conclude this section by noting that various advances in spectroscopic techniques, particularly nuclear magnetic resonance spectroscopy, have allowed a much more detailed analysis of chain structure than that available even 10 years ago, but there are still many problems that remain to be solved.

D. STUDY QUESTIONS

1. What is the difference between a *thermoplastic* and a *thermoset* ?

2. Polyethylenes are often characterized as being of high density (HDPE) or low density (LDPE). What causes the *difference in density* between these polyethylenes?

3. What is "*linear low density*" polyethylene (LLDPE) and how does it get its name?

4. What *configurational isomers* are possible following the polymerization of butadiene, $CH_2=CH-CH=CH_2$?

5. Show an example of *sequence isomerism*.

6. Different polypropylenes may be described as *atactic or isotactic*. What does this mean ?

7. What is the difference between a *block* and a *graft* copolymer ?

8. A chemist synthesizes a copolymer containing 90 mole % isoprene and 10 mole % methyl methacrylate. In a *systematic manner*, describe the polymer chain microstructures that are theoretically possible. (Ignore limitations due to chemistry, which we have not considered yet.) Be careful! There are many possibilities.

9. A sample of polystyrene is described as having a polydispersity of 1.03. What does this tell you ?

10. Consider the following distribution of polymer chains:

> 5 chains of degree of polymerization 10
> 25 chains of degree of polymerization 100
> 50 chains of degree of polymerization 500
> 30 chains of degree of polymerization 1000
> 10 chains of degree of polymerization 5000
> 5 chains of degree of polymerization 50000

(a) Calculate the number and weight average degree of polymerization and polydispersity of this collection of polymer chains.

(b) If these chains were poly(methyl methacrylate)s what would be the corresponding number and weight average molecular weights (ignore end group effects) ?

(c) Add five more chains of degree of polymerization 10 to the above distribution. Comment of the relative effect that this has on the calculated number and weight average molecular weights.

E.   SUGGESTIONS FOR FURTHER READING

P. J. Flory,   *Principles of Polymer Chemistry*,
Cornell University Press, Ithaca, New York, 1953.

J. L. Koenig   *Chemical Microstructure of Polymer Chains*,
Wiley, New York, 1982.

F. APPENDIX: CHEMICAL STRUCTURES OF SOME COMMON POLYMERS

Polymer	Chemical Structure of the Polymer Repeat
Polyethylene (PE)	$\left[\text{—CH}_2\text{—CH}_2\text{—}\right]_n$
Polypropylene (PP)	$\left[\text{—CH}_2\text{—CH(CH}_3)\text{—}\right]_n$
Polybutene-1	$\left[\text{—CH}_2\text{—CH(CH}_2\text{CH}_3)\text{—}\right]_n$
Polyisobutylene (PIB)	$\left[\text{—CH}_2\text{—C(CH}_3)_2\text{—}\right]_n$
Polybutadiene (PBD)	$\left[\text{—CH}_2\text{—CH=CH—CH}_2\text{—}\right]_n$
cis-Polyisoprene (Natural Rubber)	$\left[\text{—CH}_2\text{—C(CH}_3)\text{=CH—CH}_2\text{—}\right]_n$
trans-Polychloroprene (Neoprene® Rubber)	$\left[\text{—CH}_2\text{—C(Cl)=CH—CH}_2\text{—}\right]_n$
Polystyrene (PS)	$\left[\text{—CH}_2\text{—CH(C}_6\text{H}_5)\text{—}\right]_n$

Polymer	Chemical Structure of the Polymer Repeat
Poly(vinyl acetate) (PVAc)	$\left[\!\!\begin{array}{c} CH_2-CH \\ \mid \\ O \\ \mid \\ C \\ O\diagup\ \diagdown CH_3 \end{array}\!\!\right]_n$
Poly(vinyl chloride) (PVC)	$\left[\!\!\begin{array}{c} Cl \\ \mid \\ CH_2-CH \end{array}\!\!\right]_n$
Poly(ethylene oxide) (PEO)	$\left[\, CH_2-CH_2-O \,\right]_n$
Poly(methyl methacrylate) (PMMA)	$\left[\!\!\begin{array}{c} CH_3 \\ \mid \\ CH_2-C \\ \mid \\ C \\ CH_3-O\diagup\ \diagdown O \end{array}\!\!\right]_n$
Poly(methyl acrylate) (PMA)	$\left[\!\!\begin{array}{c} CH_2-CH \\ \mid \\ C \\ CH_3-O\diagup\ \diagdown O \end{array}\!\!\right]_n$
Polytetrafluoroethylene (PTFE)	$\left[\, CF_2-CF_2 \,\right]_n$
Polycaprolactam (Polyamide - Nylon 6)	$\left[\!\!\begin{array}{c} \quad\quad O \\ \quad\quad \| \\ (CH_2)_5-N-C \\ \mid \\ H \end{array}\!\!\right]_n$
Polycaprolactone (PCL)	$\left[\!\!\begin{array}{c} O \\ \| \\ (CH_2)_5-C-O \end{array}\!\!\right]_n$
Poly(ethylene terephthalate) (PET)	$\left[\!\!\begin{array}{c} O \quad\quad\quad\quad O \\ \| \quad\quad\quad\quad \| \\ O-C-\bigcirc-C-O-CH_2-CH_2 \end{array}\!\!\right]_n$
Poly(dimethyl siloxane) (PDMS)	$\left[\!\!\begin{array}{c} CH_3 \\ \mid \\ Si-O \\ \mid \\ CH_3 \end{array}\!\!\right]_n$

CHAPTER 2

Polymer Synthesis

> *"I am inclined to think that the development of polymerization
> is perhaps the biggest thing that chemistry has done,
> where it has had the biggest effect on everyday life"*
> —Lord Todd, 1980

A. INTRODUCTION

In the next few chapters of this book we will be considering some of the fundamentals of polymer synthetic chemistry. This will include a description of the basic mechanisms by which practically all synthetic polymers are formed, followed by a discussion of the kinetics and statistics of some of these reactions. Our aim will be to show how the mechanism of polymerization affects the molecular weight and microstructure of the polymer or copolymer that is being formed. This part of the book will culminate in Chapter 6, where spectroscopic methods for characterizing polymer chain microstructure will be described.

It is convenient to classify polymerization reactions into two or three basic types. Carothers*, for example, suggested that polymers could be broadly regarded as belonging to one of two types: *condensation* or *addition*. In the former the chemical repeat unit of the polymer has a different molecular formula to that of the monomers from which it is produced, as a result of the elimination or "condensation" of certain groups. A typical example of this type of reaction is the formation of an ester from ethyl alcohol (the active component of beer) with acetic acid (the active component of vinegar), where a molecule of water is eliminated (hence the name condensation). In this case no further condensations are possible and polymers are not formed.

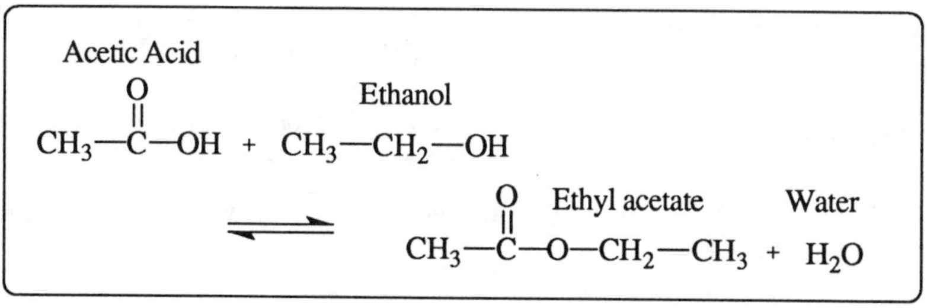

Acetic Acid

Ethanol

$$CH_3-\overset{\overset{O}{\|}}{C}-OH \ + \ CH_3-CH_2-OH$$

Ethyl acetate   Water

$$\rightleftharpoons \qquad CH_3-\overset{\overset{O}{\|}}{C}-O-CH_2-CH_3 \ + \ H_2O$$

---

* Wallace Hume Carothers was a brilliant scientist who made seminal contributions to polymer science. We will consider some of these shortly.

The structural units of addition polymers, in contrast, have the same molecular formula as their monomers, although the arrangement of bonds is different, as in ethylene and polyethylene:

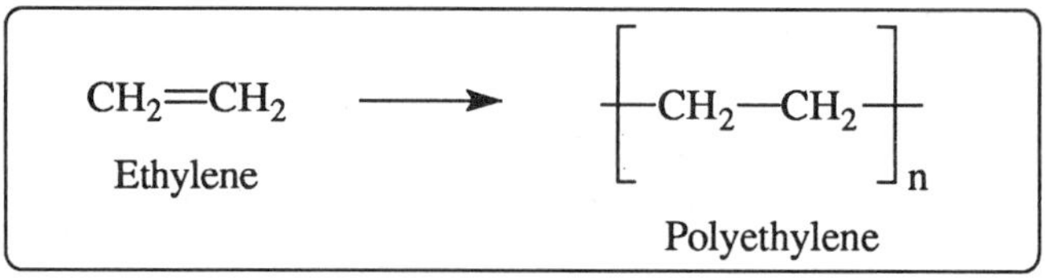

$$CH_2{=}CH_2 \longrightarrow \left[CH_2{-}CH_2 \right]_n$$

Ethylene

Polyethylene

There are exceptions, however, and some polymerizations fit uneasily into these two categories. For example, polyurethanes, which are formed by the reaction of isocyanates and alcohols, are considered condensation polymers, even though water is not eliminated during the reaction:

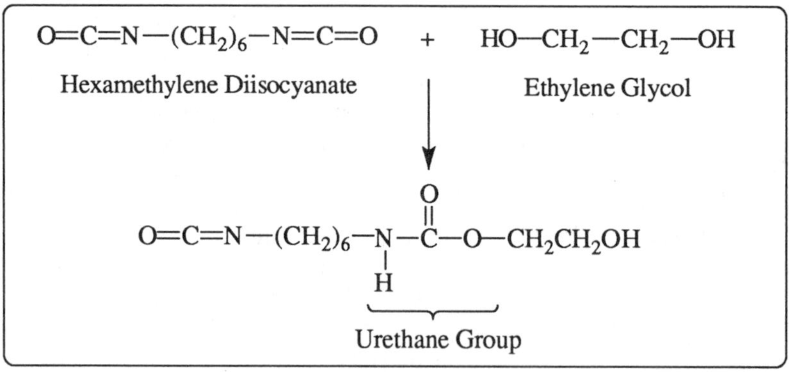

$$O{=}C{=}N{-}(CH_2)_6{-}N{=}C{=}O \quad + \quad HO{-}CH_2{-}CH_2{-}OH$$

Hexamethylene Diisocyanate

Ethylene Glycol

$$O{=}C{=}N{-}(CH_2)_6{-}\underset{H}{N}{-}\overset{\overset{O}{\|}}{C}{-}O{-}CH_2CH_2OH$$

Urethane Group

Furthermore, ring opening polymerizations must be regarded as addition polymerizations, but they form polymers that could also be synthesized by an appropriate condensation reaction, as in the polymerization of caprolactam:

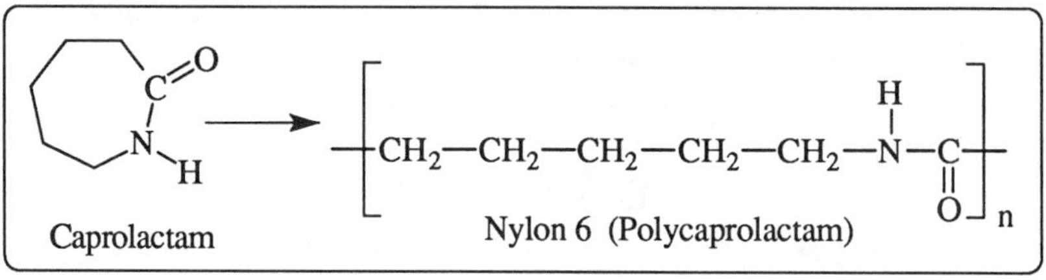

Caprolactam

$$\left[CH_2{-}CH_2{-}CH_2{-}CH_2{-}CH_2{-}\underset{}{\overset{H}{N}}{-}\overset{}{\underset{\underset{O}{\|}}{C}} \right]_n$$

Nylon 6 (Polycaprolactam)

where the resulting polyamide could also be produced by polymerization of the appropriate amino acid:

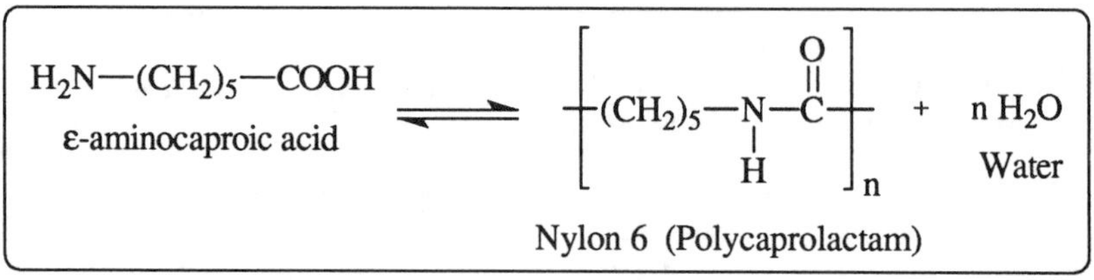

$$H_2N{-}(CH_2)_5{-}COOH \rightleftharpoons \left[(CH_2)_5{-}\underset{H}{N}{-}\overset{\overset{O}{\|}}{C} \right]_n + \ n\,H_2O$$

ε-aminocaproic acid

Water

Nylon 6 (Polycaprolactam)

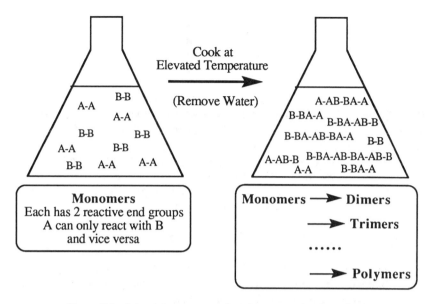

Figure 2.1 *Schematic representation of step-growth polymerization.*

Accordingly, we prefer to adopt the classification given by Rempp and Merrill (*Polymer Synthesis*, 1986), where polymerization reactions are considered to be either step-growth (this category includes polycondensations) or chain polymerizations (which include ring opening polymerizations).

One final thing needs to be kept in mind before we proceed to discuss the specifics of polymerization. In the following chapter we will discuss aspects of the kinetics of these reactions that will inherently assume that we are dealing with a *batch process*. In other words, we will consider a reaction "pot" into which we place monomer, which then reacts to form polymer and is therefore used up proportionally. In certain cases we might also have solvent, initiator, and perhaps one or two other additives present. These conditions are characteristic of most laboratory studies of polymerization that we will be concerned with here, as these provide the fundamental insights into mechanism and its relationship to the nature of the product. On the industrial scale of production continuous processes are widely used, but we will only qualitatively discuss some of these large-scale systems, at the end of this chapter.

B. STEP-GROWTH POLYMERIZATION

The central feature of this type of polymerization is the slow building of chains in a systematic stepwise fashion as depicted schematically in figure 2.1. Monomers combine with one another to give dimers:

$$M_1 + M_1 \rightarrow M_2$$

dimers and trimers can also combine with themselves or each other to give higher oligomers:

$$M_1 + M_2 \rightarrow M_3$$
$$M_2 + M_2 \rightarrow M_4$$
$$M_2 + M_3 \rightarrow M_5$$
$$M_3 + M_3 \rightarrow M_6$$

and so on. As a result of this step-growth mechanism, high molecular weight polymer is only produced at the end of the polymerization. More on this later.

There are a number of important types of polymers that are produced by this type of polymerization, including the polyamides (nylons), polyesters, polyurethanes, polycarbonates, phenolic resins, etc. We will give examples of each of these, but will start with a brief description of the discovery of nylon 6,6. This not only serves to introduce features that are characteristic of all polycondensations, but is a great and ultimately tragic story.

Historical Background—The Discovery of Nylon

Near the end of 1926 Charles Stine, Chemical Department director at Du-Pont, submitted a short proposal to his Executive Committee to initiate a program of "pure" or fundamental scientific research. In those days this was unusual (one could argue that the cycle has gone full circle and we are once again in the same unfortunate state). Nevertheless, after some persuasion the committee agreed to this proposal which included, amongst other things, fundamental work on polymerizations.

Stine believed that the success of his program would depend upon his ability to hire outstanding scientists. Because established people (predominantly academics) proved difficult to move, he decided to hire "men of exceptional scientific promise but no established reputation" (clearly a Politically Incorrect statement!). One such was Wallace Hume Carothers, then an instructor at Harvard, who was described as brilliant but mercurial.

At that time (about 1928) there was still widespread disagreement concerning the nature of what we now call polymer materials. The established view had been that these were loosely bound associations or aggregates of small molecules, but this position was becoming untenable. The champion of a different hypothesis, that these materials consisted of "giant" or macromolecules, whose units were linked by ordinary covalent bonds, was Hermann Staudinger*.

Carothers accepted Staudinger's view and his initial aim was a piece of pure science, to prove the existence of macromolecules by systematically building large molecules from smaller ones, using simple well-understood reactions. He would do this in such a way that the structure of the final product could not be questioned. He initially chose to examine the condensation of alcohols and acids to give esters, an example of which was given above. The reactants, ethanol and acetic acid are *monofunctional*, however (i.e. each has a single group capable of participating in a condensation; after this is complete no further chain forming reactions can occur). With typical insight Carothers realized that he would require *bifunctional* monomers (see below, for examples) if he were to obtain

---

* Staudinger was not awarded a Nobel Prize until 1953, perhaps because of the controversy his views initially generated; sometimes you have to outlive your enemies!

polymers (this seems simple and obvious with the benefit of hindsight, but must be placed in the context of the times, when many did not accept the existence of macromolecules at all, let alone designing reactions to produce them). By the end of 1929 Carothers' group had synthesized large molecules with molecular weights in the range 1,500–4,000, from dicarboxylic acids and dialcohols (glycols) in the presence of an acid catalyst:

$$HO-\overset{\overset{\displaystyle O}{\|}}{C}-(CH_2)_n-\overset{\overset{\displaystyle O}{\|}}{C}-OH \quad + \quad HO-(CH_2)_m-OH$$

$$\rightleftharpoons \quad HO-\overset{\overset{\displaystyle O}{\|}}{C}-(CH_2)_n-\overset{\overset{\displaystyle O}{\|}}{C}-O-(CH_2)_m-OH \quad + \quad H_2O$$

Because functional groups remain at the end of the newly formed ester, further condensations can continue, producing long chain molecules:

$$X\left[\overset{\overset{\displaystyle O}{\|}}{C}-O-(CH_2)_m-O-\overset{\overset{\displaystyle O}{\|}}{C}-(CH_2)_n\right]_n Y$$

$$X = HO-\overset{\overset{\displaystyle O}{\|}}{C}-(CH_2)_n- \qquad Y = -\overset{\overset{\displaystyle O}{\|}}{C}-O-(CH_2)_m-OH$$

The success of this approach helped to establish the macromolecular hypothesis, but initially a molecular weight of 6,000 appeared to be the upper limit for these reactions. At that time it was the consensus of opinion, apparently shared by Carothers, that the reactivity of end group decreased as the chain length got larger, so this result was perhaps not that surprising[*]. But Carothers also realized that because of the reversible nature of the reaction, water could hydrolyze the ester group back to an acid and alcohol, thus breaking up the chains. Carothers and his colleague, Julian Hill, constructed a molecular still that allowed the removal of water and polymerized a so-called 3-16 polyester, from a three carbon chain glycol and a 16 carbon chain diacid ($m = 3$ and $n = 14$ in the equation given above). Hill noticed that the molten polymer could not only initially be "drawn", but further stretched or "cold drawn" after cooling to form strong fibers, an observation of immense subsequent importance in the processing of fibers. The molecular weight of the product was found to be over 12,000, far higher than any previously prepared condensation polymer.

At this point you are probably mystified as to why we are discussing the first polyesters in a section that is supposed to be about nylons. The reason is that these polyesters, although they appeared to have useful mechanical properties (they had an elasticity that at that time could only be matched by silk), melted

---

[*] This view was challenged and defeated by Paul Flory. This is another story, however, and one to which we will return.

below 100°C, were soluble or partially soluble in dry cleaning solvents and were sensitive to water. There is not much of a market for polyester shirts made out of such materials. Accordingly, Carothers and Hill turned their attention to the chemically analogous polyamides, synthesized from diacids and diamines and known to have higher melting points:

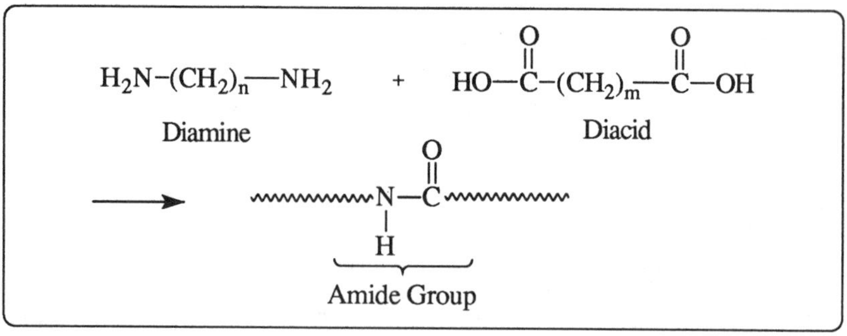

The products were designated by two numbers, the number of carbon atoms in the monomeric diamine and the number in the monomeric dicarboxylic acid, so that the polymer represented schematically above would be called nylon n, m+2. A series of different nylons were produced and initially Carothers decided that nylon 5,10 was the most suitable for forming fibers. The monomers for forming nylon 6,6, hexamethylene diamine and adipic acid, proved to be a more convenient and economic choice, however, and it is this polymer that eventually went into production.

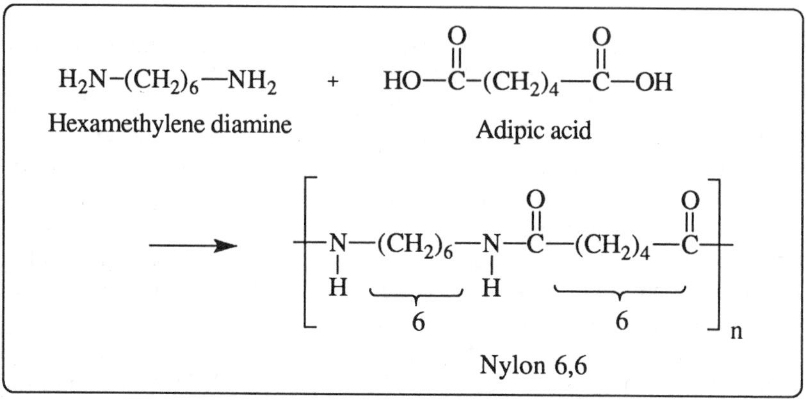

Nylon 6,6

In 1938 at the New York Herald Tribune's Eighth Annual Forum on Current Problems, Charles Stine, by then vice-president for research at DuPont, made the following statement:

> I am making the first announcement of a brand new chemical textile fiber . . . the first man-made organic textile fiber prepared wholly from materials from the mineral kingdom. . . . [Nylon] is the generic name for all materials defined scientifically as synthetic fiber-forming polymeric amides having a protein chemical structure; derivable from coal, air and water, or other

substances; and characterized by extreme toughness and strength and the peculiar property to be formed into fibers and into various shapes, such as bristles and sheets. . . .

Though wholly fabricated from such common raw materials as coal, water and air, nylon can be fashioned into filaments as strong as steel, as fine as a spider's web, yet more elastic than any of the common natural fibers and possessing a beautiful luster. In its physical and chemical properties, it differs radically from all other synthetic fibers.

This had enormous impact. In 1939 nylon stockings were first sold to DuPont employees who were residents of Wilmington, Delaware only. On May 15, 1940 there was sufficient polymer being produced that sales were extended to the whole country. *Four million* pairs of nylon stockings were sold in New York City alone in the first few hours.

The tragedy is that Carothers never lived to see this success. By 1934 he was suffering from increasing bouts of depression and on April 29, 1937 he locked himself in a Philadelphia hotel room and drank lemon juice laced with potassium cyanide, convinced that as a scientist he had failed.

Some Examples of Condensation Polymers—Linear Polymers

Polyamides

The synthesis of nylon 6,6 from adipic acid and hexamethylene diamine has been described above. The synthesis of other polyamides from diacids and diamines (e.g., nylon 6,10) should be obvious. There are three aspects of the synthesis of nylons that should be mentioned here, however. First, if high molecular weight polymer is to be obtained then it is necessary to have precise equivalence of the reactants. The reason for this will become apparent later when we discuss the statistics of linear polycondensation. This stoichiometric equivalence can be achieved by first making a salt from adipic acid and hexamethylene diamine, where the 1:1 complex is precipitated. Polymerization then occurs upon heating to appropriate temperatures and removing water.

$$
\begin{bmatrix}
{}^-OOC-(CH_2)_4-COO^- \\
{}^+NH_3-(CH_2)_6-NH_3{}^+
\end{bmatrix}
$$

Nylon Salt

The second point we wish to make is that nylons can also be made using an acid chloride monomer. In an *interfacial polycondensation* the acid chloride is dissolved in an organic solvent (such as chloroform) and the diamine is dissolved in water. When added to one another the two solutions form separate phases and a polymer skin (polyamide) is formed at the interface.

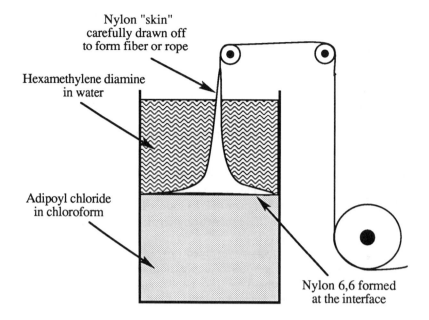

Figure 2.2 *Schematic representation of interfacial polymerization: the "nylon rope trick".*

This is illustrated in figure 2.2. In this case hydrochloric acid (HCl) is eliminated rather than water and a base is normally added to the aqueous phase to neutralize the acid formed.

$$H_2N-(CH_2)_6-NH_2 \quad + \quad Cl-\overset{\overset{O}{\|}}{C}-(CH_2)_4-\overset{\overset{O}{\|}}{C}-Cl$$

Hexamethylene diamine   Adipoyl chloride

$$\longrightarrow \quad H_2N-(CH_2)_6-\underset{\underset{H}{|}}{N}-\overset{\overset{O}{\|}}{C}-(CH_2)_4-\overset{\overset{O}{\|}}{C}-Cl \quad + \quad HCl$$

and so on.

If the nylon is carefully lifted from the interface, new polymer is immediately formed and the material can be removed in the form of a fiber or "rope". This is often demonstrated in the classroom and is referred to as the *nylon rope trick*.

Finally, polyamides such as nylon 6 are not usually synthesized by a condensation reaction, but by a ring-opening polymerization, which we will describe later.

Polyesters

General aspects of the synthesis of polyesters from diacids and dialcohols were mentioned above in our discussion of the work of Carothers. Polyesters can also be made from the acid chloride:

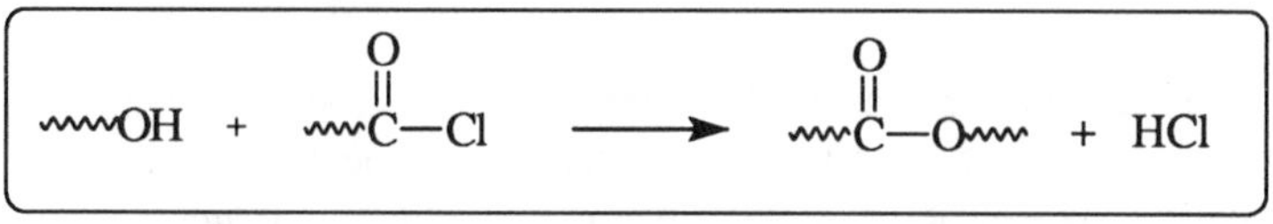

and by *transesterification*, which is used to produce the commercially important polymer, poly(ethylene terephthalate) or PET (if you buy plastic coke bottles or clothing that contains "polyester", it is usually PET).

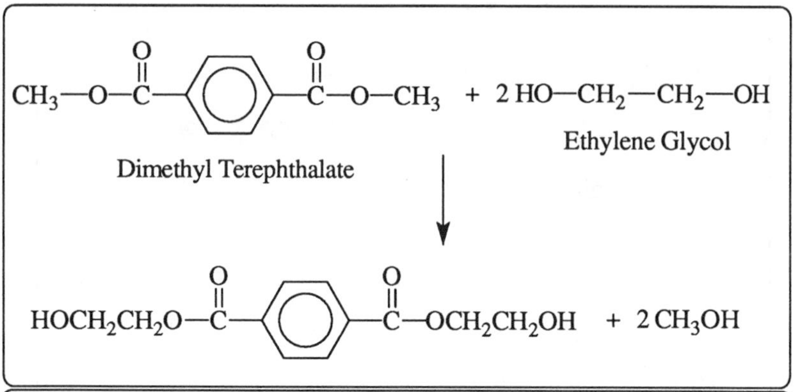

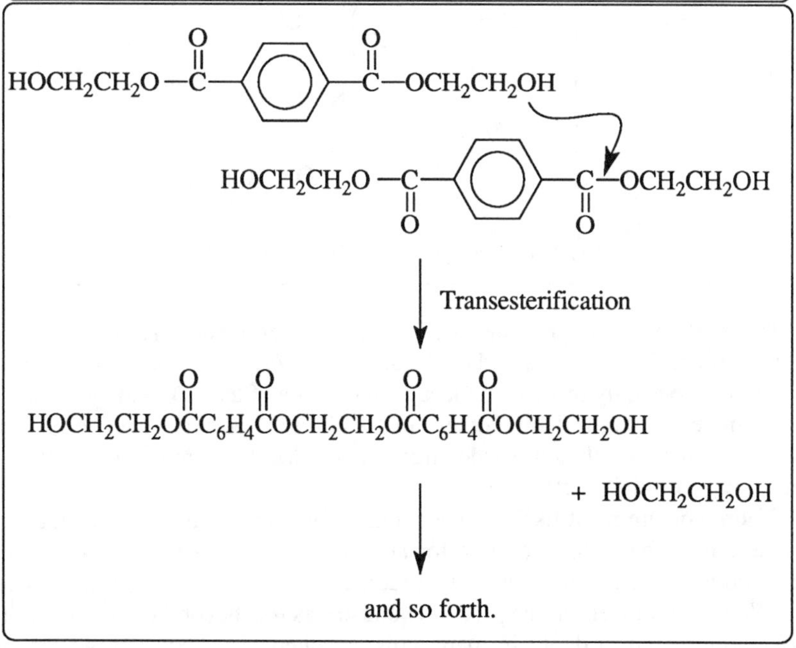

The process of transesterification is (in this case) the reaction of a terminal alcohol group with *any* ester group. The resulting shuffling of the groups is a random process which eventually leads to a most probable distribution of species, as described by Flory (who else).

Polyurethanes

Polyurethanes are synthesized by reaction of a dialcohol with a diisocyanate. A typical example of a polyurethane that would be formed from toluene diisocyanate (TDI—usually a mixture of 2,4- and 2,6- isomers) and hexamethylene diol is shown below:

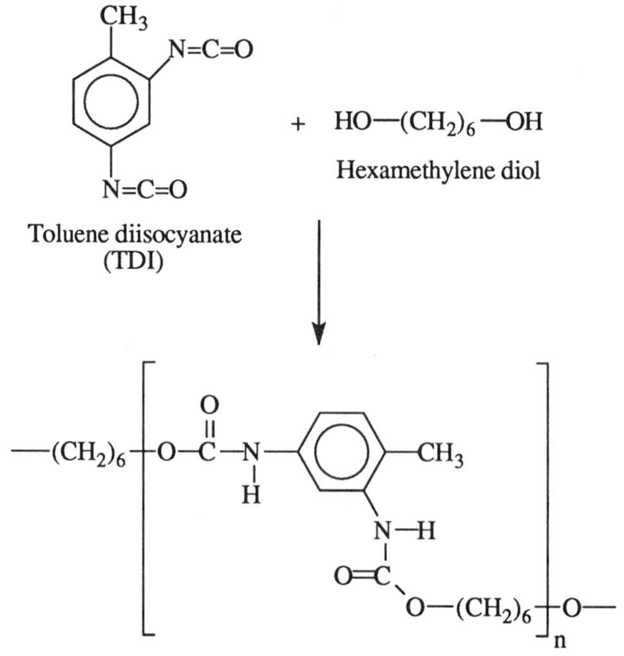

A typical aromatic / aliphatic polyurethane

This particular polyurethane is a glassy polymer at room temperature and might find application in varnishes, lacquers etc. Polyurethanes that are more flexible can be readily made by replacing all or some of the TDI with an aliphatic diisocyanate or by introducing ether linkages into the polymer backbone using hydroxy terminated ethylene oxide oligomers (molecules of moderate molecular weight, typically 1,000 g/mole.)

Where you are most likely to encounter polyurethanes is in the form of the structural hard foams that are used for thermal insulation or the soft foams that are ubiquitous in car seats, cushions, bedding, packaging and lining for clothes etc. We will not dwell on the precise chemistry as it is beyond the scope of this book and covered in depth in many polymer chemistry texts. However, we would be remiss if we did not mention that isocyanates react with water to form CO_2, an innocuous gas which is used as an *in situ* blowing agent in the manufacture of polyurethane foams.

$$R-N=C=O + H_2O \longrightarrow R-NH_2 + CO_2 \uparrow$$

Isocyanate   Amine

By varying the chemical structure, stoichiometry, temperature and amount of water present, a myriad of polyurethane foams with vastly different physical properties can be produced.

Polycarbonates

Polycarbonates are formed by transesterification or by the reaction of diphenols and phosgene. The most commonly encountered polycarbonate, lamentably (for nomenclature purists) usually called just that, polycarbonate, is derived from the diphenol, bisphenol A.

Bisphenol A

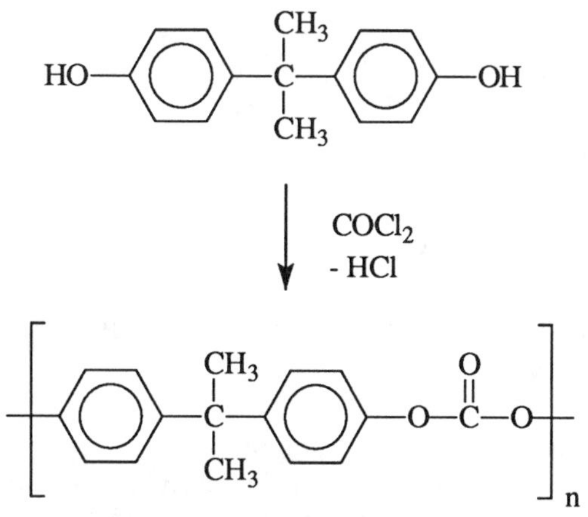

Bisphenol A Polycarbonate

Bisphenol A polycarbonate is an optically clear, glassy polymer, that is often used in the transparent covers seen in street lighting and the now ubiquitous compact discs. It is a very tough material and has found many applications in products that have to withstand mechanical abuse, such as football helmets.

Formation of Networks by Polycondensation Polymers

It should now be self-evident that if bifunctional monomers can be used to make linear polymers, then tri or higher functional monomers can be used to make network or thermosetting polymers. Two commercially important examples of these are phenol-formaldehyde and urea-formaldehyde resins. Reacting phenol with formaldehyde, for example, leads to complex mixtures of linear and branched molecules and finally various types of network structures, depending upon the catalyst used, the relative proportions of reactants etc. A typical so-called novolac prepolymer that may be formed under acid catalysized conditions is shown below:

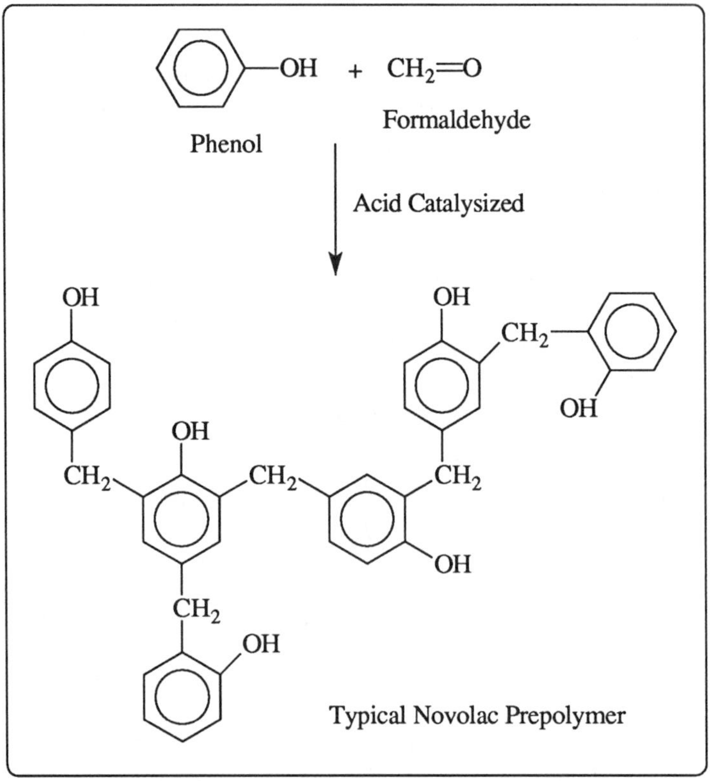

Other examples of classes of polymers that can form thermosets by condensation reactions are polyesters, polyurethanes and epoxy resins, but the details of the chemistry of these reactions is beyond the scope of the material we wish to cover. The important point that should be grasped is the use of multifunctional monomers (where the number of functional groups >2) to make networks, a general process illustrated schematically in Chapter 1 (figure 1.3).

C. CHAIN OR ADDITION POLYMERIZATION

In the preceding sections we have described the key feature of condensation polymerizations: that chains are built step by step, from monomers to oligomers to (eventually) polymers. All of the species present are involved in this process throughout the course of the reaction. In contrast, in addition or chain polymerizations there is an active site at the end of a growing chain where monomers are added sequentially, one by one. In this process only a few species

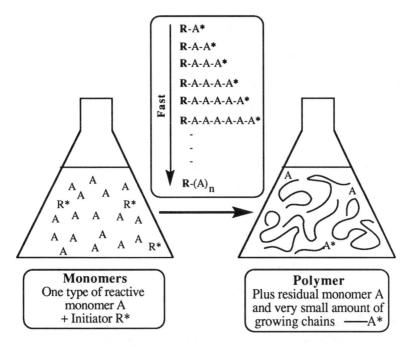

Figure 2.3 *Schematic representation of chain or addition polymerization.*

are active and at any instant of time the distribution of species present usually consists of (i) fully formed and no longer reacting polymer chains, (ii) unreacted monomer and (iii) a very small number (relatively) of growing chains, as depicted schematically in figure 2.3. These chains grow very quickly. The growing end (or sometimes ends) of the chain has an active site that obviously must be shifted to the new end after the addition of a monomer:

$$\text{wwwwwwwwM* } + \text{ M} \longrightarrow \text{ wwwwwwwwwM}-\text{M*}$$

This active site can take a number of different forms, as we will see, allowing the synthesis of a range of polymers and copolymers, as well as a variety of chain architectures or microstructures. In spite of the differences in the nature of the active site, addition polymerizations usually have most (but not always all) of the following features or steps in common:

a) The polymerization must be *initiated*, by which we mean an active site must be generated on a monomer.

b) The chains *propagate*: monomers add to the active site and the active site is simultaneously transferred to the newly added monomer.

c) The polymerization *terminates* due to destruction of the active site.

d) *Chain transfer* can occur. In this process the active site is transferred to another molecule (monomer, solvent, a specially added reagent, another polymer chain, etc.). Sometimes the active site can be transferred to a different part of the same chain. This is not, strictly speaking, "chain transfer", but just ordinary common or garden transfer.

As in step-growth polymerizations, chain polymerizations require monomers with a certain type of chemical structure. These monomers need to have either unsaturated bonds, some examples being:

$$CH_2 = CHX \qquad \text{various olefins}$$
$$CH_2 = CX - CH = CH_2 \qquad \text{various dienes}$$
$$CH \equiv CH \qquad \text{acetylene}$$

or they need to have a ring structure, usually containing at least one hetero-atom (e.g. oxygen). Examples of this type of monomer are:

ethylene oxide

caprolactam

We will discuss the key features of addition or chain polymerization by focusing our initial attention on so-called free radical polymerizations of olefin type monomers. As in our description of condensation polymerization, we will accomplish this by picking a specific polymer that has an interesting history, in this case polyethylene, and working our definitions into the narrative.

Free Radical Polymerization—The Development of Polyethylene

Polyethylene (or polymethylene) was polymerized by the decomposition of diazomethane around the turn of the century,

$$CH_2N_2 \longrightarrow -(CH_2)_n- \ + \ n \, N_2$$
$$\text{Diazomethane} \qquad \text{Polymethylene}$$

but the unstable character of the monomer (it had a propensity to explode) did not make it suitable for industrial production. This work was in any event ahead of its time (the concept of a polymer was not yet established) and it wasn't until 1920, when Staudinger published his paper "On Polymerization" that more serious attention was paid to the formation of polymer molecules by what was then described as an undefined process of "assembly".[*] By 1930 the synthesis of polystyrene, poly(vinyl chloride) (PVC) and poly(methyl methacrylate) (PMMA or Plexiglas®) had been reported. At this point, however, the mechanism of such vinyl polymerizations was still not well understood and it is in this context that the initial work on polyethylene should be placed.

---

[*] Berthelot had actually described polymerizations some sixty years earlier, but Staudinger was apparently unaware of this (see H. Morawetz, *Polymers. The Origins and Growth of a Science*, John Wiley and Sons, New York, 1985).

In 1932 a research program involving chemical reactions under high pressure was initiated in the laboratories of Imperial Chemical Industries (ICI), as a result of the introduction of previously unavailable high pressure equipment. One of many experiments that were tested involved attempts to force ethylene and benzaldehyde to react. On opening the pressure vessel after one such reaction it was observed that the benzaldehyde was apparently unchanged, but a white, waxy solid was found on the walls of the reactor. It was immediately recognized that this was probably some form of polyethylene (this observation involved a good deal of insight and was not a trivial thing), but what was not realized for some time was that a peroxide impurity in the benzaldehyde had probably played a key role. Peroxides are today routinely used to initiate polymerizations, as they readily break down to give peroxy *radicals*:

$$R-O-O-R \longrightarrow 2R-O^{\cdot}$$
Alkyl Peroxide   Peroxy Radical

This radical, with its unshared electron, is highly reactive and can then add a monomer by pairing up with an electron "stolen" from the double bond of a vinyl molecule.

$$R-O^{\cdot} + CH_2{=}CH_2 \longrightarrow R-O-CH_2-CH_2^{\cdot}$$
Initiation

or, replacing the "sticks" representing bonds by dots representing electrons.

$$R{:}O^{\cdot} + CH_2{::}CH_2 \longrightarrow R{:}O{:}CH_2{:}CH_2^{\cdot}$$

where each pair of electrons, :, represents a single covalent bond. This whole process is called *initiation*.

The next step is the *propagation* of the chain, which usually proceeds rapidly:

$$\text{wwwww}CH_2^{\cdot} + CH_2{=}CH_2$$
$$\longrightarrow \text{wwwww}CH_2-CH_2-CH_2^{\cdot}$$
Propagation

This reaction terminates when two radicals meet through the process of ordinary random collisions. We will describe termination later, but it turns out that in the free radical polymerization of ethylene at ordinary pressures the rate of propagation is much smaller than the rate of termination, so chains do not get the chance to form. At high pressures (and higher temperatures), the rate of propagation is much greater, relative to termination, and polymer is produced.

This is getting ahead of ourselves, however, and the point here is that at the time of these first experiments the process of initiation was not well understood*.

As a result, the next set of experiments attempted the polymerization of ethylene alone, resulting in a "violent decomposition" (a chemist's way of saying it blew up) that burst pipes, tubes and gauges. New equipment (and barricades) were built and in 1935 ICI again got lucky (if we define luck as the residue of hard work and intelligent persistence). The new reactor leaked and the presence of small amounts of oxygen were sufficient to initiate polymerization (there had presumably also been trace amounts of oxygen in the ethylene used in the experiment that had such explosive results.) This took months of additional work to establish, however. Nevertheless, production was eventually achieved and the new material immediately found an important use. Because it has an extremely low dielectric loss it was used to insulate radar cables. Sir Robert Watson Watt, the discoverer of radar, pointed out that:

> The availability of polythene [polyethylene] transformed the design, production, installation and maintenance of airborne radar from the almost insoluble to the comfortably manageable. . . . And so polythene played an indispensable part in the long series of victories in the air, on the sea and land, which were made possible by radar.

Up until the early 1950's polyethylene was made by the high pressure process. It was at first thought that the chains were reasonably linear, but in 1940 it was found (by infrared spectroscopy) that there were more methyl groups present than could be accounted for by chain ends alone. We now know that this is due to intramolecular transfer of the radical, or backbiting.

---

* Early polymerizations of methyl rubber, for example, involved placing monomer in big glass bottles and exposing them to sunlight. As a result, U.V. radiation induced formation of a radical, but the process was not easily controlled and it depended on the weather. Sometimes nothing happened for hours, sometimes the vessel exploded!

Because of the relative stability of a six-membered ring transition state, butyl branches can be formed. Ethyl branches may also be formed in a similar backbiting process.

A few long chain branches are also formed, but it is the short chain branches that play the major role in disrupting crystallinity, hence resulting in a material that has a lower melting point and is weaker and softer than that made from "straighter" chains. Branching could be reduced by going to still higher pressures, but it was the advent of organometallic and metal oxide catalysts in the early 1950's that allowed the synthesis of linear or high density polyethylene (at low pressures). We will come back to discuss the types of polymers produced by these catalysts in a while, but it is appropriate to leave history hanging at this point.

Examples of Polymers Synthesized by Free Radical Methods

Although many vinyl, acrylic and diene monomers are polymerized free radically, not all can be and for various reasons it is advantageous to use other methods to obtain certain polymers. This book is aimed at giving a broad overview, so we will neglect a detailed consideration of many specific systems and simply conclude this section with some brief notes on the suitability of certain monomers or types of monomers for polymerization by a free radical process.

Polyolefins

The two most important polyolefins (in terms of commercial production) are polyethylene and polypropylene. As we have seen, the former can be poly-merized free radically at high pressures and elevated temperatures.

$$R^{\bullet} + CH_2{=}CH_2 \longrightarrow R{-}CH_2{-}CH_2^{\bullet} \longrightarrow \text{Polyethylene}$$

Ethylene

$$CH_2{=}CH_2$$

$$R^{\bullet} + CH_2{=}\underset{\underset{CH_3}{|}}{CH} \longrightarrow R{-}CH_2{-}\underset{\underset{CH_3}{|}}{CH}^{\bullet}$$

Propylene

$$CH_2{=}\underset{\underset{CH_3}{|}}{CH}$$

$$R{-}CH_2{-}CH_2{-}CH_3 + CH_2{=}CH{-}CH_2^{\bullet}$$

etc.

Low molecular weight oligomers

Propylene and other olefins with a hydrogen atom on the carbon adjacent to the double bond cannot be polymerized free radically, however, because a hydrogen is transferred to propagating centers, so that only low molecular weight oligomers are produced.

Polystyrene, Polydienes, Polyacrylics etc.

Table 2.1 lists some common monomers that can be polymerized by free radical methods. Industrially significant polymeric materials with diverse physical properties such as polystyrene and poly(methyl methacrylate), which are transparent glasses at ambient temperature, or polybutadiene, polychloroprene, and styrene / butadiene copolymers, which are elastomers, are all made by free radical polymerization. Further examples include the homo- and copolymers produced from alkyl acrylates, alkyl methacrylates, acrylic and methacrylic acids used in the surface coating industry; poly(vinyl acetate) and its derivatives employed as adhesives; poly(vinyl chloride) used as siding, pipes, fixtures and flooring for dwellings and numerous other applications; poly(vinylidine chloride) and poly(vinyl alcohol),* used as barrier films in packaging; and polyacrylonitrile and acrylonitrile copolymers, which are employed as fibers.

Monomers that are not suitable for free radical polymerizations include those with bulky substituents, allyl monomers (because of chain transfer and the stability of the radicals so generated), cyclics and symmetrically substituted monomers such as maleic anhydride. Some of these latter monomers can be copolymerized free radically with other monomers, however (see later).

Ionic and Coordination Polymerization

We have discussed aspects of free radical and condensation polymerizations in some detail and by doing so we have followed the general history of the development of polymer chemistry. Although other methods, such as anionic polymerizations, were actually used in some of the earliest reported polymerizations, the nature of the active site was not well understood until the 1940's. Ionic and coordination polymerizations are now central to the synthesis of many commercially important polymers, so we will describe them here. We will not consider these in as much detail, however, when we go on to consider the kinetics and statistics of polymerization. Our purpose in this book is to establish general principles and many of the procedures and fundamental assumptions, such as the reactivity of the active site being independent of chain length, remain the same in dealing with these other types of polymerizations. We will point out qualitatively where certain important differences occur, however, and their consequences in terms of the characteristics of the polymers produced.

Anionic Polymerization

The year 1956 was a very good one for anionic polymerizations. In quick succession Szwarc announced that this mechanism can result in a "living" poly-

---

* Poly(vinyl alcohol) is not synthesized directly from vinyl alcohol, but from the hydrolysis of poly(vinyl acetate), which is produced by free radical polymerization.

Table 2.1 *Common Monomers that Polymerize Free Radically.*

Monomer	Chemical Structure
Ethylene	$CH_2{=}CH_2$
Tetrafluoroethylene	$CF_2{=}CF_2$
Butadiene	$CH_2{=}CH{-}CH{=}CH_2$
Isoprene	CH_3 $CH_2{=}C{-}CH{=}CH_2$
Chloroprene	Cl $CH_2{=}C{-}CH{=}CH_2$
Styrene	$CH_2{=}CH$ (phenyl ring)
Vinyl chloride	Cl $CH_2{=}CH$
Vinylidine chloride	Cl $CH_2{=}C$ Cl
Vinyl acetate	$OCOCH_3$ $CH_2{=}CH$
Acrylonitrile	CN $CH_2{=}CH$
Acrylic acid	$COOH$ $CH_2{=}CH$
Methyl methacrylate	$COOCH_3$ $CH_2{=}C$ CH_3
Methyl acrylate	$COOCH_3$ $CH_2{=}CH$

mer, one in which there is no termination step, while Stavely and co-workers used an alkyl lithium initiator to polymerize an isoprene containing greater than 90% cis-1,4 units. These developments had immediate consequences in producing almost monodisperse polymers, block copolymers and polymers that are more stereoregular than those obtained by free radical processes.

An anionic polymerization is one where the active site on the end of a growing chain is *negatively charged*:

$$\text{wwwww}CH_2-\overset{\overset{\displaystyle R}{|}}{C}H^{\ominus} \ + \ CH_2=\overset{\overset{\displaystyle R}{|}}{C}H$$

$$\longrightarrow \ \text{wwwww}CH_2-\overset{\overset{\displaystyle R}{|}}{C}H-CH_2-\overset{\overset{\displaystyle R}{|}}{C}H^{\ominus}$$

The charged species is in this case a *carbanion*. Certain types of ring-opening polymerizations can also proceed by this mechanism:

$$\text{wwwww}CH_2-CH_2-O^{\ominus} \ + \ H_2C\overset{\diagdown \diagup}{\underset{O}{-}}CH_2$$

Ethylene oxide

$$\longrightarrow \ \text{wwwww}CH_2-CH_2-O-CH_2-CH_2-O^{\ominus}$$

The active site is in this case an *oxanion*.

Clearly, there must be a counterion hanging around somewhere and it turns out that the separation of the end group and the counterion is a key factor in determining the stereochemistry of the propagation reaction. This depends not only upon the nature of the anion and counterion involved, but also on the solvent. The nature of the solvent is also crucial in terms of creating the conditions necessary for continuous growth and we have to distinguish between "protic" solvents, which are those that can donate a proton to the growing chain (i.e. chain transfer to solvent), and those that do not, which are "aprotic". An example of the former is the polymerization of styrene in liquid ammonia using sodium amide as an initiator. As in free radical polymerization there are initiation and propagation steps:

$$NaNH_2 \ \rightleftharpoons \ Na^{\oplus} + NH_2^{\ominus}$$

$$NH_2^{\ominus} + CH_2=\underset{\underset{\bigcirc}{|}}{CH} \ \longrightarrow \ H_2N-CH_2-\underset{\underset{\bigcirc}{|}}{C}H^{\ominus}$$

and:

$$M_x^{\ominus} + M \longrightarrow M_{x+1}^{\ominus}$$

In these reactions there is no termination of the type that occurs when free radicals collide. In "protic" solvents, however, chain growth can be stopped by transfer to solvent.

$$M_x^{\ominus} + NH_3 \longrightarrow M_xH + NH_2^{\ominus}$$

As a result, the number average degree of polymerization will depend upon the ratio of the rate of propagation relative to the rate of transfer.

If an inert solvent is used and there are no contaminants (e.g., those containing an "active" hydrogen), then it is possible to obtain a system where carbanion end groups are always present, because of the absence of termination reactions. As an example we can consider a polymerization initiated by metallic sodium:

$$CH_2{=}CH + Na \longrightarrow {}^{\cdot}CH_2{-}CH^{\ominus} \ Na^{\oplus}$$
$$\underset{R}{|} \qquad\qquad\qquad \underset{R}{|}$$

Anion radical

this gives an anion-radical that quickly forms a dimer

$$2 \ {}^{\cdot}CH_2{-}CH^{\ominus} \ Na^{\oplus} \longrightarrow$$
$$\underset{R}{|}$$

$$Na^{\oplus} \ {}^{\ominus}CH{-}CH_2{-}CH_2{-}CH^{\ominus} \ Na^{\oplus}$$
$$\qquad\quad \underset{R}{|} \qquad\qquad\qquad \underset{R}{|}$$

Dianion

which can then propagate from both ends.

One immediate consequence of the "living" nature of these polymerizations is that it allows the synthesis of block copolymers. Because the active site stays "alive", it is possible to first polymerize styrene, for example, until this monomer is effectively used up, then restart the polymerization using, say, butadiene. A third block of styrene units can subsequently be added on. Such triblock copolymers can be used as thermoplastic elastomers. The morphologies of these block copolymers are both intriguing and beautiful and are something you should look forward to studying in more advanced treatments. In practice, not all of the monomers that polymerize anionically can be used to make block copolymers and those students interested in this subject should consult the books listed at the end of this chapter that deal with polymer chemistry in more detail.

Certain initiators are practically completely dissociated before the polymerization has time to start, so that all active centers are effectively introduced at the same time, at the start of the reaction. If all of these active centers are equally susceptible to monomer addition throughout the polymerization (i.e., if the reaction medium is kept well-stirred so that there are no large concentration fluctuations), then the result is a polymer with an extremely narrow molecular weight distribution. (Any type of polymerization where initiation occurs entirely before propagation, and termination is absent, gives this result. For the most part it is just easier to achieve these conditions with an anionic polymerization.) We won't go into the details of this, but the result is a Poisson distribution where the ratio of the weight to number average degree of polymerization is given by:

$$\frac{\overline{N_w}}{\overline{N_x}} = 1 + \frac{\upsilon}{(\upsilon + 1)^2} \qquad (2.1)$$

where υ is the number of monomers reacted per initiator. Clearly as υ gets larger the polydispersity approaches 1.

Finally, the types of vinyl monomers that can be polymerized anionically are usually those where there are electron withdrawing substituent groups, which act to stabilize the carbanion. Typical examples are given in table 2.2.

Table 2.2 Common Monomers that Polymerize Anionically.

Monomer	Chemical Structure
Butadiene	$CH_2{=}CH{-}CH{=}CH_2$
Styrene	$CH_2{=}CH$
Acrylonitrile	$\overset{\displaystyle CN}{\underset{\displaystyle CH_2{=}CH}{\vert}}$
Methyl methacrylate	$\overset{\displaystyle COOCH_3}{\underset{\displaystyle \underset{CH_3}{\overset{\vert}{C}}}{CH_2{=}\overset{\vert}{}}}$
Ethylene oxide	$H_2C{-}CH_2$ with O bridging
Tetrahydrofuran	
Caprolactam	

Cationic Polymerization

Cationic polymerizations are those where the active site has a *positive* charge (i.e., a carbonium ion):

$$\text{wwww}CH_2-\overset{\overset{\displaystyle R}{|}}{C}H^{\oplus} + CH_2=\overset{\overset{\displaystyle R}{|}}{C}H$$

$$\longrightarrow \text{wwww}CH_2-\overset{\overset{\displaystyle R}{|}}{C}H-CH_2-\overset{\overset{\displaystyle R}{|}}{C}H^{\oplus}$$

and, accordingly, those monomers where R is an electron donating group are most appropriate for these types of polymerizations (e.g., isobutylene, alkyl vinyl ethers, vinyl acetals, para-substituted styrenes). The distinguishing feature of cationic polymerization is that there is a tendency for the growing chains to deactivate, because of various side reactions that can occur, although these can be suppressed by lowering the temperature. Initiation can be achieved using protonic acids and Lewis acids, with the latter apparently requiring a co-catalyst such as water or methanol:

$$H^{\oplus}A^{\ominus} + CH_2=\overset{\overset{\displaystyle R}{|}}{C}H \longrightarrow CH_3-\overset{\overset{\displaystyle R}{|}}{C}H^{\oplus} + A^{\ominus}$$

Propagation occurs in the usual fashion:

$$\text{wwww}CH_2-\overset{\overset{\displaystyle R}{|}}{C}H^{\oplus} A^{\ominus} + CH_2=\overset{\overset{\displaystyle R}{|}}{C}H$$

$$\longrightarrow \text{wwww}CH_2-\overset{\overset{\displaystyle R}{|}}{C}H-CH_2-\overset{\overset{\displaystyle R}{|}}{C}H^{\oplus} A^{\ominus}$$

Unlike anionic polymerizations, termination can occur by anion-cation recombination, as in:

$$\text{wwww}CH_2-\overset{\overset{\displaystyle R}{|}}{C}H^{\oplus} + CF_3COO^{\ominus}$$

$$\longrightarrow \text{wwww}CH_2-\overset{\overset{\displaystyle R}{|}}{C}H-O-\overset{\overset{\displaystyle O}{||}}{C}-CF_3$$

where an ester group forms. Termination can also occur by anion splitting:

$$\text{wwwwCH}_2\!-\!\overset{\overset{\displaystyle R}{|}}{C}H^{\oplus}\!+\ BF_3OH^{\ominus}$$

$$\longrightarrow\ \text{wwwwCH}_2\!-\!\overset{\overset{\displaystyle R}{|}}{C}H\!-\!OH\ +\ BF_3$$

or by reaction with trace amounts of water

$$\text{wwwwCH}_2\!-\!\overset{\overset{\displaystyle R}{|}}{C}H^{\oplus}\ A^{\ominus}\ +\ H_2O$$

$$\longrightarrow\ \text{wwwwCH}_2\!-\!\overset{\overset{\displaystyle R}{|}}{C}H\!-\!CH_2\!-\!OH\ +\ AH$$

Chain transfer to monomer can also occur by various mechanisms including:

$$\text{wwwwCH}_2\!-\!\overset{\overset{\displaystyle R}{|}}{C}H^{\oplus}\ A^{\ominus}+\ CH_2\!=\!\overset{\overset{\displaystyle R}{|}}{C}H$$

$$\longrightarrow\ \text{wwwwCH}\!=\!\overset{\overset{\displaystyle R}{|}}{C}H\ +\ CH_3\!-\!\overset{\overset{\displaystyle R}{|}}{C}H^{\oplus}\ A^{\ominus}$$

Finally, as mentioned above the types of vinyl monomers that can be polymerized cationically are usually those where there are electron donating substituent groups, which act to stabilize the cation. Typical examples are given in table 2.3.

Table 2.3 Common Monomers that Polymerize Cationically.

Monomer	Chemical Structure
Isobutylene	$CH_2\!=\!C$ with CH_3 above and CH_3 below
Vinyl methyl ether	$CH_2\!=\!CH$ with OCH_3
Styrene	$CH_2\!=\!CH$ with phenyl ring

Coordination   Polymerization

In tracing the history of the development of polyethylene we left the story at a most interesting point. Up until 1950 polyethylene could only be produced by a high pressure process. Then, in the course of a two to three year period, the use of various catalysts to produce high density polyethylene was reported by groups at Standard Oil of Indiana and Phillips Petroleum in the United States and a group directed by Ziegler in Germany. The details of these discoveries make for fascinating reading, but it is clear that Ziegler immediately recognized that the band corresponding to branches that is usually observed in the infrared spectrum of high pressure polyethylene could not be detected in his catalytically prepared material. The greater strength and higher melting point of this polymer also led Ziegler to the conclusion that he had produced "a really new type of a really straight-chain high molecular weight polyethylene". Ziegler usually drafted his own license agreements and patents and in this instance made what the benefit of hindsight clearly shows to be a major error, by writing the patent so narrowly that it only covered polyethylene. At that time he did not believe that other olefins could be polymerized with his catalyst.

In 1952 Giulio Natta had heard a lecture by Ziegler on the polymerization of ethylene and on his return to Milano had started his own program along these lines. In 1954, using what he called Ziegler catalysts, Natta's group polymerized propylene to give a product that could be fractionated into rubbery and crystalline components. The latter was immediately identified, and the fundamental characteristics of its structure established, because of the expertise of Natta's group in x-ray diffraction methods. By this time Ziegler's group had also succeeded in polymerizing propylene, but because of various delays their patent applications were ten days behind the Italians. Ziegler and Natta had been friends and collaborators, but that ended in disagreement and recrimination. This story is reconstructed in "*The Chain Straighteners*"*, for those interested in how these things can occur. They were jointly awarded the 1963 Nobel Prize in Chemistry. A number of other groups in the U.S. had also polymerized propylene with various catalysts, either by design or inadvertently. This resulted in patent litigation that has been amongst the longest and most complex in U.S. history, but Hogan and Banks of Phillips Petroleum were finally granted the patent for the production of crystalline polypropylene in 1983.

As we have noted earlier, polypropylene cannot be polymerized free radically. The Ziegler catalysts are heterogeneous and actually produce a mixture of atactic and isotactic polypropylene (from different sites on the catalyst), which was demonstrated by Natta's systematic study of the structure of this and other stereoregular polymers synthesized during this period. The nature of these catalysts appears to be still a matter of debate, but the mechanism is believed to involve complexes formed between a transition metal and the π electrons of the monomer. This *coordination* and orientation of the monomer relative to the growing chain end allows the insertion of the former in a specific manner, illustrated schematically below:

---

* F. M. McMillan, *The Chain Straighteners*, McMillan Press, London, 1979.

$$
\boxed{
\begin{array}{c}
 \qquad \text{R} \\[-2pt]
 \qquad | \\[-2pt]
\text{\small\textasciitilde\textasciitilde\textasciitilde\textasciitilde}\text{CH}_2\!-\!\text{CH} \ \text{\tiny''''''} \ \text{Catalyst} \\[2pt]
\vdots \qquad \text{\small/} \\[4pt]
\text{CH}_2\!=\!\text{CH} \\[-2pt]
| \\[-2pt]
\text{R}
\end{array}
}
$$

It is beyond the scope of this book to go into the chemistry of these systems in more detail. Common monomers that can be polymerized using Ziegler-Natta catalysts are listed in table 2.4. All are hydrocarbons and do not contain strong polar groups. The crucial point is how the development of these catalysts allowed the synthesis of high density polyethylene, linear low density polyethylene, stereoregular polyolefins (polypropylene being the most important), diene polymers and ethylene/propylene/diene (EPDM) elastomers.

One final comment before we close this section. The cost of research, development, scale up and introduction of a new polymer into the marketplace is enormous and it has been argued by many that essentially all the large scale economically viable (i.e. cheap) polymers have now been discovered. This, in large part, is why the commercial and academic interest in polymer blends is at an all time high and unlikely to abate in the foreseeable future. It follows logically that if desired properties can be attained by simply mixing two or more existing commercial polymers, surely this will be more cost effective than developing a new single polymeric material from scratch. While this is a persuasive argument, those of us who are approaching senior citizen status well remember similar arguments being expressed immediately prior to the discovery of Ziegler-Natta and metal oxide catalysts, which resulted in the introduction of isotactic polypropylene (PP), high density polyethylene (HDPE), ethylene-co-propylene elastomers (EPDM), etc.! And there appears to be no limit to the ingenuity of polymer chemists. Recently introduced metallocene and other catalysts promise "designer" homo- and copolymers, where desired structural properties of the polymeric material are, in principle, "dialed in". This part of the subject is still very much alive and we eagerly anticipate developments that in some ways will no doubt surprize us.

D. POLYMERIZATION PROCESSES

In the preceding sections we have described the principal synthetic mechanisms used in the synthesis of polymers. Making these polymers in the laboratory can usually be carried out by straightforward methods using the established art and craft of the organic chemist. At this scale, polymerizations need not be driven to high conversions and subsequent separations etc., are not usually a big problem (you just tell your graduate student or postdoc to get on with it and try not to blow up the lab!). At the industrial scale, however, there are all sorts of difficulties. The viscosity of the polymerization mass can become unmanageably large, heat transfer (practically all polymerizations are exothermic) is often a problem, the kinetics of polymerization may effectively preclude one

Table 2.4 *Common Monomers that Polymerize Using*
Ziegler-Natta Type Catalysts.

Monomer	Chemical Structure
Ethylene	$CH_2{=}CH_2$
Propylene	CH_3 $\vert$ $CH_2{=}CH$
1-Butene	C_2H_5 $\vert$ $CH_2{=}CH$
Butadiene	$CH_2{=}CH{-}CH{=}CH_2$
Isoprene	CH_3 $\vert$ $CH_2{=}C{-}CH{=}CH_2$
Styrene	$CH_2{=}CH$ $\vert$ ⬡

polymerization method, but allow a different one, and so on. In addition, the process used to polymerize a monomer can often have a profound effect on factors such as molecular weight and microstructure. It is therefore important to have at least a passing familiarity with polymerization processes, even at this introductory level, so we will briefly describe the main features of methods used in large scale polymerizations.

The first distinction that is usually made by engineers is between *batch* and *continuous* processes. The latter is preferred for large scale production, but cannot be applied to every type of polymerization. As we will see in the following chapter, step-growth polymerizations are generally slow* (hours), so that continuous reactors with inordinately long residence times would be required. For such polymerizations batch processes are employed.

The next distinction that is generally made is between *single phase* processes and *multiphase* processes. Single phase processes include *bulk* (or mass), *melt* and *solution* polymerizations. In a bulk polymerization, as the name suggests, only reactants (and added catalyst, if necessary) are included in the reaction vessel. This type of process is widely used for step-growth polymerizations, because as we will show in the following chapters, high molecular weight polymer is only produced in the last stages of the polymerization, so that the viscosity of the reaction medium stays low throughout most of the course of the

---

* Certain polyurethanes, such as those used in reaction injection molding (RIM), are formed very rapidly by step-growth polymerization and are exceptions to this general rule of thumb.

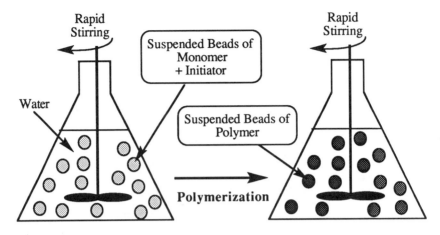

Figure 2.4 *Schematic representation of suspension polymerization.*

reaction. To prevent crystallization (hence the appearance of a solid phase in the reaction vessel) bulk polymerizations are carried out at temperatures higher than the melting point of the polymer (e.g., ≈ 250°C for nylon 6,6) and are often called melt polymerizations.

Chain polymerizations are less often performed in the bulk, because of problems with the control of the reaction*. There is a tendency for the reaction mass to form a gel (i.e., have an extraodinarily high viscosity) and "hot spots" can develop. At the extreme, the reaction rate can accelerate to "runaway" proportions (for reasons we will discuss when we consider kinetics) with potentially disastrous (explosive) consequences. Viscosity and heat control can be achieved, if necessary, by carrying out the polymerizations to a relatively low conversion, with the unreacted monomer being separated and recycled.

Another way to control the viscosity and heat transfer problems of chain polymerizations is to perform the polymerization in solution. A major concern with this method is that "chain transfer" to solvent can occur (i.e., the solvent can take part in the reaction (see next chapter). In addition, the problem of removing and recycling the solvent is introduced, which is expensive and not a trivial factor in today's enviromentally conscious society. Even trace amounts of potentially toxic or carcinogenic solvents can pose a major predicament.

Obviously, one solvent, ideal from many points of view, is water. Nobody cares about trace amounts of water in polymer films that might be used to wrap food, for example. Trace amounts of benzene (a grade A carcinogen) would be unacceptable, however. The problem is, of course, that most polymers (or monomers for that matter) do not dissolve in water. This brings us to the topic of mutiphase processes where polymerizations are performed in water with the

---

* An interesting exception, nonetheless, is poly(methyl methacrylate), a polymer that is soluble in its own monomer (not all polymers are), and which is synthesized commercially by chain (free radical) polymerization very slowly in bulk. The resulting polymer has outstanding optical properties (clarity) because there are very few impurities.

monomer or polymer suspended in the form of droplets or dispersed in the form of an emulsion.

With the exception of a few polar molecules, such as acrylic acid, ethylene oxide, vinyl pyrrolidone and the like, most monomers are largely insoluble* in water. A mixture of the two will phase separate into layers, usually a less dense "oily" layer of monomer sitting on top of the water layer. Polymerization in this state would obviously not have much of an advantage over bulk polymerization. Continuous rapid stirring can produce smaller globs or spherical beads of monomer suspended in the water, however, and each bead then becomes a miniature reaction vessel, as depicted schematically in figure 2.4. The initiator used must also be essentially water insoluble so that it prefers to reside in the monomer phase where it can initiate polymerization. Problems of heat transfer and viscosity are overcome by this method. In addition, the final product is in the form of "beads", which is convenient for subsequent processing. There are difficulties involving coalescence of particles, however, so a variety of additives (protective colloids etc.) are used to stabilize the droplets. The beads or monomer droplets in such a *suspension* polymerization (sometimes also called bead or pearl polymerization) are usually about 5 μm in diameter and require the mechanical energy of stirring to maintain their integrity. If the stirring is stopped, a gross phase separation into two layers occurs.

There is a way to suspend even smaller monomer particles in water such that the monomer droplets are stable and do not aggregate to form a separate layer. Essentially, a surfactant (soap) is used to form an emulsion. Surfactant molecules consist of a polar head (hydrophilic) group attached to a non-polar (hydrophobic) tail, such that it looks something like a tadpole, as depicted schematically in figure 2.5.

In water, soap molecules arrange themselves so as to keep the polar groups in contact with water molecules, but the non-polar tails as far from the water as possible (at concentrations above a certain level, called the critical micelle concentration). One way of doing this is to form a *micelle*, which usually has a spherical or rod-like shape, as also illustrated in figure 2.5. In this micelle, which is about 10^{-3}-10^{-4} μm in diameter, the polar groups are on the outside surface while the non-polar tails are hidden away inside. The non-polar groups are compatible with non-polar monomer, however, so that if monomer is now added to the water and dispersed by stirring, the very small amounts of monomer that dissolve in the water can diffuse to the micelles and enter the interior, non-polar hydrocarbon part (then more monomer enters the aqueous phase to replace that which has departed). In the same way, surfactants molecules can diffuse to the dispersed monomer droplets (whose size depends upon the stirring rate, but is usually in the range of 1-10 μm), where they are absorbed onto their surface, thus stabilizing them, again as illustrated in figure 2.5.

Unlike suspension polymerization, where a water *insoluble* initiator is used, in *emulsion* polymerization a water *soluble* initiator is added. The polymerization, for the most part, occurs in the swollen micelles, which can be thought

---

* It is important to emphasize the word *largely*, because even when you see a grossly phase separated system, like oil and water, a small amount of oil has actually dissolved in the water and *vice versa*. This will be important when we go on to consider emulsion polymerization.

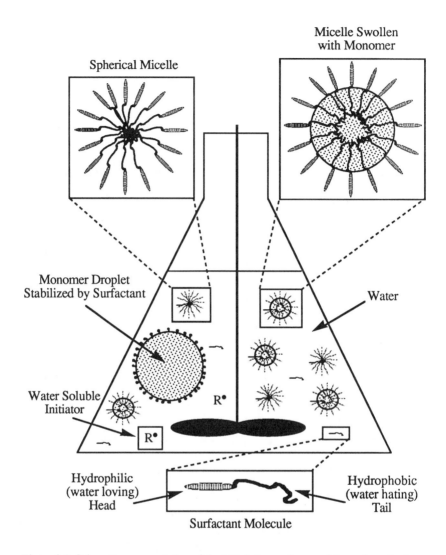

Spherical Micelle

Micelle Swollen
with Monomer

Monomer Droplet
Stabilized by Surfactant

Water

Water Soluble
Initiator

R•

R•

Hydrophilic
(water loving)
Head

Hydrophobic
(water hating)
Tail

Surfactant Molecule

Figure 2.5 *Schematic representation of the initial stages of an emulsion polymerization.*

of as a meeting place for the water soluble initiator and the (largely) water insoluble monomer*. As the polymerization proceeds, the micelles grow by addition of new monomer diffusing in from the aqueous phase. At the same time, the size of the monomer droplets shrinks, as monomer diffuses out into the aqueous phase. The micelles where the polymerization is occuring grow to about 0.5 μm in diameter and at this stage are called polymer particles. After a while ($\approx$ 15% conversion of monomer to polymer) all the micelles become polymer

---

* A small amount of polymerization sometimes occurs in the monomer droplets and almost certainly in solution, but the latter does not contribute significantly because of the low monomer concentration in the water phase.

particles and a while later (40-60% monomer conversion) the monomer droplet phase finally disappears. It is all quite remarkable.

Termination occurs when a radical (usually from the initiator) diffuses in from the aqueous phase. This is a major advantage of the emulsion technique for the polymerization of monomers such as butadiene, which cannot be polymerized easily by free radical means using homogeneous (single phase) polymerization, because it has a fairly high rate of termination. The termination step in an emulsion polymerization is controlled by the rate of arrival of radicals at the polymer particles, which depends only on the concentration of initiator and the concentration of surfactant (hence initial micelle concentration).

There are some obvious additional advantages to emulsion polymerization; as in suspension polymerization, the viscosity is always low and heat control is relatively straightforward. Also, the final product is an emulsion of $\approx 0.1 \ \mu m$ diameter polymer particles that is typically 50% by volume polymer and 50% water. This makes it almost immediately applicable as a surface coating (paint). The major disadvantage is the presence of the surfactant, which is difficult to completely remove even if the polymer product is precipitated and washed.

There are various types of multiphase processes that are widely used in the mass production of polymers. The two phases can both be liquids, as in suspension and emulsion polymerization, or can be a gas/solid, gas/melt (liquid) or liquid/solid system. In the *interfacial polymerization* of nylon 6,6, for example, the two monomers are initially dissolved in different solvents, hexamethylene diamine in water and adipoyl chloride in chloroform, as we mentioned previously on page 36. In this system, the aqueous phase sits on the top of the chloroform solution and a solid polymer is formed at the interface. Polymerizations of an initial miscible mixture of, say, monomer + initiator or monomer + solvent + initiator also become multiphase if the polymer formed is insoluble in its own monomer (e.g. poly(vinyl chloride) or polyacrylonitrile), or the monomer solvent mixture (e.g. the cationic polymerization of isobutylene in methylene chloride).

Finally, multiphase processes where the monomer is a gas are important methods for polymerizing monomers such as ethylene and propylene. Some low density polyethylenes, as well as several well-known ethylene copolymers such as ethylene-co-vinyl acetate copolymers, are still made by high pressure free radical polymerization methods. The process is usually continuous, with compressed ethylene, for example, at about 200°C flowing through a heavy walled tubular reactor. Initiator is injected into the ethylene stream as it enters the vessel. About 15% of the entering gas is converted to molten polymer which is carried out in the gas stream (there is often a polymer buildup on the lower surface of the reactor which must be periodically removed by a process called "blowing down"). These days, however, polymers like high density polyethylene are often produced in a gas/solid multiphase process. Catalyst can be suspended in an upward flowing stream of monomer gas in a fluidized bed reactor, for example, and crystalline polyethylene grows on the catalyst surface. Polyethylene and polypropylene are also polymerized by forming a slurry of monomer, insoluble catalyst and a solvent and again the polymer "grows" on the catalyst surface. But this is enough, you should have the general picture by now. There is a range of industrially important methods for synthesizing polymers and if your interest is in the engineering side of producing these materials you will have to study the methods we have mentioned (and others) in much greater depth.

E. STUDY QUESTIONS

1. Write down a typical chemical structure for the following polymers together with the starting materials (monomers) involved in their synthesis:
   (a) Polyisobutylene
   (b) Poly(ethylene terephthalate)
   (c) Nylon 6,12
   (d) Polyacrylonitrile
   (e) Poly(ethylene oxide)
   (f) Poly(vinyl alcohol)

2. Give a typical example of a commercial application for each of the six polymers listed above in question 1.

3. What are the major differences between step growth and chain (or addition) polymerization ?

4. Nylon 6 and nylon 6,6 are two polyamides that are produced commercially. How are they synthesized and how do their chemical repeat units differ ?

5. What is transesterification and why is it important in the production of high molecular weight poly(ethylene terephthalate) ?

6. Describe the chemistry involved in the formation of a urethane group and briefly explain how flexible polyurethane foams are prepared.

7. Describe the four primary features that characterize most addition polymerizations.

8. Under conditions of high pressure and temperature and in the presence of a free radical initiator ethylene polymerizes to produce a high molecular weight polymer which is referred to as low density polyethylene (LDPE).
   (a) Why is LDPE formed and what chemical reactions are responsible for the formations of this particular polymer ?
   (b) If a 50:50 mixture of ethylene and propylene is substituted for ethylene under the same reaction conditions no high molecular weight polymers are formed. Why do you think this is so ?
   (c) High density polyethylene (HDPE) also cannot be produced by the free radical reaction conditions mentioned above. How is HDPE synthesized and how does it differ in terms of polymer chain structure from that of LDPE ?

9. Describe the unique feature that distinguishes "living" polymerization from that of ordinary addition polymerization and show how the former can be used to synthesize styrene-butadiene-styrene block copolymers.

10. Compare and contrast suspension and emulsion polymerization processes that have been used to produce polystyrene beads of very different dimensions.

F. SUGGESTIONS FOR FURTHER READING

(1)   K. J. Saunders, *Organic Polymer Chemistry*,
       Chapman and Hall, London, 1973.

(2)   G. Odian, *Principles of Polymerization*,
       3rd. Edition, Wiley, New York, 1991.

(3)   H. R. Allcock and F. W. Lampe, *Contemporary Polymer Chemistry*,
       Prentice Hall, New Jersey, 1981.

(4)   P. Rempp and E. W. Merrill, *Polymer Synthesis*,
       Heuthig and Wepf, Basel, 1986.

CHAPTER 3

Kinetics of Step-Growth and Addition Polymerization

"For some cry 'Quick' and some cry 'Slow'
But, while the hills remain,
Up hill 'Too-slow' will need the whip,
'Down hill 'Too quick', the chain."
—Alfred, Lord Tennyson

A. INTRODUCTION

Thermodynamics can tell us whether or not a process or reaction is capable of occurring, but nothing about its rate, which is often all important. For example, ordinary sugar (sucrose) can react with oxygen in the atmosphere with a free energy change of about - 5.7 kJ per mole. This is the release of a lot of energy, but we have never seen the sugar bowl lying around on our kitchen table suddenly burst into flames*, because the rate at which this reaction occurs is so slow. In chemical synthesis a particular product might be thermodynamically preferred (i.e., have a lower free energy), but be kinetically unfavorable, in that a different product is formed at a much faster rate. We will find in our studies of polymers that such considerations apply not only to chemical synthesis, but also to physical processes such as crystallization (see Chapter 8).

The application of chemical kinetics is crucial, not only to an understanding of the fundamentals of polymerization, but also to its practical implementation by process engineers. First, it provides information on how long a reaction takes. We will see that step-growth polymerizations generally require long times, but some addition polymerizations occur so rapidly that they could be considered explosive. The relative rates at which different reactions occur (e.g., the rate at which chains are terminated relative to the rate at which they propagate), can have a profound effect on molecular weight and microstructure (e.g., in copolymerization, where the composition and sequence distributions of the copolymer will depend upon the relative rates at which the monomers add to the growing chain end). In this chapter we will discuss the kinetics of step-growth (condensation) and free radical polymerizations, as this will serve to illustrate the general principles, leaving the kinetics of other addition polymerizations to more advanced treatments. Copolymerization will be treated in a separate chapter.

We will assume that you have a basic familiarity with the fundamentals of

---

* We stole this analogy from D. Eisenburg and D. Crothers, *Physical Chemistry with Applications to the Life Sciences*, Benjamin/Cummings Publishing Co., Menlo Park, CA, 1979.

kinetics, essentially the relating of reaction rates (i.e. the rate of disappearance of reactants or appearance of products) to the concentration of the various species present and a *rate constant*. This is the approach we will take in this chapter, but there is another way of looking at some of the same problems through the use of probability theory. We will consider this latter approach separately in the following chapter, but it is interesting to observe here that the first of many major contributions made by Paul Flory to polymer physical chemistry (which would culminate in the 1974 Nobel Prize for Chemistry) was the theoretical determination of the distribution of the molecular weights of chains formed in condensation polymerizations using probability arguments. Flory had been hired by DuPont in 1934 and assigned to Carothers' group. Carothers had mentioned that he believed that mathematics could be used extensively in the study of polymers and as early as 1935 he presented some of Flory's initial work on molecular weight distributions at a Faraday Society meeting. In deriving his results Flory made the fundamental assumption that the reaction rate constants of the functional groups are independent of the length of the chain to which they are attached. As we have already mentioned, at that time many found this assumption implausible, but Flory subsequently proved his point with a detailed study of the kinetics of polyesterification. We will discuss these latter results first, reversing the flow of historical events.

B. THE KINETICS OF STEP-GROWTH POLYMERIZATION

The assumption that the reaction rate is independent of chain length can, at first, be confusing. First, it is not exactly true, in that for very small molecules the rate can be appreciably different. The rate constant quickly approaches an asymptotic value as chain length increases, however, as in the example of the reaction of ethanol with various carboxylic acids, shown in figure 3.1. Even with these experimental results, many students still feel that there is something not quite right about Flory's assumption or that it is counter-intuitive. The problem relates to how we define the concentration terms in the usual rate equations. Clearly, the overall rate of reaction of acid groups with alcohol groups (in the example given above) will decrease as the length of the alkyl group attached to the acid functionality increases, as there are fewer reacting groups per unit volume.

Flory's assumption[*] is that the inherent reaction rate per functional group is independent of chain length in a situation where there are *equal numbers of reacting groups per unit volume*. In other words, we could take an acid with N=3 (i.e., $3CH_2$ groups) and react it with an alcohol in the presence of a given amount of an inert (non-reacting) diluent, say hexane (assuming all these molecules mix with one another). The amount of diluent would be chosen so that the concentration of functional groups (acids and alcohols) present in this first mixture would be exactly the same as in a second mixture, where the acid chain length is longer, say N=7. Flory's assumption means that the rate of reaction (esterification) in these two mixtures would be the same.

---

[*] P. J. Flory, *Principles of Polymer Chemistry*, Cornell University Press, 1953.

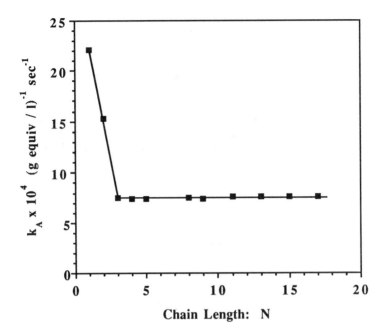

Figure 3.1 *Plot of esterification rate constant, k_A, vs chain length, N for CH_3CH_2OH + $H(CH_2)_NCOOH$. Redrawn from the data of P. J. Flory,* Principles of Polymer Chemistry, *Cornell University Press, 1953, p. 71.*

It is important to note that if we now go on to consider polymerizations, this assumption *does not mean that the overall reaction rate remains constant throughout the course of chain growth.* Clearly, after the reaction has proceeded a while there is a smaller number of unreacted functional groups per unit volume. As we will see, these groups are attached to chains of different length. The meaning of the Flory assumption is that in this reaction mass the probability of reaction of a group attached to a dimer is exactly the same as that of a group attached to a 20-mer, and this probability depends only upon the number of groups remaining, which are assumed to be randomly distributed in the mixture.

There are other aspects to the Flory assumption that should be considered, but we will neglect them here in favor of proceeding to an experimental demonstration of the validity of this assumption and refer the interested student to Flory's book. We will simply make the additional observation that in polymerizations where the reaction rate is fast and the viscosity of the medium is high, there may be a non-equilibrium distribution of reacting groups and the rate of reaction becomes diffusion controlled, so that the relationships we will now describe become invalid.

Flory investigated polyesterifications of the type:

$$A—A + B—B \rightleftharpoons A—AB—B$$

where A—A is a dicarboxylic acid or diacid and B—B a dialcohol or diol. The kinetic equations for such bimolecular reactions are often of the form:

$$\text{Reaction Rate} = -\frac{d[A]}{dt} = k_2[A][B] \qquad (3.1)$$

where k_2 is the rate constant. It is important to grasp that [A] and [B] are the concentrations of *functional groups* not molecules (there are two functional groups per molecule in the example we are considering here). Esterifications are catalyzed by acids, however, and in the absence of any added strong acid a second molecule of the carboxylic acid can act as a catalyst in each reaction step and the kinetics are now third order

$$-\frac{d[A]}{dt} = k_3[A]^2[B] \qquad (3.2)$$

If the concentrations of carboxylic acid and alcohol groups are exactly equal we can let:

$$c = [A] = [B] \qquad (3.3)$$

hence,

$$-\frac{dc}{dt} = k_3 c^3 \qquad (3.4)$$

If c_0 is the initial concentration of acid and alcohol groups (i.e., at t = 0), then we can integrate as follows:

$$-\int_{c_0}^{c} \frac{dc}{c^3} = k_3 \int_{t=0}^{t} dt \qquad (3.5)$$

to obtain:

$$2k_3 t = \frac{1}{c^2} - \frac{1}{c_0^2} \qquad (3.6)$$

We will define p as the extent of reaction, which is equal to the fraction of functional groups that have reacted after a time t:

$$p = \frac{\text{Number of COOH groups reacted}}{\text{Number of COOH groups originally present}} \qquad (3.7)$$

In his studies of polyesterification Flory determined p from the results of titration (i.e., by measuring the fraction of remaining or unreacted acid groups, which is equal to 1 - p). It follows that c, the concentration of groups at time t, is related to c_0 by:

$$c = c_0 (1 - p) \qquad (3.8)$$

Substituting in equation 3.6:

$$2c_0^2 k_3 t = \frac{1}{(1 - p)^2} - 1 \qquad (3.9)$$

and plots of $1/(1 - p)^2$ as a function of time should be linear. A plot of some of Flory's data is shown in figure 3.2.

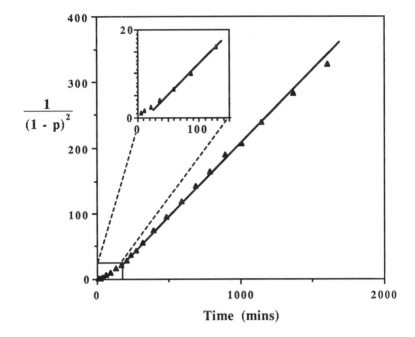

Figure 3.2 *Plot of $1/(1 - p)^2$ vs time, for an uncatalysed polyesterification. Redrawn from the data of P. J. Flory, JACS, 61, 3334 (1939).*

It can be seen from the insert in figure 3.2 that data for the initial stages of the reaction are not linear, but at higher conversions the predicted straight line relationship is obtained. This typically also happens in simple esterifications (i.e. between monofunctional reactants). In ion sensitive reactions of this type this behavior can be attributed to the change in the character of the medium as the reaction proceeds and the acids and alcohols are converted to esters.

Further evidence supporting Flory's assumption concerning the reaction rates of functional groups was obtained from polyesterifications catalyzed by the addition of small amounts of strong acid (p-toluene-sulfonic acid). The concentration of this added catalyst remains constant throughout the reaction and can therefore be included in the rate constant k' $\{= k_2[\text{Acid}]\}$:

$$- \frac{d[A]}{dt} = k'[A][B] \tag{3.10}$$

Again assuming equal amounts of reacting functional groups

$$-\frac{dc}{dt} = k'c^2 \tag{3.11}$$

and:

$$c_0 k't = \frac{1}{(1 - p)} - 1 \tag{3.12}$$

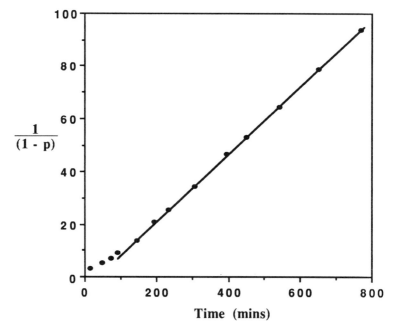

Figure 3.3 *Plot of 1/(1 - p) vs time, for a catalysed polyesterification. Redrawn from the data of P. J. Flory,* JACS, **61**, *3334 (1939).*

A plot of $1/(1 - p)$ against t, shown in figure 3.3, again displays the predicted linear relationship after a certain degree of conversion has been reached. Obviously, if the intrinsic reactivity of functional groups decreased with increasing chain length a linear relationship would not have been obtained.

It can be seen from both figures 3.1 and 3.2 that polyesterifications take a long time to reach high degrees of conversions (i.e., $p \rightarrow 1$, hence $1/(1 - p) \rightarrow$ very large). This is a general characteristic of step-growth polymerizations or polycondensations and is important because it is relatively easy to show that the number average degree of polymerization is equal to $1/(1 - p)$, so that polymer is only obtained at very high conversions near the end of the reaction. We will not derive this equation here, because we think it makes for an easier introduction into the use of probability arguments for obtaining molecular weight distributions, a topic which seems to give many students a hard time, so we will consider it at the beginning of the next chapter. The crucial point is that this behavior stands in marked contrast to addition polymerizations, where some polymer is obtained almost immediately. This brings us to the kinetics of free radical polymerization.

C. KINETICS OF FREE RADICAL POLYMERIZATION

We saw in Chapter 2 that the processes occurring in free radical polymerization can be broken down into a number of steps:

a) Initiation
b) Propagation
c) Chain transfer
d) Termination

It is convenient to consider a, b and d first and for the moment defer our discussion of chain transfer.

Initiation of free radical polymerization can be achieved by a number of methods, such as the irradiation of monomers with high energy radiation or thermal initiation *without* added initiator. Styrene and methyl methacrylate can be polymerized in this way. Most often, however, free radicals are generated by adding initiators which form radicals when heated or irradiated. There are several groups of such initiators and two common examples are benzoyl peroxide:

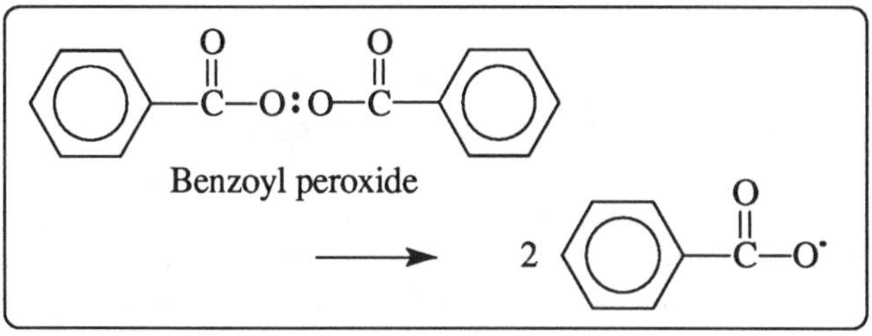

Benzoyl peroxide

and azobisisobutyronitrile (AIBN).

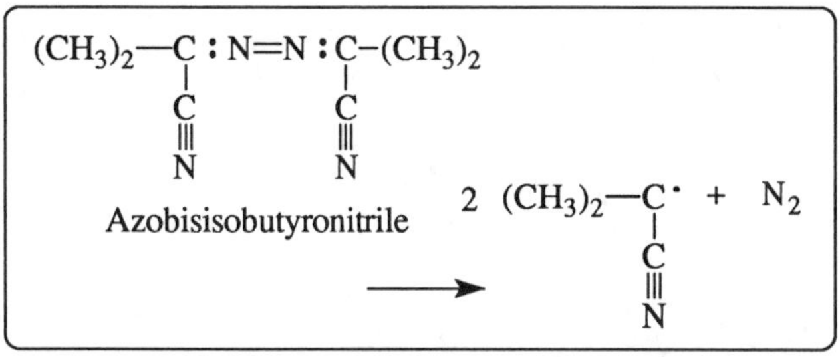

Azobisisobutyronitrile

These types of initiators usually give two identical radicals upon decomposition, which we can represent as:

$$I_2 \xrightarrow{k_d} 2\,R^\bullet \tag{3.13}$$

where k_d is the first order rate constant describing this process.

This radical can now "attack" the double bond of a monomer:

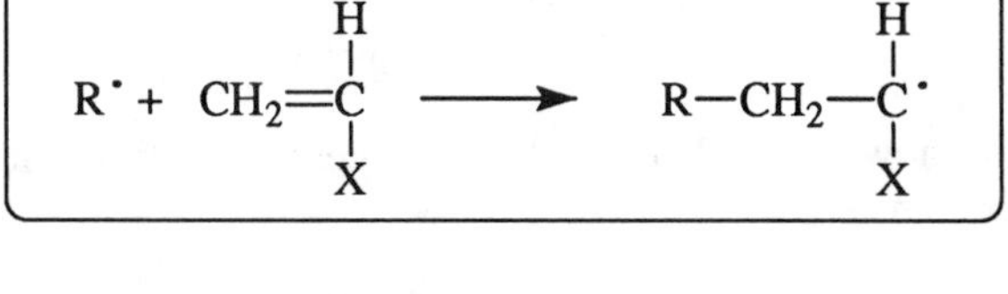

We will use the abbreviation M to represent monomer and let the rate constant for this process be k_i, so that we can write:

$$R^{\cdot} + M \xrightarrow{k_i} M_1^{\cdot} \tag{3.14}$$

These two reactions are together considered to be the process of initiation. In our development of the kinetics of these polymerizations we can usually assume that the first of these two reactions (decomposition of initiator) is the rate determining step (i.e., it is much slower than the addition to the first monomer), so that we can write for the rate of initiation, r_i:

$$r_i = \frac{d[M_1^{\cdot}]}{dt} = 2k_d[I] \tag{3.15}$$

where the 2 is obtained because:

$$-\frac{d[I]}{dt} = \frac{1}{2}\frac{d[M_1^{\cdot}]}{dt} = k_d[I] \tag{3.16}$$

The factor 2 could be included in the rate constant, if one was of a mind to do so. However, not all primary radicals formed by the decomposition of initiator will react with the first monomer. In the case of benzoyl peroxide, for example, several other reactions occur:

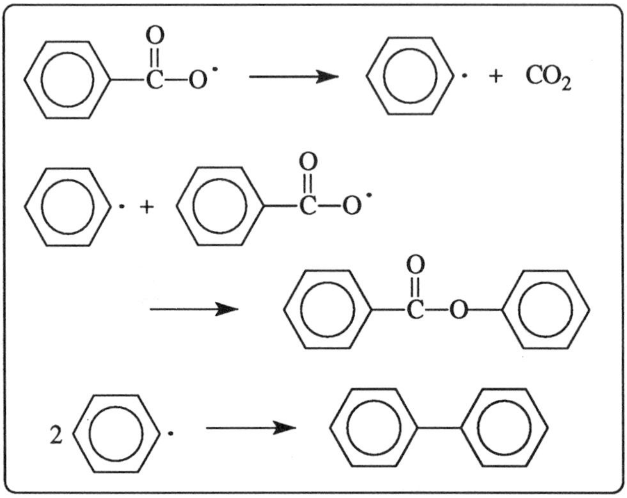

If we let f be the fraction of initially formed radicals that actually start chain growth, then the rate of initiation is given by:

$$r_i = \frac{d[M_1^{\cdot}]}{dt} = 2fk_d[I] \tag{3.17}$$

Propagation now proceeds through the successive addition of monomers, which can be written as:

$$M_1^\bullet + M \xrightarrow{\ k_p\ } M_2^\bullet \tag{3.18}$$

or, in general:

$$M_x^\bullet + M \xrightarrow{\ k_p\ } M_{x+1}^\bullet \tag{3.19}$$

Note that just as in polycondensation we have assumed that reactivity is independent of chain length by using the same rate constant, k_p, for each step. The rate of propagation, r_p, or monomer removal, is given by

$$r_p = -\frac{d[M]}{dt} = k_p [M^\bullet][M] \tag{3.20}$$

Termination always involves the reaction of two radicals, but this can go in one of two ways. The first is the simple formation of a bond between two radicals:

$$\text{wwwCH}_2-\underset{\underset{X}{|}}{\overset{\overset{H}{|}}{C}}{}^\bullet + {}^\bullet\underset{\underset{X}{|}}{\overset{\overset{H}{|}}{C}}-\text{CH}_2\text{www}$$

$$\xrightarrow{\ k_{tc}\ } \text{wwwCH}_2-\underset{\underset{X}{|}}{\overset{\overset{H}{|}}{C}}-\underset{\underset{X}{|}}{\overset{\overset{H}{|}}{C}}-\text{CH}_2\text{www}$$

This is called, obviously enough, *combination*. The second termination mechanism is called *disproportionation*, where a proton is transferred and a double bond is formed:

$$\text{wwwCH}_2-\underset{\underset{X}{|}}{\overset{\overset{H}{|}}{C}}{}^\bullet + {}^\bullet\underset{\underset{X}{|}}{\overset{\overset{H}{|}}{C}}-\text{CH}_2\text{www}$$

$$\xrightarrow{\ k_{td}\ } \text{wwwCH}{=}\underset{\underset{X}{|}}{\overset{\overset{H}{|}}{C}} + H-\underset{\underset{X}{|}}{\overset{\overset{H}{|}}{C}}-\text{CH}_2\text{www}$$

These reactions can be represented schematically as:

$$M_x^\bullet + M_y^\bullet \xrightarrow{\ k_{tc}\ } M_{x+y} \quad \text{(combination)} \tag{3.21}$$

$$M_x^\bullet + M_y^\bullet \xrightarrow{\ k_{td}\ } M_x + M_y \quad \text{(disproportionation)} \tag{3.22}$$

Both these reactions involve two radicals and are kinetically identical, so we can write for the rate of termination, r_t

$$r_t = -\frac{d[M^\bullet]}{dt} = 2k_t[M^\bullet]^2 \tag{3.23}$$

where $k_t = k_{tc} + k_{td}$ and the factor 2 is included because two radicals are consumed in each termination reaction. The relative proportions of disproportionation and combination depend upon the type of monomer involved and the temperature.

We can now obtain expressions for the rate of polymerization, R_p, and then the degree of conversion as a function of polymerization time—things that are nice to know. The first expression we will obtain is an approximation based on the often used assumption in kinetics of a *steady state* concentration of a transient species, in this case $M^\bullet$. For this to be so, radicals must be generated at the same rate as they are consumed, or the rate of initiation, r_i, must equal the rate of termination, r_t. From equations 3.17 and 3.23:

$$2fk_d[I] = 2k_t[M^\bullet]^2 \tag{3.24}$$

which allows us to obtain an expression for $M^\bullet$:

$$[M^\bullet] = \left(\frac{fk_d[I]}{k_t}\right)^{1/2} \tag{3.25}$$

This is necessary in all the equations that follow, because the concentration of radicals during a polymerization would be a difficult thing to measure experimentally. We can then immediately obtain an expression for the rate of polymerization (which is equal to the rate of propagation) by substituting in equation 3.20:

$$R_p = r_p = -\frac{d[M]}{dt} = k_p\left(\frac{fk_d[I]}{k_t}\right)^{1/2}[M] \tag{3.26}$$

This tells us that the rate of polymerization is directly proportional to monomer concentration (or, in other words, is first order in monomer concentration), but proportional to the square root of the concentration of initiator (i.e., one-half order).

At low conversions it is reasonable to assume that the concentration of initiator, [I], is constant, but a more precise equation can be obtained by accounting for the consumption of initiator. If we reexamine equation 3.16:

$$-\frac{d[I]}{dt} = k_d[I] \tag{3.27}$$

We can integrate to obtain:

$$[I] = [I]_0\, e^{-k_d t} \tag{3.28}$$

and after substituting in equation 3.26 the rate of polymerization becomes:

$$R_p = \left\{ k_p \left(\frac{f k_d}{k_t} \right)^{1/2} \right\} \left\{ [I]_0^{1/2} [M] \right\} e^{-k_d t / 2} \qquad (3.29)$$

where we have broken the equation into three parts for the purpose of our discussion.

First, we see that the rate of polymerization is still proportional to $[M][I]_0^{1/2}$, although now we are talking about the initial initiator concentration. Consequently, if we are dealing with a solution polymerization and we wish to speed up the rate of polymerization, we simply increase the monomer concentration and the rate of reaction should increase proportionally. Polymerizations in the bulk (i.e., just monomer and initiator and maybe some inconsequential additives) would require significantly increasing the initiator concentration (because of the square root dependence). This is not always practical or desirable as this will *reduce* the molecular weight of the product, as we will show shortly.

The second point is that the polymerization will slow down in an exponential fashion as the initiator is used up (final term). More initiator would have to be added to keep the polymerization going (or we start with the right amount for our purposes and recover polymer product before the reaction slows down too much).

Third, we see that the rate of polymerization is proportional to $k_p / k_t^{1/2}$ (first term in brackets). We have already mentioned one consequence of this for the specific case of ethylene polymerization. At 130°C and a pressure of 1 bar this ratio has a value of 0.05, but this increases to 0.7 at 2500 bar. Increasing the temperature to 200°C at this latter pressure increases the ratio to 3. To reiterate, in the absence of catalyst the rate of propagation is too small relative to the rate of termination for ethylene to polymerize at ordinary pressures.

There is another consequence of the dependence of the rate of polymerization on $k_p / k_t^{1/2}$. Various kinetic experiments performed upon polymers in solution have yielded the expected first order dependence on monomer concentration. In concentrated solutions or in the polymerization of an undiluted monomer there is often a marked acceleration of the polymerization rate at some point in the reaction, however. For methyl acrylate this can occur after a conversion of only 1 percent and can result in an explosion. The reason for this is that the viscosity of the medium increases as polymer is formed. This does not affect the rate of propagation, which depends upon the barely affected diffusion of small molecules. Termination involves the much slower diffusion of larger macromolecular species, however, and this increase in viscosity can result in a large decrease in the rate of termination. The "reactivity" of the radicals is not altered, but the ability of the radicals to find each other is, so that the apparent value of k_t is reduced. This has a dramatic effect on the rate of polymerization, because of its dependence on $k_p / k_t^{1/2}$. The rate of heat evolved from these exothermic reactions also increases and in situations where this is not quickly dissipated there can be catastrophic consequences. This phenomenon of *autoacceleration* is often called the *Trommsdorff* effect (although it was originally discovered by Norrish and Smith, see Morawetz, cited earlier).

We can now obtain an expression for the degree of conversion as a function of time by simply noting that:

$$R_p = -\frac{d[M]}{dt} \tag{3.30}$$

Substituting the expression for R_p given in equation 3.29 and integrating, we obtain:

$$\ln\frac{[M]_0}{[M]} = 2k_p\left(\frac{f}{k_d k_t}\right)^{1/2}[I]_0^{1/2}\left(1 - e^{-k_d t/2}\right) \tag{3.31}$$

The degree of conversion is equal to $([M]_0 - [M])/[M]_0$, the fraction of monomer that has been reacted (recall that [M] is the concentration that is left or unreacted, so that $([M]_0 - [M])$ where $[M]_0$ is the initial monomer concentration, is the concentration of monomer that has reacted). We can then obtain:

$$\text{Conversion} = \frac{[M]_0 - [M]}{[M]_0} = 1 - \frac{[M]}{[M]_0} \tag{3.32}$$

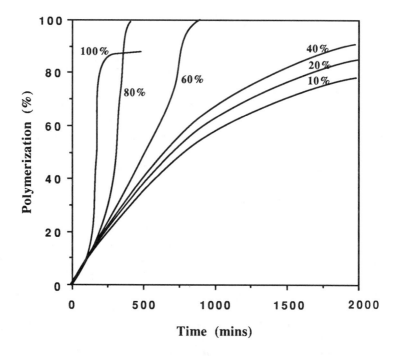

Figure 3.4 *Plot of conversion vs time, for the free radical polymerization of methyl methacrylate at 50°C using benzoyl peroxide at various concentrations in benzene. Redrawn from the data of G.V. Schulz and G. Harborth, Macromol. Chem., 1 106 (1967).*

or:

$$\text{Conversion} = 1 - \exp\left\{-\left\{2k_p\left(\frac{f}{k_d k_t}\right)^{1/2}[I]_0^{1/2}\left(1 - e^{-k_d t/2}\right)\right\}\right\} \quad (3.33)$$

Obviously, conversion is never complete as it is always less than 1 by an amount that is given by the exponential term. If we let t go to infinity we obtain an expression for the maximum conversion that is less than 1 by an amount that depends upon the initial initiator concentration:

$$\text{Maximum conversion} = 1 - \exp\left\{-\left\{2k_p\left(\frac{f}{k_d k_t}\right)^{1/2}[I]_0^{1/2}\right\}\right\} \quad (3.34)$$

If we had assumed a steady state concentration of initiator, then the maximum conversion would approach 1 after long periods of time. Schematic plots of conversion (or % polymerization) vs time illustrating this are shown in figure 3.4. This figure also shows the autoacceleration effect that can occur at high monomer concentrations.

Finally, in this section we have so far ignored the effects of chain transfer upon the rate of polymerization. When the active site is transferred to monomer, solvent, initiator, polymer itself, etc., another radical is generated and the total number of active sites is not altered. If these new sites add monomer, then the rate of polymerization is barely affected. The degree of polymerization is altered, however, and we will discuss that in the next section. We conclude here by noting that chain transfer to certain types of molecules called *inhibitors* can retard or prevent the onset of polymerization for a while, if the radicals so generated are unable to react with monomer. Obviously, it is a good idea to add such inhibitors to monomers such as methyl acrylate, styrene, etc. that are being stored and then remove them (by distillation or by passing through an appropriate packed column) when they are to be polymerized.

Average Degree of Polymerization in Free Radical Polymerizations

In discussing condensation polymers we can use (next chapter) a statistical method to obtain expressions for the molecular weight distribution, because of the stepwise nature of the polymerization and the homogeneous nature of the reaction mass. There is a continuous distribution of species present, ranging from monomer through polymer and all species take part in the reaction. Chain polymerizations are entirely different. We do not generally have a situation where shorter chains can latch on to one another to form longer chains. Instead, in free radical polymerizations an active site is formed by an initiator, monomer adds very quickly and a complete polymer is formed, often in a matter of seconds. The length of this polymer chain depends on the random radical encounters that lead to termination, so there will again be a distribution of chain lengths. But unlike condensation polymers, there is no continuous distribution of species, and at a given degree of conversion we have a mixture of polymer chains and monomers (although there will obviously be some oligomeric species from chains that by chance terminated quickly). In a condensation polymerization the average degree of polymerization increases continuously during the reaction and when we calculate say $\bar{x}_n$, the number average degree of polymerization, it reflects *all the species present in the reaction vessel* and is a function of the extent of reaction only. This is something we want to know, so that we can judge when we have polymer of the required molecular weight. In contrast, in a free radical polymerization we have polymer of the required molecular weight immediately, providing that we have started with the correct proportions of monomer, initiator, chain transfer agents, etc. (note that the degree of polymerization obtained in a free radical polymerization depends upon far more variables than a condensation polymerization, as we will shortly demonstrate more explicitly). This initially formed polymer then usually sits around inertly while other polymers are formed. These could have a different molecular weight, however, as initiator is depleted and monomer is used up. In this situation what we want to know (and actually all that we can determine) is the degree of polymerization of the product, not the average of the polymer *and* monomer in the reaction vessel at a given time in the reaction. We can obtain this by the definition of a *kinetic chain length*. Because of the wide variety of side reactions that can occur in free radical polymerizations the theoretical calculation of distributions is difficult, but it is a fairly straightforward procedure to calculate a number average chain length, which is therefore all we will attempt. We will start by neglecting transfer reactions, then subsequently modify our result to account for these.

It should be intuitively obvious that in a reaction vessel containing monomer and initiator (and perhaps some inert solvent) the average length of the chains we will obtain will be inversely proportional to the initiator concentration (or some power of the initiator concentration). If we generate a smaller number of initial radicals then on average the polymer chains will grow longer before their active ends meet and terminate (i.e., there is a smaller number of radicals per unit volume). It also follows that if we do not have steady state conditions, then as a polymerization proceeds and initiator is used up the number of radicals present decreases and the average polymer chain length of the product will be greater.

If for a given period of time we know how many growing polymer chains have been started, and also how many monomer units have been polymerized, then we could simply obtain the number average degree of polymerization of these chains by dividing one by the other (i.e., if 1,000,000 monomers have been used up and 100 chains formed in this time period, then the number average length of each chain would be 10,000). For the moment we are neglecting what happens upon termination, so we can consider these chains to be radicals (i.e., have active sites on one end). Obviously, the longer the time period we choose, the more things will be different at the start than near the end, as monomer is used up, etc. We get a more accurate representation of what is occurring if we consider shorter and shorter time periods, and this is in effect what we do by defining a kinetic chain length, ν, as the *rate* of propagation divided by the *rate* of initiation at some instant of time:

$$\nu = \frac{r_p}{r_i} \qquad (3.35)$$

If we use the steady state assumption, we can first substitute from equations 3.17 and 3.20 to obtain:

$$v = \frac{k_p[M][M^\bullet]}{2fk_d[I]} \qquad (3.36)$$

then substituting for $[M^\bullet]$ from equation 3.25:

$$v = \left\{ \frac{k_p}{2(fk_dk_t)^{1/2}} \right\} \left(\frac{[M]}{[I]^{1/2}} \right) \qquad (3.37)$$

where we see that there is a first order dependence on monomer concentration, but that the average chain length varies as the inverse of the square root of the initiator concentration.

The kinetic chain length is clearly equal to the number average degree of polymerization of the radicals formed at the particular time in the polymerization that we have chosen. This must also be equal to the number average degree of polymerization of the polymer, if termination is by disproportionation, as this does not change the number of chains nor their length. If termination were exclusively by combination however, the number of chains would halve and their length would (on average) double.

i.e. $\bar{x}_n = v$ (termination by disproportionation)
         $\bar{x}_n = 2v$ (termination by combination)

Since both mechanisms of termination can occur in certain polymerizations, it is useful to define a parameter ξ, which is equal to the average number of dead chains formed per termination. We can obtain an expression for ξ by again using a kinetic argument and noting that this must be equal to the *rate* of dead chain formation divided by the *rate* of termination. The rate of dead chain formation is given by:

$$-\frac{d[M^\bullet]}{dt} = 2k_{td}[M^\bullet]^2 + k_{tc}[M^\bullet]^2 \qquad (3.38)$$

where the factor 2 is included because disproportionation results in *two* dead chains. The rate of termination reactions is simply given by:

$$k_t[M^\bullet]^2 = (k_{tc} + k_{td})[M^\bullet]^2 \qquad (3.39)$$

Hence:

$$\xi = \frac{k_{tc} + 2k_{td}}{k_{tc} + k_{td}} = \frac{k_{tc} + 2k_{td}}{k_t} \qquad (3.40)$$

Accordingly, the instantaneous number average chain length can now be obtained by dividing the rate of addition of monomer units to the rate of "dead" polymer formation, i.e.

$$\bar{x}_n = \frac{k_p[M][M^\bullet]}{(2k_{td} + k_{tc})[M^\bullet]^2} = \left(\frac{k_p}{\xi(fk_dk_t)^{1/2}} \right) \left(\frac{[M]}{[I]^{1/2}} \right) \qquad (3.41)$$

It is important to note that the rate of polymerization is proportional to $[I]^{1/2}$, while $\bar{x}_n$ is proportional to $1/[I]^{1/2}$. In other words, if we try to speed up the polymerization by adding more initiator we end up getting shorter chains, which is not always desirable. We have to strike a balance for the product we are making.

The Effect of Chain Transfer

As we noted at the start of our discussion of free radical polymerization, there is also the possibility of chain transfer reactions, where a growing chain radical is terminated and a new one is initiated in its place. Such reactions are common and generally take the form:

$$M_x^{\cdot} + R{-}H \longrightarrow M_x + R^{\cdot}$$

$$R^{\cdot} + M \longrightarrow R{-}M^{\cdot} \quad \text{etc.} \tag{3.42}$$

[Note: it is not always a proton that is transferred.]

In general, these reactions do not affect the rate of polymerization (although in certain circumstances, they can), but obviously affect the molecular weight, because on average shorter chains are produced. A lot of components in a polymerization can act as chain transfer agents; monomer, the solvent, dead or terminated polymer (producing long chain branches), and so on.

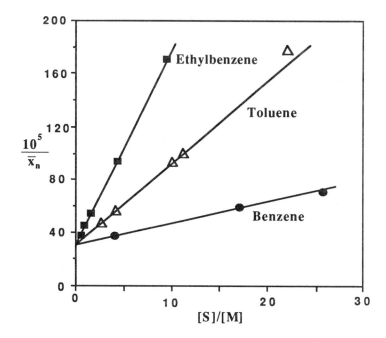

Figure 3.5 *Effect of solvent on* $\bar{x}_n$ *on polymerization of styrene at 100°C. Redrawn from the data of R. A. Gregg and F. R. Mayo,* Faraday Soc. Discussions, **2**, *328 (1947).*

Sometimes a chain transfer agent, such as a mercaptan, is deliberately added to control molecular weight. The effect of chain transfer on the instantaneous number average molecular weight is easily accounted for by adding the effect of dead chain formation by this mechanism to equation 3.38. Then:

$$-\frac{d[M^\bullet]}{dt} = 2k_{td}[M^\bullet]^2 + k_{tc}[M^\bullet]^2 + k_{tr}[T][M^\bullet] \tag{3.43}$$

where [T] is the concentration of the chain transfer reagent. It follows directly that:

$$\bar{x}_n = \frac{k_p[M]}{\xi\,(fk_dk_t[I])^{1/2} + k_{tr}[T]} \tag{3.44}$$

Rearranging:

$$\frac{1}{\bar{x}_n} = \frac{1}{(\bar{x}_n)_0} + C\frac{[T]}{[M]} \tag{3.45}$$

where $(\bar{x}_n)_0$ is the average length of the chain in the absence of chain transfer agent (cf. equation 3.41) and C is simply k_{tr}/k_p. A plot of $1/\bar{x}_n$ against [T]/[M] should then be linear with a slope of k_{tr}/k_p and an intercept $1/(\bar{x}_n)_0$. This equation has been applied to the polymerization of styrene in various solvents, which act as chain transfer agents, and the results are shown schematically in figure 3.5. There is very good agreement between theory and experiment (in this case).

D. STUDY QUESTIONS

1. Show that the kinetics of polyesterification catalysed by a strong acid may be represented by:

$$1/(1-p) = c_0\, k't + \text{constant}$$

2. A large scale polycondensation reaction of a diol and a diacid is performed at 200°C. At various intervals small samples are removed, introduced into tared conical flasks, quenched in a dri ice/acetone, weighed, dissolved and titrated with a standard 0.1N sodium hydroxide solution. From the results given below determine whether this was a catalysed or non-catalysed reaction.

Time (min)	Sample Weight (g)	Titration (cm³ 0.1N NaOH)
0	0.3402	46.58
25	0.4501	30.82
50	0.4931	16.88
100	0.6005	10.28
150	1.1014	11.62
200	1.5113	10.35
300	2.1377	8.78
400	4.2978	12.95
500	5.5118	11.92
600	5.9007	11.32
700	6.1387	10.09

3. Define all symbols employed and show that the rate of polymerization for a free radical process is given by:

$$R_p = k_p \left[\frac{fk_d}{k_t} \right]^{1/2} [I]^{1/2} [M]$$

4. What effect would increasing the initiator concentration have on the degree of polymerization in the bulk free radical polymerization of styrene ?

5. Ethylene does not polymerize at ordinary pressures under conventional free radical processes. Why is this ?

6. Describe the following giving appropriate examples:
    (a) Chain transfer
    (b) Retardation
    (c) The Trommsdorff effect
    (d) Inhibition
    (e) Termination by disproportionation

7. The rate of many free radical polymerizations can be followed fairly accurately by dilatometry, which is a technique that measures volume changes. Furthermore, it has been shown that in a number of systems the rate of polymerization, R_p, is approximately constant for conversions of up to 10 - 20%. Assume that this holds for the bulk polymerization of styrene (initial concentration, $[M]_o = 8.3$ moles/l) at 70°C, initiated by AIBN (molecular weight = 164; $k_d = 3.5 \times 10^{-5}$ sec^{-1}). The following *initial* rates of polymerization, measured in terms of % contraction per hour, and the number average degrees of polymerization, $\bar{x}_n$, of the initial product were obtained as a function of initiator concentration.

[I] (g / 100ml)	Contraction (% / h)	$\bar{x}_n$
0.061	0.84	860
0.100	1.20	625
0.185	1.48	500
0.352	2.09	363

Assume that termination occurs solely by disproportionation and that chain transfer takes place only to monomer. Given that there is an 18% volume contraction when styrene is completely (100%) polymerized, calculate:

    (a) the value of k_t / k_p^2
    (b) the efficiency of AIBN as an initiator
    (c) the value of $C (= k_{tr} / k_p)$

State explicitly what additional assumptions you make in performing these calculations. [Hint: first derive an expression for $\bar{x}_n$ in terms of R_p and other constants.]

8. In this chapter we only considered the initiation of polymerization by chemical methods. Given that for a certain monomer thermal generation of free

radicals involves two molecules of monomer (i.e. second order in monomer concentration) such that:

$$M + M \xrightarrow{\;k_i\;} R^{\bullet} + \text{inert products}$$

(a) Show that the rate of polymerization varies as $[M]^2$.

(b) Obtain an expression for the kinetic chain length and comment on the dependence of $\bar{x}_n$ on $[M]$.

E. SUGGESTIONS FOR FURTHER READING

(1)   P. J. Flory, *Principles of Polymer Chemistry*,
        Cornell University Press, Ithaca, New York, 1953.

(2)   R. W. Lenz, *Organic Chemistry of Synthetic High Polymers*,
        Interscience, J. Wiley & Sons, New York, 1967.

CHAPTER 4

Statistics of Step-Growth Polymerization

*"The true logic of this world
is the calculus of probabilities"*
—James Clerk Maxwell

A. INTRODUCTION

Probability theory has a long and disreputable history (which is why it appeals to us), having its origins in the desire to work out the odds in games of chance and thus get rich without having to work for a living. Even the parts of it whose origins are not apparently connected to gambling can seem a bit weird. For example, Watson and Galton (in 1874[*]) considered the problem of the extinction of particular family names, which (in those days of moral and intellectual darkness when women took the names of their husbands) was related to average fertility and the probability of having sons. They came to the conclusion that all such family names would quickly die out. In a subsequent study of the reproductive rates of the British Aristocracy (whose records go back a long way and therefore provide a good data base), Galton found evidence to support this. However, the original analysis contained an algebraic error, but we also suspect that they did not account for the unfortunate effects of in-breeding. In any event, the result of their efforts was the development of the theory of branching processes (or Markov chains). We will use the simplest part of this theory in the next chapter, when we discuss copolymerization. Here we will confine ourselves to very simple probability arguments and about the only tool you will need to start with, apart from a modicum of common sense, is the definition that the probability of an event E, P{E} is given by:

$$P\{E\} = \frac{N_E}{N} \qquad (4.1)$$

where N_E is the number of E events occurring and N is the total number of events. For example, if you toss an unbiased coin 100 times and get 50 heads, you would calculate that the probability of getting a head in any coin toss is 50/100 or 0.5. (In truth, if you actually tossed a coin 100 times you would be unlikely to get exactly 50 heads, because of fluctuations.   We will consider this

[*] See T. E. Harris, *The Theory of Branching Processes*, Springer-Verlag, Berlin, 1963.

some more when we talk about chain conformations, but to get the expected result of 0.5 you would have to perform the 100 coin tosses a large number of times and average all of the results. We don't have to worry about that here, because if we make a mole of polymer we have *billions* and *billions* of chains, so the probabilities all work out, more or less.) Also, if there are only two possible outcomes, like in a coin toss, then the probability of the other event must equal $1 - P\{E\}$. Note that the probability of an event is equal to the *number fraction* of these events and is always between 0 (never occurs) and 1 (always occurs).

We will start by examining molecular weight averages and distributions in step-growth polymerizations, as these are particularly well-suited to a statistical treatment (because all the species present are involved in the polymerization throughout the course of the reaction). The nature of addition polymerization is such that the statistical approach cannot be readily applied to the analysis of molecular weight distributions, but it is a powerful tool in the study of such aspects of microstructure as copolymer sequence distributions and various types of isomerism. We will discuss these latter applications of probability theory in Chapters 5 and 6.

B. THE STATISTICS OF LINEAR POLYCONDENSATION

We will start by considering the number average degree of polymerization in a linear step-growth reaction. This really involves practically no statistics at all and in most texts (e.g., Flory's) is included in the discussion of kinetics. Over the years, however, we have found it useful pedagogically to start our discussion of statistics with this topic by noting that the extent of reaction, p, defined in the previous chapter as equal to the fraction of functional groups of a particular type that have reacted (e.g. acid groups in polycondensation), is also equal to the probability that one such group, taken at random from the reaction vessel, has reacted. This follows directly from the definition of a probability given in equation 4.1. This is not always grasped immediately by students who think about probabilities in a sloppy manner (like, what is the probability that I will drink too much and do something stupid if I go out on Friday night). Remember, when we are talking about reactions in a vessel containing *billions* and *billions* of molecules, the probability that any one of these molecules has reacted after a certain time is simply equal to the fraction that have reacted in the same time period. (Of course, this assumes that reactivity does not change with chain length.)

We can consider two types of polycondensations, the first involving bifunctional monomers that have different (but complementary) functional groups:

$$\text{A—B} + \text{A—B} \rightleftharpoons \text{A—BA—B} \quad \text{(Type I)}$$

where, for example, the A group could be an acid, the B an alcohol and the BA product an ester.

The second type of reaction involves monomers that each contain only one type of functional group:

$$A-A + B-B \rightleftharpoons A-AB-B \quad \text{(Type II)}$$

In reactions of Type I we can "count" the number of molecules present if we can measure the concentration of one of the end groups (the acid group, in the example we have used, by titration or spectroscopic methods). Students whose brains have not yet engaged in forward gear at the time of reading this section can convince themselves of this by examining figure 4.1 and counting the number of molecules and the number of (say) A end groups. The same holds for polycondensations of Type II, *providing that we start with exactly equal equivalents of the two reactants.*

This counting of the number of molecules present can be used to provide a measure of the *number average degree of polymerization,* $\bar{x}_n$, which is simply defined as the total number of molecules originally present (monomers), equal to N_0, divided by the total number of molecules in the system after the polymerization has been stopped, N:

$$\bar{x}_n = \frac{N_0}{N} \tag{4.2}$$

This definition of an average degree of polymerization is easily grasped. If, for example, we started with 20 monomers and ended up with 4 (short) chain molecules, the average length of each chain would be 20/4 or 5. This says nothing about the actual distribution of the chain lengths. We could indeed have four chains of equal length, but a distribution where one of the chains contained 17 units while the other 3 units remained monomers would also give the same average. A feel for the breadth of the distribution can be obtained by also

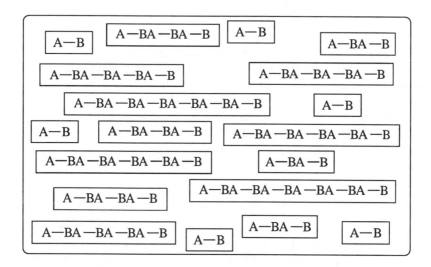

Figure 4.1 *Schematic representation of n-mers.*

measuring the weight average molecular weight, but it is often advantageous to have a more precise knowledge of the actual distribution (many physical properties can be affected by this). We will come back to this shortly. First, we can express $\bar{x}_n$ in terms of the conversion p by recalling that:

a) the concentration of one of the functional groups, c, present after a fraction p has been reacted, is related to the initial concentration of such groups by:

$$c = c_0(1 - p) \qquad (4.3)$$

b) the number of molecules (N, N_0) can be simply converted to concentrations (c, c_0) so that:

$$\bar{x}_n = \frac{N_0}{N} = \frac{c_0}{c} = \frac{1}{1 - p} \qquad (4.4)$$

where the conversion constants cancel. If we wish to obtain a number average molecular weight, $\bar{M}_n$, instead of a degree of polymerization, we can use:

$$\bar{M}_n = M_0\,\bar{x}_n = \frac{M_0}{(1 - p)} \qquad (4.5)$$

Be careful here. For condensations of Type II the *chemical repeat* unit contains bits of both monomers. In nylon 6,6, for example, the chemical repeat is:

Chemical repeat of Nylon 6,6

In terms of a degree of polymerization as we have defined it here however, this chemical repeat unit contains two structural repeat units (it has been made by joining two monomers). Accordingly, we must define M_0 as the *mean* molecular weight of a structural unit (in the above example this can be obtained by dividing the molecular weight of the chemical repeat unit by 2).

The result obtained in equation 4.4 is an important one, so we will write it out again and put it in a box so you will hopefully remember it:

$$\boxed{\bar{x}_n = \frac{1}{(1 - p)}}$$

It tells us that high molecular weight is only achieved at high degree of conversions. In their laboratory experiments many organic chemists would be happy with a 90% conversion (p = 0.9). This would give us a polymer with a number average degree of polymerization of only 10 (and this would not be a particularly useful material). A conversion of 95%, which would make our hypothetical chemist hop up and down with glee, is not much better ($\bar{x}_n = 20$). To obtain a degree of polymerization of the order of 200, the reaction has to be driven almost to completion (p = 0.995). Not only that, but it is necessary to do this reproducibly on an industrial scale!

These results are summarized in figure 4.2, which displays a plot of $\bar{x}_n$ as a function of conversion (p). As it turns out, for nylon 6,6 it is desirable from the point of view of subsequent processing to limit the molecular weight to the 10,000–15,000 range ($\bar{x}_n \approx 100\text{-}150$). Although this could be achieved by stopping the reaction at the appropriate degree of conversion by simply cooling, polymerization would recommence when the product is reheated during processing. For nylon 6,6 a "chain stopper" (a monofunctional reagent, acetic acid) is added to the polymerization and this brings us to the general subject of the control of molecular weight.

We will start our discussion of this topic by considering the effect of a non-stoichiometric equivalence of the bifunctional monomers. Our intention is to obtain an expression for the number average degree of polymerization. This requires that we know two things, the number of monomers we started with (N_0) and the number of chains after a fraction of p groups has been reacted (N). We will designate the groups that are present in excess by the letter B, so that if N_A is the number of A groups and N_B the number of B groups initially present, then the total number of *monomers* initially present is equal to:

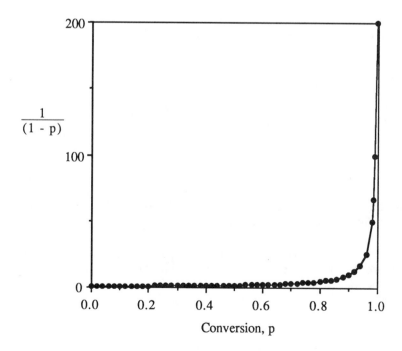

Figure 4.2 *Plot of the number average degree of polymerization, $\bar{x}_n$, versus conversion, p.*

$$N_0 = \frac{N_A + N_B}{2} = \frac{N_A}{2}\left(1 + \frac{1}{r}\right) = \frac{N_A}{2}\left(\frac{1 + r}{r}\right) \qquad (4.6)$$

where:

$$r = \frac{N_A}{N_B} \qquad (4.7)$$

and can only take fractional values between 0 and 1. The factor 2 in equation 4.5 is there because the number of monomers is equal to half the number of (bi)functional groups. After a fraction of p groups has been reacted the number of *chain ends* (not molecules) remaining (i.e. not reacted) is equal to:

$$N_A(1 - p) + (N_B - pN_A) \qquad (4.8)$$

where the origin of the first term should be self evident. The second term is equally straightforward. Because N_A molecule reacts with exactly one N_B unit, the number of B units that have reacted must be equal to pN_A. Therefore the number of B units unreacted (equal to the number of chain ends that are B units) is given by the second term. Substituting $r = N_A/N_B$ and dividing by two to obtain the number of molecules:

$$N = \frac{1}{2}\left[N_A(1 - p) + N_B(1 - rp)\right]$$
$$= \frac{N_A}{2}\left[(1 - p) + \frac{(1 - rp)}{r}\right] \qquad (4.9)$$

Substituting and rearranging:

$$\bar{x}_n = \frac{N_0}{N} = \frac{1 + r}{1 + r - 2rp} \qquad (4.10)$$

We can now let $p = 1$ (i.e. complete reaction) to show that there is a theoretical upper limit to the molecular weight that cannot be exceeded if there is an excess of one component:

$$\bar{x}_n \rightarrow \left[\frac{1 + r}{1 - r}\right] \text{ as } p \rightarrow 1 \qquad (4.11)$$

If, for example, there is an excess of just 1% of B units at the start of the reaction (by which we mean $N_A/N_B = 99/100$), then the upper limit for $\bar{x}_n$ is 199.

The same equation can be applied to a polymerization involving equal numbers of bifunctional reagents and a small amount of a "chain stopper" [e.g., a monofunctional reactant such as acetic acid (CH_3COOH)], providing that r is now defined by the following equation (see Flory[*]):

---

[*] P. J. Flory, *Principles of Polymer Chemistry*, Cornell University Press, 1953.

$$r = \frac{N_A}{N_B + 2N_B^M} \qquad (4.12)$$

where N_B^M is the number of monofunctional groups (note that $N_A = N_B$).

Clearly, if it is desired to obtain high molecular weight polymer great care must be taken to make sure that there are equal amounts of monomer at the start of the reaction, that these monomers are pure (i.e., do not have monofunctional contaminants) and that one of these monomers does not get lost or escape (e.g., by evaporation) in preference to the other.

C. MOLECULAR WEIGHT DISTRIBUTIONS IN LINEAR CONDENSATION POLYMERS

In this section we really start using probabilities and we will derive expressions for the molecular weight distributions in Type I condensation polymerizations. Equivalent expressions can be obtained for Type II, but the algebra can get much more complicated when the number of A–A molecules does not equal the number of B–B molecules. It is sufficient for our discussion of general principles to limit our attention to polymerizations of A–B type molecules where, of course, the number of A functional groups is always equal to the number of B groups.

Following Flory, we determine the probability that a molecule selected at random from a mixture of polymerizing chains, where the extent of reaction is equal to p, is an x-mer (i.e., has x units in its chain). We focus our attention on the first B unit in the chain and ask "What is the probability that this has reacted?" This must, of course, be equal to p (to reiterate, the probability that a group has reacted is just equal to the fraction that have reacted). The probability that the second group in the chain has reacted is also equal to p, and so on:

```
A—BA—BA—BA—BA —————————————B
  └┬┘ └┬┘ └┬┘ └┬┘            └┬┘
   p   p   p   p             1-p
```

There are (x - 1) AB linkages in an x-mer, so the probability that all these groups have reacted is just the product of each term, equal to p^{x-1}. We must also include a term that gives the probability that the end B group has *not reacted*, which is equal to 1 - p. The probability that the molecule has exactly x units, P_x, is therefore:

$$P_x = (1 - p)\, p^{x-1} \qquad (4.13)$$

The probability of finding an x-mer at random must, by definition, be equal to the mole fraction of x-mers present, so that we can write:

$$\frac{N_x}{N} = X_x = (1 - p)\, p^{x-1} \qquad (4.14)$$

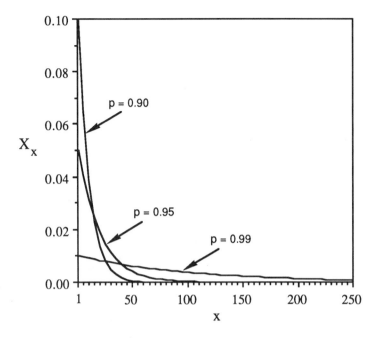

Figure 4.3 Mole fraction distribution of x-mers.

where N_x is the number of x-mers present in our polymerization pot, N is the total number of molecules present when the extent of reaction is equal to p and X_x is the mole fraction of x-mers.

Rather than use the mole fraction of x-mers, some authors prefer the number distribution N_x, which can be obtained from equation 4.14 using:

$$\bar{x}_n = \frac{N_0}{N} = \frac{1}{(1-p)} \qquad (4.15)$$

hence:

$$N = N_0 (1-p) \qquad (4.16)$$

and:

$$N_x = N_0 (1-p)^2 p^{x-1} \qquad (4.17)$$

We can now use equation 4.14 to calculate the mole fraction of x-mers present for various values of p. The results are shown graphically in figure 4.3. It can be seen that there is always a larger number of monomers present than any other species and this is true of all stages of the reaction (i.e. all values of p).

By weight, however, the amount of monomer can be very small (obviously a single 100 mer is much heavier than 5 or 10 monomers). This can be shown more explicitly by calculating the weight fraction of x-mers present, w_x, which is given by:

$$w_x = \frac{x N_x M_0}{N_0 M_0} = \frac{\text{weight of all x-mers}}{\text{weight of all units}} \qquad (4.18)$$

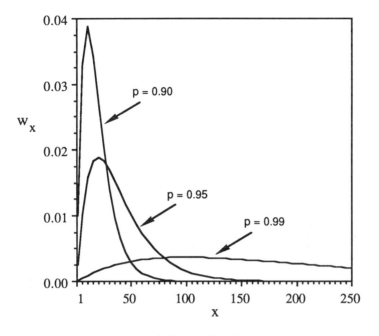

Figure 4.4 *Weight fraction distribution of x-mers.*

where we have assumed that the average molecular weight of the structural repeat unit is the same as the monomer average (this introduces a small error through neglect of end protons, OH groups etc.) Substituting from equation 4.17, the weight fraction of x-mers is then:

$$w_x = x \, (1 - p)^2 \, p^{x-1} \tag{4.19}$$

Values of w_x as a function of x are plotted in figure 4.4 for the same values of p that were used to calculate the mole fraction distribution. It can be seen that there is a maximum in this distribution that shifts to higher values as p increases. The distribution also broadens with increasing extent of reaction.

The distribution described by equations 4.14 and 4.19 is called the *most probable* or Schulz-Flory distribution. It rests on the fundamental assumption that the reactivity of a functional group is independent of the length of a chain to which it is attached. The validity of this assumption has never been contradicted by experiment, but until the advent of gel permeation chromatography (GPC - see later) there were few studies of the molecular weight distribution. This is because of the difficulty of cleanly separating a polymer into its fractions by methods such as precipitation from solution (you tend to get fractions that also contain a distribution of molecular weights). An unusually sharp fractionation was obtained many years ago by Taylor, however, and these results are shown in figure 4.5. It can be seen that the agreement of theory with experiment is within error.

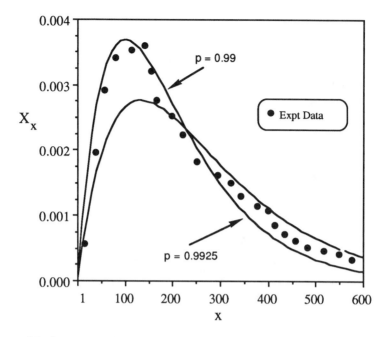

Figure 4.5 *Comparison of calculated weight fraction distributions (p = 0.9900 and p = 0.9925) to experimental nylon 6,6 data. Redrawn from the data of G. B. Taylor, JACS, 69, 638 (1947).*

Finally, the equations for the most probable distribution given above can be used to obtain expressions for the number and weight average degree of polymerization. The former has been given already, but it is usually interesting and informative to obtain fundamental results by different methods.

The number average degree of polymerization is defined as:

$$\bar{x}_n = \frac{\sum x\, N_x}{\sum N_x} \qquad (4.20)$$

which in its expanded form is:

$$\bar{x}_n = \frac{N_1}{\sum N_x} + \frac{2N_2}{\sum N_x} + \frac{3N_3}{\sum N_x} + \text{---------------} \qquad (4.21)$$

As above, if we let X_x be the mole fraction of x-mer ($N_x/\sum N_x$), it follows that:

$$\bar{x}_n = \sum x\, X_x \qquad (4.22)$$

substituting from equation 4.14:

$$\bar{x}_n = \sum x\, p^{x-1}\, (1-p) \qquad (4.23)$$

Because $p < 1$ we can use the following equation for a series convergence:

$$\sum x\, p^{x-1} = \frac{1}{(1-p)^2} \qquad (4.24)$$

to obtain:

$$\overline{x}_n = \frac{1}{(1-p)} \qquad (4.25)$$

The weight average degree of polymerization is given by:

$$\overline{x}_w = \sum x\, w_x = \sum x^2\, p^{x-1}\,(1-p)^2 \qquad (4.26)$$

Using:

$$\sum x^2\, p^{x-1} = \frac{(1+p)}{(1-p)^3} \qquad (4.27)$$

we obtain:

$$\overline{x}_w = \frac{(1+p)}{(1-p)} \qquad (4.28)$$

The weight average molecular weight (degree of polymerization) is always larger than the number average (they are only equal in a *monodisperse* sample, where all the chains have exactly the same length). As mentioned in Chapter 1, the ratio of the two is called the *polydispersity* and is a measure of the breadth of the distribution. Substituting from equations 4.25 and 4.28:

$$\frac{\overline{x}_w}{\overline{x}_n} = (1+p) \qquad (4.29)$$

and as $p \rightarrow 1$

$$\frac{\overline{x}_w}{\overline{x}_n} = 2 \qquad (4.30)$$

D. MULTICHAIN CONDENSATION POLYMERS

We have seen that for an essentially complete reaction (i.e. $p \approx 1$) the polydispersity of polymers synthesized by linear polycondensation is predicted to be 2. Thus, if we were to submit a commercial nylon 6,6 polymer for molecular weight analysis by say both osmometry and light scattering (from which number and weight average molecular weights, respectively, may be determined—as we will see later in Chapter 10), we would expect, and commonly find, results showing $\overline{M}_w$ to be roughly twice that of $\overline{M}_n$. If branching reactions occur, from chemical side reactions, impurities or by design using multifunctional monomers, the polydispersity generally broadens. There is an interesting polycondensation case, however, where the polydispersity *narrows*. Consider the condensation polymerization of A–B with a *small* amount of a multifunctional monomer, R-A$_f$. As the polymerization reaction approaches completion, linear species will disappear in favor of multichain polymer molecules. The polymer formed upon complete reaction will consist of f chains attached to a central unit R.

$$R\left[A-(BA)_{y-1}-B-A\right]_f$$

For example: ε-aminocaproic acid (A–B) plus a tetrabasic acid (RA$_4$)would yield schematically:

$$A-B \ + \ \underset{\overset{|}{A}}{\overset{\overset{A}{|}}{A-R-A}}$$

Large excess

Small amount
Tetrafunctional monomer

$$\longrightarrow \ A-(BA)\!_{\overline{y}_4}\!-BA-\underset{\overset{|}{AB-(AB)\!_{\overline{y}_2}\!-A}}{\overset{\overset{AB-(AB)\!_{\overline{y}_3}\!-A}{|}}{R-AB-(AB)\!_{\overline{y}_1}\!-A}}$$

Complete reaction

Tetrafunctional star

The total number of molecules will equal the number of tetrafunctional units in the system, while the total number of units in the entire molecule, x, depends upon the sum of the y-values of the individual chains. A size, x, much larger or smaller than average is only likely to occur if several (or all) of the chains in the molecule are abnormally large or small. Accordingly, the distribution will be *narrower* than conventional linear polycondensation (i.e., < 2). If this is not immediately obvious, look at it this way. Say we initially perform a linear polycondensation to essentially complete conversion. In our reaction pot is a distribution of linear chains with a polydispersity of 2. Now, holding an RA$_f$ molecule in one hand, we reach into the pot with our molecular tweezers and withdraw a molecule at random and attach it to one of the A groups of the RA$_f$ molecule. We now repeat this until all the A groups of the RA$_f$ molecule have attached chains. As the selection of the chain lengths was performed randomly, it is very unlikely that we would select, for example, only small (or large) molecules. In fact, such cooperation among statistically uncoordinated compounds will be comparatively rare and it is much more likely that the collection of molecules we select at random will contain molecules of average, large and small sizes. While the average molecular weight of the fully reacted star shaped molecules increases *vis-a-vis* the original linear molecules, the polydispersity decreases.

As before, we will define p as the probability that an A group has reacted. The probability that a particular chain contains y units is $p^y(1 - p)$. (This is similar to equation 4.13, but note that one more A group has reacted in this case, because of attachment to the RA$_f$ molecule.)

The probability that the f chains have lengths y$_1$, y$_2$, y$_3$ ......y$_f$, respectively is:

$$p^{y_1}p^{y_2}p^{y_3} \text{.............} p^{y_f} (1 - p)^f \tag{4.31}$$

Now, if the central unit, R, is counted as one of the total x:

$$y_1 + y_2 + y_3 + \text{----------} + y_f = (x - 1) \qquad (4.32)$$

then the probability of obtaining a f-armed x-mer may be expressed as:

$$P_{x,f} = p^{x-1} (1 - p)^f \qquad (4.33)$$

If you think about it, this result is straightforward. It is exactly the same as equation 4.13 describing the probability of reacting x - 1 groups to form a linear chain, but instead of a final term of (1 - p) we have $(1 - p)^f$, because there are f end groups in the f-armed star.

So far we have been able to derive all our results from the simplest type of probability argument. Now, however, we must take a step up in "degree of difficulty". Unlike the linear case, where P_x could be equated directly to mole fraction of x-mer species, X_x, here we must also include the condition that the lengths of the individual chains, $y_1, y_2, y_3 - - - - - y_f$, can vary from x-mer to x-mer. In other words, different x-mers can have their units distributed amongst the f arms in many different ways (e.g., all the same length, (f - 1) really short chains and one really long one, and so on). We don't care how the units are distributed, we are just counting those star molecules whose total number of units is equal to x. The probability that the molecule contains exactly x units distributed over f chains *in any manner whatsoever* is therefore equal to equation 4.33 multiplied by the total number of combinations of y's that fulfills the condition expressed by equation 4.32. This can be worked out by imagining the f blocks arranged in a line (of x - 1 units) instead of in a star-like fashion. To link them together would require f - 1 bonds. The problem is therefore reduced to the number of ways of distributing f - 1 bonds amongst the x - 1 units, i.e., the number of combinations of (x - 1) + (f - 1) things taken at a time. From simple combinatorics this is equal to:

$$\left[\frac{(x + f - 2)!}{(f - 1)! \, (x - 1)!} \right] \qquad (4.34)$$

The mole fraction of x-mers thus becomes:

$$X_{x,f} = \left[\frac{(x + f - 2)!}{(f - 1)! \, (x - 1)!} \right] p^{x-1} (1 - p)^f \qquad (4.35)$$

If you are not used to combinatorics, this is difficult to grasp at first, so we won't discuss the derivation of the corresponding equation for the weight fraction of x-mers and we will just present the result derived by Stockmayer[*].

$$w_{x,f} = \left[\frac{x(x + f - 2)!}{(f - 1)! \, (x - 1)!} \right] \left[\frac{(1 - p)^{f+1}}{fp + 1 - p} \right] p^{x-1} \qquad (4.36)$$

---

[*] W. Stockmayer, *J. Chem. Phys.*, **12**, 125 (1944)

This looks truly awful, but fortunately, using a number of simplifying approximations it was shown by Stockmayer that:

$$\frac{\overline{x}_w}{\overline{x}_n} \approx \left(1 + \frac{1}{f}\right) \tag{4.37}$$

This simple equation illustrates nicely the point we wish to make, that as f increases the distribution sharpens. For example, the polydispersities predicted for A–B polycondensation systems containing small amounts of RA_f molecules with functionalities of 4, 5 and 10, are 1.25, 1.20 and 1.10, respectively. For f = 2, which is equivalent to the case of combining two independent chains into one molecule,

$$A—(BA)_n—B + A—R—A + B—(AB)_m—A$$

$$\longrightarrow \quad A-(BA)_n—BA—R—AB-(AB)_m—A$$

the polydispersity is predicted to be 1.5. (Incidentally, an analogous situation is present in free radical polymerization when chain termination is exclusively by combination.)

It should be emphasized that in this section we have considered a rather special case of the polycondensation of A–B molecules in the presence of multifunctional molecules that contain only A functional groups. If we were to add B–B molecules to this system, network structures are possible, which lead us to our next major topic.

E. THEORY OF GELATION

The presence of polyfunctional units nearly always presents the possibility of forming chemical structures of macroscopic dimensions or *infinite networks*. Examples include phenolic resins, polyurethanes, gel formation in diene polymers etc. In this section we will be concerned with defining the critical conditions for the formation of infinite networks and molecular weight distributions for non-linear polymers.

Consider a polycondensation reaction involving two difunctional molecules and one trifunctional molecule, as illustrated below:

$$A—A + \quad \begin{array}{c} A \\ \diagdown \\ \diagup \\ A \end{array} \!\!\!\! {>}—A \quad + \quad B—B$$

We again impose the restriction that reaction can only occur between A and B. After some time, this would lead to typical branched structures like:

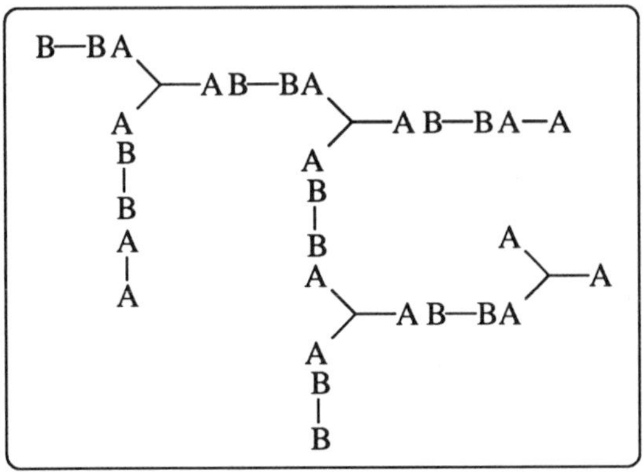

Our purpose will be to define the conditions under which an infinitely large chemical structure or infinite network will occur. Flory, in his analysis, invoked two major assumptions: first the principle of *equal reactivity*, where it is assumed that all functional groups, regardless of where they are situated, have the same reactivity; and second, that *intramolecular condensation* does not occur. The former is a reasonable assumption, although it is known that the secondary hydroxyl group in glycerol, $HO-CH_2-CH(OH)-CH_2-OH$, has a somewhat different reactivity than the primary OH groups. The latter assumption is not as reasonable, however, and we will see that intramolecular cyclization does lead to significant errors.

Flory defined a parameter, which he termed α, which is the probability that a given functional group of a branch unit leads, via a chain of bifunctional units, to another branch unit. This may appear to be a mouthful, but is actually a simple concept as we illustrate below:

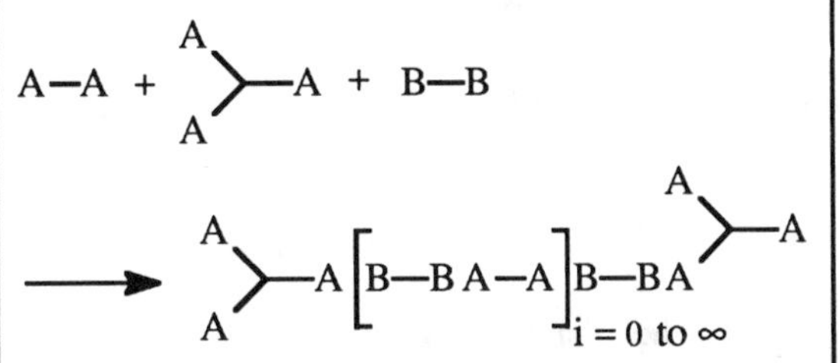

Definition of α

It may be helpful for some students to visualize a reaction pot in which we put known amounts of A–A, B–B and RA_3. After a certain reaction time let us pretend that we have a pair of molecular tweezers and can randomly pull out a molecule. We now ask the question, "What is the probability that the molecule selected at random has the structure shown above?" This may seem a rather strange question to ask, but bear with us as this is an example of the type of conceptual leap made by Flory that separates the extraordinary scientist who shakes up the field from the mundane scientist who merely extrapolates. Enough philosophy, we now need to define some terms.

Let:

p_A be the probability that an A group has reacted.

p_B be the probability that a B group has reacted.

ρ be the ratio of A's (reacted and unreacted) belonging to branch units to the total number of A's.

Hence:

The probability that a B group has reacted with a branch unit is $p_B\rho$.

The probability that a B group is connected to an A–A unit is $p_B(1 - \rho)$.

Now examine the chain linking two branch units:

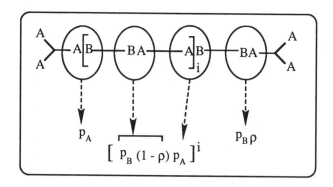

The probability that an A group is connected to a sequence of units as depicted above is thus $p_A[p_B(1 - \rho)p_A]^i p_B\rho$.

The parameter α is therefore given by:

$$\alpha = \sum_{i=0}^{\infty} \left[p_A p_B(1 - \rho) \right]^i p_A p_B \rho \tag{4.38}$$

This is another of those summations that converge to give:

$$\alpha = \frac{p_A p_B \rho}{\left[1 - p_A p_B(1 - \rho) \right]} \tag{4.39}$$

(For those of you who wish to work this out, note that;

$$\sum_{i=0}^{\infty} x^i = \sum_{i=1}^{\infty} x^{i-1} = \frac{1}{1 - x}$$

when $x < 1$.)

As in the case of linear polycondensation we will let r equal the ratio of A to B groups which leads to:

$$p_B = rp_A \tag{4.40}$$

[Note that if there are more A groups than B, then it follows that the probability that a B group has reacted is greater than that of the A group.]

Hence:

$$\alpha = \frac{rp_A^2 \rho}{\left[1 - rp_A^2(1 - \rho)\right]} \qquad (4.41)$$

or:

$$\alpha = \frac{p_B^2 \rho}{\left[r - p_B^2(1 - \rho)\right]} \qquad (4.42)$$

It is important to recognize that it is a relatively straightforward task to calculate α as a function of conversion, as both r and ρ are determined by the concentrations of the initial ingredients and the unreacted end groups, A or B, may be determined analytically by a variety of techniques at various stages of the reaction.

There are a two special cases that deserve mention as they further simplify equations 4.41 and 4.42. First, when there are no A–A groups present, $\rho = 1$ and α is given by:

$$\alpha = rp_A^2 = \frac{p_B^2}{r} \qquad (4.43)$$

Second, when A and B groups are present in equivalent amounts, $r = 1$ and $p_A = p_B = p$, and α is given by:

$$\alpha = \frac{p^2 \rho}{\left[1 - p^2(1 - \rho)\right]} \qquad (4.44)$$

Note that the above treatment is not completely rigorous. For example, the polycondensation of R–A_f and R'–B_g cannot be handled. In addition, differing reactivities such as the secondary hydroxyl in glycerol, which is less reactive than the primary hydroxyl groups, are not taken into account.

The Critical Value of α

Now we come to the clever bit. Flory reasoned that at some critical value of α, incipient formation of an infinite network* will occur and this will depend upon the functionality of the branching unit. Let us first consider a *trifunctional* branching unit. Each chain which reacts with this unit is succeeded by *two* more chains. Similarly, if both of these react with the branching unit, *four* chains are produced and then *eight* and so forth. Schematically:

---

* For those students who have never observed the onset of gelation, which is a truly spectacular phenomenon, we recommend that a rising bubble experiment be performed. A simple polycondensation involving say ethylene glycol, adipic acid and a trifunctional acid is carried out in a glass reaction pot containing a fine capillary tube that bleeds nitrogen bubbles into the material. One can observe the nitrogen bubbles rising monotonously for a long period of time (typically 2 hours) and then there is an remarkably abrupt loss of fluidity at the incipient gel point, when the bubbles stop rising and appear frozen in place. The precision and reproducibility of this measurement is quite amazing.

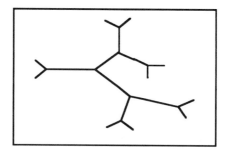

Now if $\alpha < 0.5$, there is a less than even chance that each chain will lead to a branch unit and thus to two more chains, etc. In fact, there is a greater than even chance that it will end at an unreacted functional group. Eventually, termination of chains must outweigh continuation of the network through branching. Thus we reach the important conclusion that when $\alpha < 0.5$, *all molecular structures must be limited.* In other words, all the molecules in the pot are of finite size.

On the other hand, if $\alpha > 0.5$, each chain has a better than even chance of reproducing two more chains. Molecular stuctures of an infinite size (networks) are now possible.

Accordingly, $\alpha = 0.5$ represents the critical condition for the *incipient formation of an infinitely large network* in a trifunctionally branched system. It is very important to recognize that when $\alpha > 0.5$ this does not mean that all the material is combined in an infinite structure. Both *gel* (infinite network) and *sol* (molecules of finite size) exist in varying amounts.

In general, if f = the functionality of the branching unit, gelation will occur when $\alpha(f - 1)$ exceeds unity. In another form:

$$\alpha_c = \frac{1}{(f - 1)} \tag{4.45}$$

where α_c is the *critical value* of α. This remarkably simple expression illustrates that α_c is dependent solely on the functionality of the branched molecule.

Experimental Testing of the Gelation Theory

Kienle and coworkers[*] studied the reaction between glycerol and *equivalent* amounts of a number of dibasic acids.

$$HO-CH_2-\underset{\underset{OH}{|}}{CH}-CH_2-OH + HOOC-R-COOH$$

Glycerol   Dibasic acid

The incipient gel point was observed to occur at $76.5 \pm 1\%$ esterification; in other words, p = 0.765. For this case $\rho = 1$ and $\alpha = p^2$. Substituting for p yields an experimentally determined value for α of 0.58. This is significantly

---

[*] R. H. Kienle et al., *JACS*, **61**, 2258 (1939); *ibid.* **61**, 2268 (1939); *ibid.* **62**, 1053 (1940); *ibid.* **63**, 481 (1941).

higher than the theoretical $\alpha_c = 0.50$. Some of the discrepancy was no doubt due to the different reactivity of the secondary OH group, but not all of it. Flory* studied a different system, the reaction between diethylene glycol with adipic or succinic acids (bifunctional) and varying amounts of a tricarballylic acid (trifunctional).

$$
\text{HO}-\text{CH}_2-\text{CH}_2-\text{OH} \;+\; \left\{
\begin{array}{l}
\text{HOOC}-\text{R}'-\text{COOH} \\[4pt]
\text{Adipic or succinic acid} \\[12pt]
\text{HOOC}-\text{R}-\text{COOH} \\
\qquad\qquad\;\; | \\
\qquad\qquad \text{COOH} \\[6pt]
\text{Tricarballylic acid}
\end{array}\right.
$$

Ethylene glycol

Recall that r and ρ are calculated from the relative amount of each component used initially and p_A can be determined by withdrawing small samples and titrating to obtain the extent of esterification of the acid groups. The gel point was determined experimentally by the loss of fluidity and the absence of rising bubbles. An accurate experimental value of p_A at the gel point was obtained by extrapolation.

The value of α at the gel point was then calculated using equation 4.41. For the four systems studied, values of $\alpha = 0.60 \pm 0.02$ were obtained which is again significantly higher than the theoretical value of $\alpha_c = 0.5$. This discrepancy is primarily caused by the failure of the theory to account for intramolecular condensation.

Flory also calculated the number average degree of polymerization from:

$$
\overline{x}_n = \frac{f\left(1 - \rho + \dfrac{1}{r}\right) + 2\rho}{f\left(1 - \rho + \dfrac{1}{r} - 2p_A\right) + 2\rho} \tag{4.46}
$$

In figure 4.6 we show a schematic diagram depicting the variation of p_A, $\overline{x}_n$, α and viscosity (η) as a function of reaction time. Note that $\overline{x}_n$ is not very large, nor is it increasing rapidly at the gel point. This means that at the gel point many molecules are present, but only a fractional amount of indefinitely large structures exist beyond the gel point. Note also the very rapid rise of the viscosity as the gel point is approached, allowing for a sharp transition which is experimentally easy to observe.

We have intimated that the discrepancy between the theoretical prediction and experimental observation of α at the gel point is primarily due to the fact that intramolecular condensation was not taken into consideration in the development of the theory. An elegant set of experiments by Stockmayer and Weil** showed

---

* P. J. Flory, *Principles of Polymer Chemistry*, Cornell University Press, 1953.
** W. H. Stockmayer and L. L. Weil, Chapter 6 in *Advancing Fronts in Chemistry*, S. B. Twiss, Ed., Reinhold, New York, 1945.

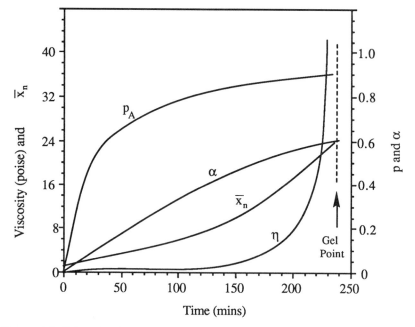

Figure 4.6 *Schematic diagram of a typical polyesterification involving a trifunctional monomer. Redrawn from the data of P. J. Flory,* Principles of Polymer Chemistry, Cornell University Press, 1953.

this indeed to be the case. These authors studied the reaction between pentaerythritol (tetrafunctional) and adipic acid (bifunctional).

$$HOOC-(CH_2)_4-COOH \;+\; HO-H_2C-\overset{\displaystyle CH_2-OH}{\underset{\displaystyle CH_2-OH}{\overset{|}{\underset{|}{C}}}}-CH_2-OH$$

Adipic acid   Pentaerythritol

As pentaerythritol is tetrafunctional the theoretical critical value of alpha is $\alpha_c = 0.333$. This corresponds to a $p_c = 0.577$. Experimentally, it was found that the gel point occurred when $p = 0.63$. This is again in line with the previous results; i.e., a result higher than predicted by theory. However, in a classic experiment the authors reasoned that the probability of intramolecular condensation would increase if the system was diluted with an inert solvent. A series of experiments were performed in the presence of different amounts of such a solvent and p was obtained at the gel point in each case. The results were then plotted against *inverse* concentration of the inert solvent in order to obtain an extrapolated value of p at *infinite concentration*; i.e., at $(c^{-1})_{c=\infty}$, where intramolecular condensation would be expected to be completely eliminated. The extrapolated value of p at $(c^{-1})_{c=\infty}$ was determined to be 0.578 ± 0.005 which is in remarkable agreement with theory.

F. RANDOM BRANCHING WITHOUT NETWORK FORMATION

This is the final system that we will consider in this chapter. It is included because it is an interesting special case of branching which currently (1996) happens to be undergoing a revival of sorts with the renewed interest in so-called hyperbranched polymers. Consider, if you will, the polycondensation of the monomer $A-R-B_{f-1}$. Again, A may only condense with B and vice versa. After some reaction has occurred typical structures such as that shown below (a 7-mer) will be present: A cursory look at this structure might lead you to the conclusion that an infinite network is inevitable. This is not the case, however. Note that each x-meric species contains $(x + 1)$ unreacted B groups (8 for the 7-mer shown above) and only one A group.

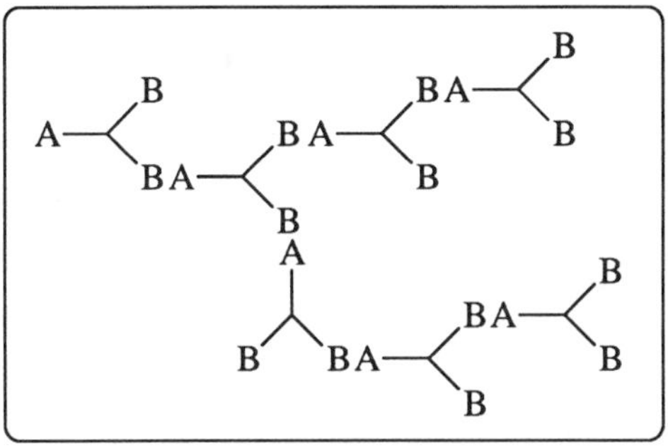

In fact, in general, analogous polymers formed from $A-R-B_{f-1}$ will always only contain one unreacted A and $(f - 2)x + 1$ unreacted B groups. Interestingly enough, if bifunctional A–B monomers are added to the $A-R-B_{f-1}$ system the essential character of the molecular structure remains identical.

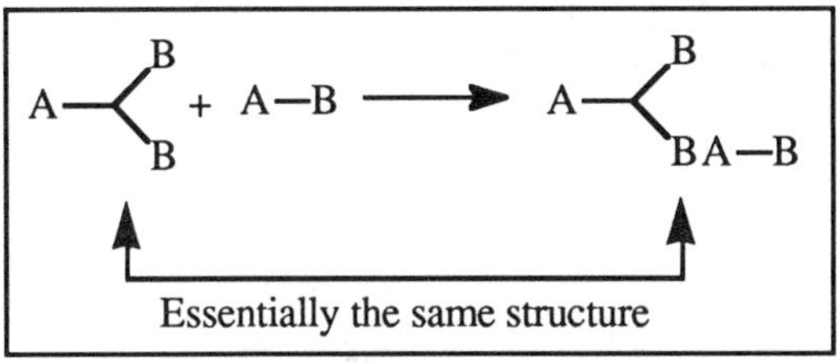

However, if A–A, or A–A and B–B molecules are added to the $A-R-B_{f-1}$ system, we immediately introduce the possibility of forming infinite structures. It is important to note that now stuctures are formed which have more than one A group and this leads to the possibility of networks.

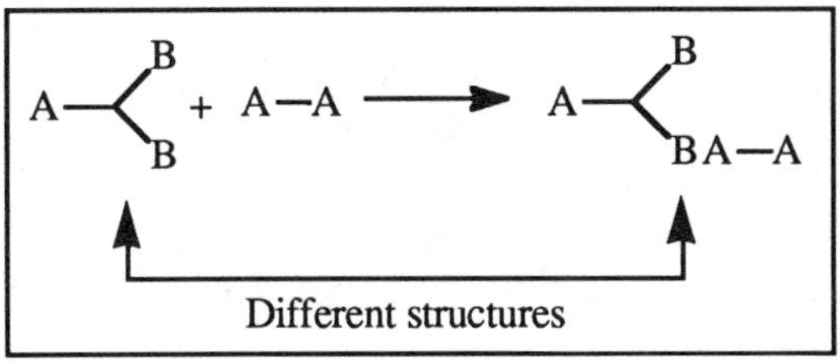

Let us now see if we can use the principles that we have developed in this chapter to explain why the A–R–B$_{f-1}$ system does not form infinite networks. For the polycondensation of A–R–B$_{f-1}$ we know from the definition of α that:

$$\alpha = p_B \tag{4.47}$$

(Remember the definition of α —the probability that a given functional group of a branch unit leads, via a chain of bifunctional units, to another branch unit—and here there are no bifunctional units and α is thus simply p_B.)

Now:

$$p_A = p_B (f - 1) \tag{4.48}$$

(This is self-evident, as there are (f - 1) more B groups than A groups.)

Substitution of equation 4.48 into 4.47 leads to:

$$\alpha = \frac{p_A}{(f - 1)} \tag{4.49}$$

We have previously seen (equation 4.45) that the critical value of α is:

$$\alpha_c = \frac{1}{(f - 1)} \tag{4.50}$$

The magnitude of α cannot equal α_c, as this requires p_A to equal 1. The extent of reaction can approach 1 but not attain it, thus the formation of network structures (gelation) is not possible! Very large, but finite molecules will be present at high extents of reaction.

Finally, for completeness we will present equations for the number and weight average degree of polymerization for the A–R–B$_{f-1}$ system, but without the derivation which is outside the scope of this book. The interested reader is referred to Flory's book for the details.

$$\bar{x}_n = \frac{1}{\left[1 - \alpha(f - 1) \right]} \tag{4.51}$$

$$\bar{x}_w = \frac{\left[1 - \alpha^2(f - 1) \right]}{\left[1 - \alpha(f - 1) \right]^2} \tag{4.52}$$

The polydispersity is thus given by:

$$\frac{\bar{x}_w}{\bar{x}_n} = \frac{1 - \alpha^2(f - 1)}{1 - \alpha(f - 1)} \tag{4.53}$$

For the polycondensation reaction of A–R–B$_2$, where f = 3 and α is calculated from equation 4.49, polydispersity values of approximately 6, 11 and 51 are calculated for p_A values of 0.90, 0.95 and 0.99, respectively. This broadening of the molecular weight distribution is thus in stark contrast to the narrowing discussed previously for the system consisting of A–B and a small amount of RA$_f$.

G. STUDY QUESTIONS

1. The extent of reaction for the linear polycondensation of A-B is defined as p. Write down the probability P_x that a molecule selected at random contains exactly x BA units.

2. The number and weight distribution functions for linear polycondensation are given below:

$$N_x = N_0 (1 - p)^2 p^{(x - 1)}$$

$$w_x = x (1 - p)^2 p^{(x - 1)}$$

   (a) Define all the symbols used in these equations
   (b) Sketch two curves showing the relationship between N_x and w_x as a function of x using two values of p = 0.95 and 0.99.
   (c) Describe the significance of the graphs you have drawn.

3. The number and weight average degrees of polymerization for linear polycondensation are given respectively by:

$$\bar{x}_n = \sum x X_x \text{ and } \bar{x}_w = \sum x w_x$$

   Define the symbols used in the above equations and show that the polydispersity of this system approaches two at high conversions.

4. Calculate the weight and number average *molecular weights* as a function of conversion for the linear polycondensation of the aminoacid:

$$NH_2 - (CH_2)_7 - COOH$$

   (a) Select appropriate values for the extent of reaction and display your results graphically.
   (b) 1g of acetic acid is added per 100g of the above aminoacid. Recalculate the number average molecular weights as a function of conversion and display your results graphically.
   (c) Comment on the significance of your results.

5. If 0.52 moles of terephthalic acid are reacted with 0.50 moles of ethylene glycol, what is the number average degree of polymerization at 90 and 100% reaction ?

6. A polycondensation reaction is performed using 46g of glycerol and 109.5g of adipic acid. A 10 ml aliquot is removed at the beginning of the reaction and titrated with a standard aqueous NaOH solution. 12.5 ml of the NaOH solution was required to neutralize the sample. The gel point was determined by the rising bubble technique. Just before the gel point, a 20 ml aliquot was removed and 5.88 ml of the standard NaOH was reqired to neutralize this sample. Calculate the value of the parameter α for this reaction and compare your result to the theoretical critical value of α. Comment on the significance of your result.

7. Calculate the extent of reaction at the gel point for a mixture of 0.8 moles of pentaerythritol, 0.5 moles of ethylene glycol and 2.0 moles of phthalic acid.

8. A chemist synthesizes a cyclic molecule containing six functional NH_2 groups that is incorporated in a polycondensation reaction with bifunctional adipic acid.

(a) What is the critical value of α for the onset of gelation ?

(b) If equivalent amounts of acid and amine functional groups are originally present, at what extent of reaction will the mixture theoretically gel ?.

(c) How would you explain experimental results that indicate a greater extent of reaction is required to reach the gel point than that predicted theoretically?

9. A polycondensation reaction involving an aminodicarboxyllic acid $(A-R-B_2)$ is performed. After a certain period of time an aliquot is removed and it is determined that the extent of reaction of carboxyllic acid groups is 0.43. Calculate the polydispersity of the polymer at this stage and comment upon the value you obtain.

10. Five different polymers, denoted [I] through [V], all based upon the bifunctional monomer ε-aminocaproic acid [$NH_2 (CH_2)_6 COOH$], were prepared by condensation polymerizations that were taken to high conversion. The feed compositions differed in the folllowing manner:

[I] No additional reagents

[II] A small amount of a monofunctional acid, RCOOH, is added

[III] A small amount of a trifunctional amine, $R(NH_2)_3$, is added

[IV] A small amount of a pentafunctional acid, $R(COOH)_5$, is added

[V] A small amount of a diaminoacid, $(NH_2)_2RCOOH$, is added

Using reasonable assumptions and with appropriate justification:

(a) Compare the relative molecular weights of [I] and [II]

(b) Compare the expected polydispersities of [I] and [II]

(c) Describe the type of polymers that are formed in [III], [IV] and [V]

(d) Compare the expected polydispersities of [III], [IV] and [V]

(e) If polymers [I] and [II] are mixed together what would be the effect on the polydispersity ?

H. SUGGESTIONS FOR FURTHER READING

(1)   P. J. Flory, *Principles of Polymer Chemistry*,
Cornell University Press, Ithaca, New York, 1953.

CHAPTER 5

Copolymerization

"See plastic Nature working to this end,
The single atoms to each other tend,
Attract, attracted to, the next in place'
Form'd and impell'd its neighbour to embrace."
—Alexander Pope

A. GENERAL OVERVIEW

At the beginning of this text we defined copolymers as those containing more than one type of chemical unit in the same chain. Copolymers come in various forms, according to the arrangements of units in the chain, and we defined four types in Chapter 1: statistical (random), alternating, block, and graft.

Copolymerization allows various monomers to be combined in such a way so as to provide materials with useful and sometimes unique properties. Linear polyethylene and isotactic polypropylene homopolymers are both semi-crystalline plastic materials (we will discuss exactly what we mean by semi-crystalline later), but a random (or more precisely, statistical) copolymer of the two, in the appropriate proportions, is an elastomer. Block copolymers of styrene and butadiene can be used to make thermoplastic rubbers or alternatively, an impact resistant rigid material, the property obtained being dependent upon the exact arrangement and the relative length of the blocks. Ionomers consist of statistical copolymers of a nonpolar monomer with (usually) a small proportion of ionizable units. Ethylene-co-methacrylic acid (containing ≈ 5% methacrylic acid) is an example. The protons on the acid are exchanged to form salts and these ionic species phase separate into domains that act to cross-link and toughen the polymer. One application of these materials was found to be as "cut proof" outer covers for golf balls, thus making them one of the most momentous discoveries of this century! Clearly, it is useful to spend some time discussing copolymerization, and we will do so here. We will initially consider some examples of the types of copolymers that can be produced by various chain and step-growth polymerizations and then examine the kinetics of free radical polymerization in more detail.

Copolymers are synthesized using the same types of chemistry as described in Chapter 2. Chain growth polymerization can be used to make statistical copolymers, for example, by including two monomers, say A and B, in the same reaction vessel. The active site, denoted * in the schematic overleaf, could be a radical, or ionic species, or a coordination complex (and could be a B rather than an A).

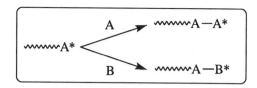

The distribution of species in the chain will depend upon the rate at which one monomer adds to an active site relative to the other. Usually this results in a statistical distribution of units in a chain that it is possible to describe in terms of rate constants (actually, their ratios). We will examine this in more detail shortly.

If instead of including the monomers in the reaction vessel at the same time, we add them sequentially, making sure that monomer is used up in each stage before adding a different batch, we can make block copolymers, *providing we use a living polymerization technique* (usually anionic). A sequential procedure can also be used to make graft copolymers by first copolymerizing a unit with an appropriate functional group, X, into a chain and then grafting onto these sites a second polymer with an end group Y that reacts with X. Alternatively, free radical sites on polymer chains swollen in monomer can be generated by irradiation. The free radical sites then add monomer.

Step or condensation polymers can also be used to make copolymers. Blocks, for example can be made by first separately polymerizing short polymer chains (macromers) with one type of functional group at each end (say an acid) and a different type of macromer with a distinct but complementary functional group at each of its ends (say an alcohol). The macromers can then be reacted to form chains (in this example, linked by ester groups). Thermoplastic elastomers, which amongst other things are used extensively in the automobile industry, are produced in a similar fashion, in this case taking advantage of the facile urethane chemical reaction.

Step copolymerization of different types of units in the same vessel usually leads to truly random copolymers which usually have the same composition as the feed (i.e., if we start with 50% A and 50% B units in a pot (the feed), then the resulting copolymer has 50% A and 50% B units). These copolymers are usually truly random because in nylons and polyesters, for example, transamidation and transesterification reactions occur, scrambling any initial distribution of species imposed by kinetic factors, so that a truly random product is formed. This may not hold true when chain growth is a rapid, irreversible, step-growth reaction, as in the *in-situ* formation of polyurethanes that are used in reaction injection molding (commonly referred to as RIM), and here the kinetics of the polymerization will be the deciding factor, as in chain growth polymerizations.

This brings us full circle. The kinetics of chain growth copolymerization, particularly free radical, have been extensively studied and we will now examine this in more detail. We will see that a knowledge of the kinetics allows the prediction of the instantaneous composition of the resulting copolymers. Then we will turn our attention to the powerful application of simple probability theory to sequence distribution and the characterization of copolymer chains.

B. THE COPOLYMER EQUATION

We will focus our attention on the types of copolymers that can be produced by copolymerizing two comonomers, which for generality we will label M_1 and M_2. It is possible to extend the treatment to copolymers with three or more units, but at the expense of a significant increase in algebra. It is the general principles of the relationship of copolymer composition to the initial concentration of monomers and the kinetics of chain addition that we are seeking to get across here, so we will neglect these complications. In the same vein, we will confine our discussion to free radical copolymerization. In principle, it should be possible to apply these or similarly derived equations to other types of polymerizations, but in doing so there are often significant complications that are beyond the scope of what we want to cover. For example, in a polymerization that uses a heterogeneous catalyst, polymers with different sequence distributions could be produced at different catalyst sites. The copolymer equation should be applicable to each type of site, but there may be no way to distinguish and separate chains produced from one site from those produced at another for the purpose of analysis. In the final product all the different types of chains are jumbled up together. The full scope of this difficulty should become more apparent as we discuss simple free radical copolymerization, because even here we will see that the composition of copolymer chains can vary dramatically throughout the course of a polymerization (i.e., as a function of conversion).

Free radical copolymerization involves the same steps as ordinary free radical polymerization; initiation, propagation and termination. With one or two exceptions, initiation and termination are much the same as in homopolymerization, and it is the propagation step that gives copolymerization its special character.

For the polymerization of a binary copolymer we can distinguish four possible propagation reactions. The first two apply to the reactions that can occur if the active site (radical) at the end of a growing chain happens to be monomer of type 1. This can either add another unit of type 1 with a rate constant given by k_{11} (i.e., reflecting the rate of adding a type 1 to a type 1 radical)

$$\text{wwwwM}_1^{\bullet} \ + \ M_1 \ \xrightarrow{\ k_{11}\ } \ \text{wwwwM}_1^{\bullet} \tag{I}$$

or a unit of type 2

$$\text{wwwwM}_1^{\bullet} \ + \ M_2 \ \xrightarrow{\ k_{12}\ } \ \text{wwwwM}_2^{\bullet} \tag{II}$$

where the rate constant k_{12} is a measure of the rate which a type 2 unit is added to a type 1 radical. Similar equations can be written for the addition to a type 2 radical:

$$\boxed{\text{\tiny www}M_2^{\bullet} + M_1 \xrightarrow{\;k_{21}\;} \text{\tiny www}M_1^{\bullet}}$$

(III)

$$\boxed{\text{\tiny www}M_2^{\bullet} + M_2 \xrightarrow{\;k_{22}\;} \text{\tiny www}M_2^{\bullet}}$$

(IV)

Here we have made a fundamental assumption, that the rate of addition of monomers depends only upon the nature of the radical species at the end of the growing chain. In other words, whether it is a

$$\text{\tiny www}M_1^{\bullet} \text{ or a } \text{\tiny www}M_2^{\bullet}$$

This is the so-called *terminal* model. It has been proposed that *penultimate* effects may be important in certain systems and a penultimate model has been constructed (see later) where there is now a dependence on the character of the final two units in the growing chain:

$$\text{\tiny www}M_1{-}M_1^{\bullet} \text{ or } \text{\tiny www}M_2{-}M_1^{\bullet} \text{ etc.}$$

This gets algebraically complicated, as we now have to consider eight rate constants instead of four. We will ignore the penultimate model for now, for three reasons:

1) The terminal model illustrates the general principles that most concern us.

2) The terminal model is adequate in describing most copolymerizations.

3) In situations where it is apparently inadequate, where there appears to be a variation in the values of the ratios of the rate constants (the reactivity ratios, defined below) with initial monomer composition, alternate explanation to penultimate effects can often (but not always) be invoked (e.g., preferential solution of polymer chain by one of the two monomers).

If we accept the general validity of the terminal model, we can immediately describe some interesting limiting conditions:

1) If $k_{11} \gg k_{12}$ and $k_{22} \gg k_{21}$, then a chain whose active site happens to be a type 1 monomer unit always prefers to add another 1, while the ends whose active sites are 2's always prefer to add other 2's. Thus there is a tendency to form block copolymers, or even homopolymers (if the tendency is so great that one type of monomer unit almost never adds a unit of the other type).

2) If $k_{12} \gg k_{11}$ and $k_{21} \gg k_{22}$ then there is a large tendency for the copolymers to be alternating, based on the same type of kinetic argument.

3) If $k_{12} = k_{11}$ and $k_{21} = k_{22}$ then the copolymers will be truly random. Clearly, the ability to calculate copolymer composition and sequence distributions is a valuable thing (as is the ability to measure them, to make sure the model is applicable to the system you happen to be interested in). We can obtain an equation for the instantaneous copolymer composition by applying our old friend, the steady state assumption, to the radical species $\text{\tiny www}M_1^{\bullet}$ and $\text{\tiny www}M_2^{\bullet}$ present in the polymerization. This states that in the (short) time frame that we will be considering (remember that a free radical polymerization can initiate, propagate and terminate all in a matter of seconds), the concentration

of the radical species is constant. This, in turn, means that the radical species ⋱⋱⋱⋱$M_1^{\bullet}$ and ⋱⋱⋱⋱$M_2^{\bullet}$ are formed and removed at equal rates. If we examine reactions I through IV we immediately see that in only two of these, II and III, are radicals of one type generated from the other (i.e., forming one and removing the other). In reactions I and IV a type 1 radical is converted to another type 1 radical and a similar thing happens to type 2 radicals. There is no net change. Applying the steady state assumption we can therefore say that the rate of generation of new species of type I (reaction III) is equal to their rate of removal (reaction II), so that:

$$k_{12}[M_1^{\bullet}][M_2] = k_{21}[M_2^{\bullet}][M_1] \tag{5.1}$$

(Obviously, we get the same equation by considering the rate of formation and disappearance of type 2 radicals.) Note that we have let the concentration of chain radical species ⋱⋱⋱⋱$M_1^{\bullet}$ and ⋱⋱⋱⋱$M_2^{\bullet}$ be represented by $[M_1^{\bullet}]$ and $[M_2^{\bullet}]$, respectively.

It will (hopefully) be recalled that in our discussion of free radical polymerization we used the steady state assumption to obtain an equation for the concentration of radical species in terms of the concentration of monomer units, because in general we do not know the former and they are very difficult to measure, while the concentration of monomers is usually known (at least at the beginning of the reaction). Here we have more than one radical species, so we obtain an equation for their ratio:

$$\frac{[M_1^{\bullet}]}{[M_2^{\bullet}]} = \frac{k_{21}[M_1]}{k_{12}[M_2]} \tag{5.2}$$

We can now write equations for the rate of consumption of monomer 1 and monomer 2:

$$-\frac{d[M_1]}{dt} = k_{11}[M_1^{\bullet}][M_1] + k_{21}[M_2^{\bullet}][M_1] \tag{5.3}$$

$$-\frac{d[M_2]}{dt} = k_{22}[M_2^{\bullet}][M_2] + k_{12}[M_1^{\bullet}][M_2] \tag{5.4}$$

The ratio of these two equations is then:

$$\frac{d[M_1]}{d[M_2]} = \frac{k_{11}[M_1]\left(\dfrac{[M_1^{\bullet}]}{[M_2^{\bullet}]}\right) + k_{21}[M_1]}{k_{22}[M_2] + k_{12}[M_2]\left(\dfrac{[M_1^{\bullet}]}{[M_2^{\bullet}]}\right)} \tag{5.5}$$

where the top and bottom parts of the equation have each been divided by $[M_2^{\bullet}]$.

A rearrangement into a more useful form can be obtained by substituting from equation 5.2 and then defining reactivity ratios:

$$r_1 = \frac{k_{11}}{k_{12}} \text{ and } r_2 = \frac{k_{22}}{k_{21}} \tag{5.6}$$

to obtain:

$$\frac{d[M_1]}{d[M_2]} = \frac{\left(r_1 \frac{[M_1]}{[M_2]}\right) + 1}{\left(r_2 \frac{[M_2]}{[M_1]}\right) + 1} \tag{5.7}$$

This equation is the amount of monomer 1 that is being polymerized relative to monomer 2 in a time dt. We also sometimes see equation 5.7 expressed simply as:

$$y = \frac{d[M_1]}{d[M_2]} = \frac{1 + r_1 x}{1 + \frac{r_2}{x}} \tag{5.8}$$

where x is the monomer feed ratio:

$$x = \frac{[M_1]}{[M_2]}$$

These monomers are not just disappearing into thin air, but are being incorporated into the polymer, so this equation must be equal to the ratio of monomers found in the polymer chains (or segments of those chains) that are being produced *in this very small time period* and is therefore called the *instantaneous copolymer composition equation.*

If, instead of the ratio of the monomers in the polymer, we wish to know the mole fraction of each, we can use a rearrangement of equation 5.7. If F_1 is the mole fraction of monomer 1 that is being incorporated into the copolymer at some instant of time, while f_1 is the mole fraction of monomer that is left in the reaction mass at that same instant, then:

$$F_1 = \frac{d[M_1]}{d[M_1] + d[M_2]} \tag{5.9}$$

from the definition of a mole fraction. Similarly:

$$f_1 = \frac{[M_1]}{[M_1] + [M_2]} \tag{5.10}$$

Combining equations 5.7, 5.9 and 5.10 we can obtain:

$$F_1 = \frac{r_1 f_1^2 + f_1 f_2}{r_1 f_1^2 + 2 f_1 f_2 + r_2 f_2^2} \tag{5.11}$$

The quantities F_2 and f_2 are, of course, simply given by $1 - F_1$ and $1 - f_1$, respectively.

The reactivity ratios that we have employed in the copolymer equations are extremely important quantities. First, they are a measure of the relative preference of a radical species for the monomers. For example, r_1 is a measure of the relative preference of $[M_1^\bullet]$ for M_1 relative to M_2 ($= k_{11} / k_{12}$, the rate of adding 1 divided by the rate of adding 2). Second, these two quantities, r_1 and r_2, represent the only two independent rate variables that we need to know, rather than the four individual rate constants k_{11} etc. Finally, experimental methods for the measurements of r_1 and r_2 have been established, so that in principle we should be able to calculate copolymer composition.

This is where we come to the next problem. Equation 5.7 tells us the ratio of monomers in the polymer *at some instant* relative to the ratio of the concentration of monomers in the reaction mass (i.e., a measure of the amount of monomers that at that time has not yet reacted). The relative proportions of unreacted monomer *after polymerization has proceeded for a while may be very different to the proportions at the very start of the polymerization* (e.g. one monomer could react far more quickly than the other and is therefore used up faster). This means that we have *composition drift,* the copolymer composition usually varies throughout the polymerization (except for a special case we will consider below) and differs from the monomer "feed" concentration. It also means that we can not employ equations 5.7 or 5.11 by simply treating $[M_1] / [M_2]$ as a constant over the course of the polymerization. To reiterate, the copolymer equations relate the proportions of monomer incorporated into the copolymer at some point in time during the polymerization *to the monomer concentrations present at the same time.* Nevertheless, the effect of the reactivity ratios on the types of copolymer produced can be calculated and we will now consider this in more detail.

C. REACTIVITY RATIOS AND COPOLYMER COMPOSITION

Before getting down to the nitty gritty of calculating copolymer composition for some real monomer pairs, it is useful to consider some limiting cases, as these provide a good "feel" for the significance of the reactivity ratios and the results we might expect from using the copolymer equation.

Special Case I: $r_1 = r_2 = 0$

This means that a radical of type 1 never wants to add to itself (i.e. $k_{11} = 0$), a radical of type two never wants to add to itself either ($k_{22} = 0$), but each can add to the other (if k_{12}, k_{21} are also not equal to zero). In this situation a perfectly alternating copolymer is produced until all of the monomers are used up (if $[M_1]_0 = [M_2]_0$ where the subscript 0 indicates the initial monomer concentration) or until one of the monomers is used up, at which point polymerization stops (i.e., if $[M_1]_0 \neq [M_2]_0$)

Special Case II: $r_1 = r_2 = \infty$

If $r_1 = \infty$ then k_{12} must be equal to zero. Similarly, for $r_2 = \infty$, $k_{21} = 0$. Accordingly, monomer 1 will only add to M_1 radicals and monomer 2 will only add to M_2 radicals. Hence two homopolymers are produced.

Special Case III: $r_1 = r_2 = 1$

This is a very special case, as will soon be apparent, and means that monomers 1 and 2 add with equal facility to either type of radical active site, M_1 and M_2 radicals.   The resulting copolymer not only has a truly random distribution of monomers, but the copolymer composition is exactly the same as that of the initial monomer concentration and stays so throughout the course of the polymerization (i.e. $F_1 = f_1 = (f_1)_{t=0}$).

Special Case IV: $r_1r_2 = 1$

This situation is called an *ideal* copolymerization.   The distribution of monomers in the chain at any point in the polymerization is truly random, but the copolymer composition is not usually the same as the composition of the monomers in the reaction mass (i.e. $F_1 \neq f_1$). To see how this comes about, we first write out the condition $r_1r_2 = 1$ in terms of the rate constants and rearrange the terms to obtain:

$$\frac{k_{11}}{k_{12}} = \frac{k_{21}}{k_{22}} \text{ (ideal copolymerization)} \qquad (5.12)$$

This means that each radical displays the same preference for one of the monomers over the other. (i.e., the *ratio* of the rate of adding an M_1 to an M_1 radical to the rate of adding an M_2 to an M_1 radical is exactly equal to the equivalent ratio of adding an M_1 to an M_2 radical relative to an M_2 to an M_2 radical).   In other words, no matter what the radical active site happens to be, the rate of adding M_1 relative to an M_2 is always the same (it could be 100:10 for k_{11}/k_{22}, for example).   This results in a random placement of the units.   This is confusing to many students, because they don't have a good grasp of probability theory (but, then again, how many people really do?).   A random placement does not necessarily mean there has to be 50% A units and 50% B units placed according to coin-toss statistics along a chain.   There might be only 10% A units in the chain, but providing that the probability that an A is located at any point along that chain is in this case 0.1, then the distribution of units is random.   In an ideal copolymerization the distribution of units in the copolymer formed at some instant is usually different than the distribution of monomers in the reaction vessel at that time, but the sequence distribution in the polymer is nevertheless random.   Furthermore, unless we have the special limiting case that $r_1 = r_2 = 1$, then there will also be composition drift (which, you will recall, means that the copolymer composition changes during the polymerization).

In practice, special case II ($r_1 = r_2 = \infty$) has not been obtained, but there are circumstances in which special case I ($r_1 = r_2 = 0$) has been approached (i.e. where both r_1 and r_2 are very small).   We will present an example of this below. In certain systems ideal copolymerization ($r_1r_2 = 1$) has been approached, but in most cases r_1r_2 is less than 1.

A more precise description of the effect of the values of the reactivity ratios can be obtained by making plots of the instantaneous copolymer composition. By this we mean a plot of F_1 against f_1 (or, equivalently, F_2 against f_2).   This

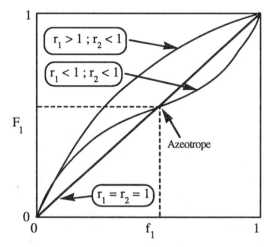

Figure 5.1 *Schematic diagram of instantaneous compositions of copolymers.*

does not tell us exactly how the copolymer composition varies as a function of conversion for a given starting monomer composition (we will consider this later), but does tell us the copolymer composition we will obtain for some given proportion of monomers. Examples are shown in figure 5.1. The special case of $r_1 = r_2 = 1$ gives the straight line along the diagonal of this plot. The curve that is above the diagonal over the whole composition range represents a situation where the copolymer formed instantaneously is always richer in units of type 1 than the monomer mixture from which it was polymerized. This happens when $r_1 > 1$ and $r_2 < 1$. If both r_1 and r_2 are less than 1, then the curve that crosses the diagonal can be obtained. The point where the curve intersects the diagonal is termed the azeotrope, by analogy to the situation where a vapour has the same composition as a liquid with which it is in equilibrium. Here the copolymer being formed has the same composition as the monomer feed, so that $d[M_1]/d[M_2] = M_1/M_2$. Under this condition, of course, there is no composition drift, because monomer is incorporated into the polymer exactly in proportion to its composition in the feed.

The Determination of Reactivity Ratios

Reactivity ratios have been determined by a number of different methods. The most frequently used depend upon algebraic rearrangement of the copolymer equation to give a form which allows linear plots. Of course, this assumes the validity of the terminal model, but if we accept this then we can start by first defining:

$$\frac{[M_1]}{[M_2]} = x \quad \text{and} \quad \frac{d[M_1]}{d[M_2]} = y \tag{5.13}$$

then the copolymer equation has the simple form shown previously in equation 5.8:

$$y = \frac{1 + r_1 x}{1 + \dfrac{r_2}{x}} \qquad (5.14)$$

Solving for r_2:

$$r_2 = \frac{x(r_1 x + 1)}{y} - x \qquad (5.15)$$

The graph obtained using this equation is called the Mayo-Lewis plot, and the equation can be solved by an intersection method by plotting values of r_2 vs r_1 for given values of x and y. This can be confusing at first sight, because we don't know r_1 and r_2! However, if (in principle) we pick any old value of r_1, then we can use equation 5.15 to calculate r_2. This is repeated for various experimental values of x and y and as a result a set of straight lines is obtained, as illustrated in figure 5.2. The intersection point defines the values of r_1 and r_2. In practice, the lines intersect over an area and some method for picking the best point within this area has to be applied.

Up to now we have said nothing about the experimental quantities x and y, the ratio of the monomer concentrations incorporated into the copolymer relative to the ratios of the monomer concentrations in the reaction mass at some instant of time. Because in most cases both compositions vary as the reaction proceeds, it is usual practice to limit the conversion to a small fraction of the monomer mixture. The quantity x is then assumed to be accurately given by the ratio of the initial monomer concentrations, while y is given by the average or the integral value of the copolymer composition over the range of conversion chosen and is usually measured by spectroscopic methods. Obviously, the smaller the degree of initial conversion, the more accurately y becomes a representation of the instantaneous copolymer composition. Each experiment, where y is determined for a polymer obtained with a chosen value of x, requires a separate synthesis

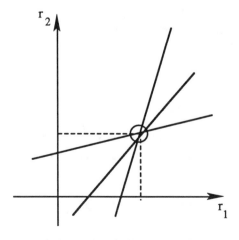

Figure 5.2 Schematic diagram of a Mayo-Lewis plot.

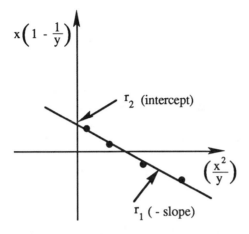

Figure 5.3 *Schematic diagram of a Fineman-Ross plot.*

and spectroscopic characterization. Accordingly, the determination of sufficient data points to determine r_1 and r_2 with a reasonable degree of accuracy is a time consuming business, but such experiments have nevertheless been performed on a wide range of monomer pairs.

An alternative graphical procedure for determining the reactivity ratios, but one which depends upon performing the same experimental measurements, is the Fineman-Ross plot. The copolymer equation written in terms of the quantities x and y, given above, can be rearranged into the following form:

$$x\left(1 - \frac{1}{y}\right) = r_1\left(\frac{x^2}{y}\right) - r_2 \tag{5.16}$$

A plot of the quantities in brackets then gives a straight line whose slope is r_1 and whose intercept is $-r_2$. A schematic example of such a plot is shown in figure 5.3.

Both the Mayo-Lewis and Fineman-Ross methods depend upon use of the instantaneous copolymer equation and as such are subject to various errors. First of all, the validity of the terminal model is assumed. In certain systems penultimate effects may play a role. Furthermore, it has been shown that these linearization methods transform the error structure in such a way that linear least squares procedures are inappropriate. An alternative method designed to overcome these shortcomings, attributed to Kelen and Tüdős,[*] has gained popularity in recent years. Their equation is written in the form:

$$\eta = r_1 \xi - \frac{r_2}{\alpha}(1 - \xi) \tag{5.17}$$

where:

---

[*] T. Kelen and F. Tüdős, *J. Macromol. Sci.-Chem.*, **A9(1)**, 1 (1975).

$$\eta = \frac{\dfrac{x(y-1)}{y}}{\alpha + \dfrac{x^2}{y}} \quad \text{and} \quad \xi = \frac{\dfrac{x^2}{y}}{\alpha + \dfrac{x^2}{y}} \tag{5.18}$$

A plot of η vs ξ calculated from the experimental data (x and y) should yield a straight line with extrapolated intercepts at $\xi = 0$ corresponding to $-r_2/\alpha$ and at $\xi = 1$ corresponding to r_1. The parameter α serves to distribute the experimental data uniformly and symmetrically between the limits of 0 and 1, and is generally determined from:

$$\alpha = \sqrt{\left(\frac{x^2}{y}\right)_{low} \cdot \left(\frac{x^2}{y}\right)_{high}} \tag{5.19}$$

where the subscripts low and high represent the lowest and highest values of x^2/y calculated from the experimental data. Figure 5.4 shows a typical Kelen-Tüdõs plot that we prepared to determine the reactivity ratios of n-decyl methacrylate and p(t-butyldimethylsilyloxy)styrene[*].

Finally, it should be mentioned that problems can also arise in the determination of reactivity ratios when the monomers are capable of association, or, for example, in the polymerization of dienes, where chain propagation can proceed in various ways (1,2; 1,4; etc).

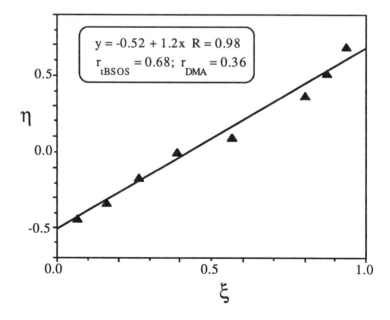

Figure 5.4 *Kelen-Tüdõs plot for* n-decyl methacrylate-co-p(t-butyldimethylsilyloxy)styrene *copolymers.*

---

[*] Y. Xu, P. C. Painter and M. M. Coleman, *Polymer*, **34**, 3010 (1993).

Factors Affecting Monomer Reactivity

There are a number of factors that affect the ability of a monomer to react in a copolymerization and hence the reactivity ratios. The obvious ones include:

a) Steric factors
b) Resonance stabilization of the radical site
c) Polarity of the double bond

Monomers with very bulky groups could clearly have trouble polymerizing. This is why styrene polymerizes (free radically) in a head-to-tail fashion, but in poly(vinylidene fluoride), $-(CH_2-CF_2)_n-$, the smaller fluorine groups allow the formation of some head-to-head and tail-to-tail sequences. In general, monomers with the common structure $CH_2=CR_2$ (R being the bulky group) will polymerize, but monomers of the type $CHR=CHR$ will not (they can add to radicals of another kind, however).

If the monomer has a structure that allows a radical site to be resonance stabilized (e.g., styrene, where several mesomeric resonance forms can be written), then in general there is a decrease in the reactivity of this monomer, relative to one where there is no resonance stabilization.

Electron donating and electron withdrawing groups can affect the polarity of the double bond. Obviously, a monomer of one type would preferentially add a monomer of the other and in certain circumstances the effect can be so marked that charge transfer complexes are produced.

Alfrey and Price attempted to account for two of these factors, resonance stabilization and polarity, with parameters labeled Q and e. These were determined relative to a chosen monomer standard (styrene) and allowed the calculation of reactivity ratios. Although this scheme ignores steric effects it nevertheless allowed reactivity ratios to be at least semi-quantitatively estimated at a time when their determination was more difficult. The interested reader is referred to standard polymer chemistry books for further details as we will not cover this here.

D. COPOLYMER SEQUENCE DISTRIBUTION AND THE APPLICATION OF PROBABILITY THEORY

The methods described above to determine reactivity ratios, although still used, have their origin in a time when even the determination of the composition of a copolymer was a difficult task, usually accomplished by elemental analysis. The advent of spectroscopic methods, particularly nuclear magnetic resonance (nmr) spectroscopy, not only facilitated the determination of overall copolymer composition, but allowed the measurement of sequence distributions. The theoretical treatment of copolymerization can be extended to the prediction of the frequency of occurrence of various sequences (e.g., $M_1M_1M_1$ "triads"). As we will see later, these and higher sequences can now be routinely measured by nmr spectroscopy.

Brace yourself, for now we are going to change from the kinetic description that we have used so far in this chapter and that chemists usually like, to one based upon probability theory, which is a far more convenient way to describe sequence distributions, but when introduced to students is often met with vacant stares. This really shouldn't worry even those who are not mathematically inclined, as the concepts that we will discuss are truly simple (and elegant) and just require us to think a little differently, practice a little and learn a new language. Regarding the latter, we will immediately start by changing the nomenclature used so far in this chapter. We would do this deliberately, even if we didn't have to, just to satisfy the sadistic streak within us! But, in truth, it is conventional and more convenient when applying probability theory to use A and B to define the two different comonomers, rather than M_1 and M_2 favored in kinetic treatments. First we will be considering *general statistical relationships* between the various experimental sequence probabilities that *do not* depend upon polymerization mechanism. Then we will consider *specific relationships* characteristic of specific polymerization mechanisms. But first a review of some definitions.

General Statistical Relationships

Probability theory is designed to tackle situations in which there is more than one possible outcome and attempts to answer questions concerning the likelihood of the various possible results. To remind you, the probability of an event E is defined as:

$$P\{E\} = \frac{N_E}{N} \tag{5.20}$$

where N_E is the number of E events occurring and N is the total number of events. Hence, the probability is equal to the *number fraction* of desirable events occurring and the terms number fraction and probability will often be used interchangeably. Probabilities are restricted to a range of 0 to 1, which represent impossibility and certainty, respectively.

Let us define:

$$P_n\{x_1 x_2 x_3 x_4 \ldots\ldots x_n\} \tag{5.21}$$

as the measured number fraction of a particular sequence of units $x_1\ x_2\ x_3\ \ldots\ x_n$. For example, P{AABA} is the probability of the sequence AABA in a copolymer chain consisting of only two monomers A and B. It should be self-evident that if there are only two events, A or B, then $P_1\{A\}$ and $P_1\{B\}$ are the *mole fractions* of A and B in the polymer, respectively, and:

$$P_1\{A\} + P_1\{B\} = 1 \tag{5.22}$$

Similarly, when considering pairs of units or *diads*:

$$P_2\{AA\} + P_2\{AB\} + P_2\{BA\} + P_2\{BB\} = 1 \tag{5.23}$$

The concentrations of the lower order placements can be expressed as sums of the two appropriate higher order sequences, since any sequence has a *successor* or *predecessor* which can only be either an A or a B.

$$P_1\{A\} = P_2\{A\underline{A}\} + P_2\{A\underline{B}\} = P_2\{\underline{A}A\} + P_2\{\underline{B}A\}$$
$$P_1\{B\} = P_2\{B\underline{B}\} + P_2\{B\underline{A}\} = P_2\{\underline{B}B\} + P_2\{\underline{A}B\} \tag{5.24}$$

$$\phantom{P_1\{A\} = }\underbrace{\phantom{P_2\{A\underline{A}\} + P_2\{A\underline{B}\}}}_{\text{successor}}\quad\underbrace{\phantom{P_2\{\underline{A}A\} + P_2\{\underline{B}A\}}}_{\text{predecessor}}$$

This leads to the important conclusion:

$$P_2\{AB\} = P_2\{BA\} \tag{5.25}$$

In other words, the number fraction of AB sequences must equal the number fraction of BA sequences.

Similarly, for the diads in terms of *triads*:

$$P_2\{AA\} = P_3\{AA\underline{A}\} + P_3\{AA\underline{B}\} = P_3\{\underline{A}AA\} + P_3\{\underline{B}AA\}$$
$$P_2\{AB\} = P_3\{AB\underline{A}\} + P_3\{AB\underline{B}\} = P_3\{\underline{A}AB\} + P_3\{\underline{B}AB\}$$
$$P_2\{BA\} = P_3\{BA\underline{A}\} + P_3\{BA\underline{B}\} = P_3\{\underline{A}BA\} + P_3\{\underline{B}BA\}$$
$$P_2\{BB\} = P_3\{BB\underline{A}\} + P_3\{BB\underline{B}\} = P_3\{\underline{A}BB\} + P_3\{\underline{B}BB\} \tag{5.26}$$

$$\underbrace{\phantom{P_3\{BB\underline{A}\} + P_3\{BB\underline{B}\}}}_{\text{successor}}\quad\underbrace{\phantom{P_3\{\underline{A}BB\} + P_3\{\underline{B}BB\}}}_{\text{predecessor}}$$

which requires that:

$$P_3\{AAB\} = P_3\{BAA\}$$
$$P_3\{BBA\} = P_3\{ABB\} \tag{5.27}$$

Note the *reversibility* of these sequences. In fact this is a general principle and it is always true that the probability or number fraction of a sequence is always equal to that of its mirror image. For example, $P_6\{ABABBA\}$ must equal $P_6\{ABBABA\}$.

Conditional Probability

We will require a more specific way of defining a particular sequence for our purposes, which is referred to as conditional probability.

We define:

$$P\{x_{n+1}/x_1x_2.......x_n\} \tag{5.28}$$

as the conditional number fraction of units of a particular polymer chain with a sequence $x_1x_2.....x_n$ and which has the particular unit x_{n+1} as the next unit. For example, $P\{A/B\}$ is the number fraction of A units given that the previous unit was a B. It follows therefore that :

$$P_2\{AA\} = P_1\{A\}\ P\{A/A\}$$
$$P_2\{BA\} = P_1\{B\}\ P\{A/B\} \tag{5.29}$$

and:

$$P_3\{AAA\} = P_1\{A\}\ P\{A/A\}\ P\{A/AA\}$$
$$P_3\{BAA\} = P_1\{B\}\ P\{A/B\}\ P\{A/BA\} \tag{5.30}$$

where $P\{A/A\}$, $P\{A/B\}$, $P\{A/AA\}$, $P\{A/BA\}$, represent the conditional number fraction that the next unit is A given that the preceding placement (or placement pair) was A, B, AA, or BA, respectively.

Rearranging these equations, the conditional probabilities can be computed as ratios of the measured probabilities:

$$P\{A/A\} = P_2\{AA\} / P_1\{A\} \qquad P\{B/A\} = P_2\{AB\} / P_1\{A\}$$
$$P\{A/B\} = P_2\{BA\} / P_1\{B\} \qquad P\{B/B\} = P_2\{BB\} / P_1\{B\}$$
$$P\{A/AA\} = P_3\{AAA\} / P_2\{AA\} \qquad P\{B/AA\} = P_3\{AAB\} / P_2\{AA\}$$
$$P\{A/BA\} = P_3\{BAA\} / P_2\{BA\} \qquad P\{B/BA\} = P_3\{BAB\} / P_2\{BA\}$$
$$P\{A/AB\} = P_3\{ABA\} / P_2\{AB\} \qquad P\{B/AB\} = P_3\{ABB\} / P_2\{AB\}$$
$$P\{A/BB\} = P_3\{BBA\} / P_2\{BB\} \qquad P\{B/BB\} = P_3\{BBB\} / P_2\{BB\}$$

$$(5.31)$$

and

$$P\{A/A\} + P\{B/A\} = 1$$
$$P\{A/B\} + P\{B/B\} = 1$$
$$P\{A/AA\} + P\{B/AA\} = 1$$
$$P\{A/BA\} + P\{B/BA\} = 1$$
$$P\{A/AB\} + P\{B/AB\} = 1$$
$$P\{A/BB\} + P\{B/BB\} = 1 \qquad (5.32)$$

At this point you may be wondering, "Why are these guys giving us all this stuff?" They are tools you will need to analyze microstructure. Unfortunately, we are not finished yet, but the equations have a simple structure and once you have used them a couple of times they become familiar. (In a sense, its a bit like having your teeth cleaned. You don't like it, but it's necessary!)

Conditional Probability of Different Orders

In considering compound events like $P_3\{AAA\}$, it is perfectly legitimate to break them down into conditional probabilities of *different orders*: For example:

$$P_3\{AAA\} = P_1\{A\} \, P\{A/A\} \, P\{A/AA\} \qquad (5.33)$$

The general equation relating the number fraction or probability of a sequence of n consecutive units, in this case all A's, is:

$$P_n\{A^n\} = P_1\{A\} \, P\{A/A\} \, P\{A/AA\}\text{-------}P\{A/A^{n-1}\} \qquad (5.34)$$

However, we will be considering polymerization systems where only *one type of conditional probability is applicable*. For example, in free radical polymerization where commonly two reactivity ratios are used to describe the instantaneous composition of a copolymer chain formed from two different monomers, first order Markovian statistics (the so-called terminal model) are applicable. The order referred to here depends upon how many preceding units are included in the conditional probability. For the general case of kth order Markovian statistics we have:

$$P_n\{A^n\} = \left(P_1\{A^k\} \, P\{A/A\}^k\right)^{n-k} \tag{5.35}$$

When $k = 0$, we have *zero order Markovian* statistics, also known as *Bernoullian* statistics, and:

$$P_n\{A^n\} = \left(P_1\{A\}\right)^n \tag{5.36}$$

When $k = 1$, we have *first order Markovian* statistics which correspond to the *terminal* model, and:

$$P_n\{A^n\} = P_1\{A\}\left(P\{A/A\}\right)^{n-1} \tag{5.37}$$

When $k = 2$, we have *second order Markovian* statistics which correspond to the *penultimate* model, and:

$$P_n\{A^n\} = P_2\{AA\}\left(P\{A/AA\}\right)^{n-2} \tag{5.38}$$

As an example, consider the sequence ABABA. In *Bernoullian* statistical terms the probability may be expressed as:

$$P_5\{ABABA\} = P_1\{A\} \, P_1\{B\} \, P_1\{A\} \, P_1\{B\} \, P_1\{A\}$$

or

$$P_5\{ABABA\} = \left(P_1\{A\}\right)^3\left(P_1\{B\}\right)^2 \tag{5.39}$$

In *1st order Markovian* statistical terms:

$$P_5\{ABABA\} = P_1\{A\} \, P\{B/A\} \, P\{A/B\} \, P\{B/A\} \, P\{A/B\}$$

or

$$P_5\{ABABA\} = P_1\{A\} \left(P\{B/A\}\right)^2 \left(P\{A/B\}\right)^2 \tag{5.40}$$

You can see that this is quite easy, you just write down all the conditional probabilities required to describe the sequence, then gather all like terms. But if you are having trouble with this stuff, try not to miss the wood for the trees. Remember that nmr now provides a way to measure sequences such as this, so these equations have direct practical importance.

In *2nd order Markovian* statistical terms:

$$P_5\{ABABA\} = P_2\{AB\} \, P\{A/AB\} \, P\{B/BA\} \, P\{A/AB\}$$

or

$$P_5\{ABABA\} = P_2\{AB\} \left(P\{A/AB\}\right)^2 P\{B/BA\} \tag{5.41}$$

Some Useful Parameters

There are a number of parameters that will be of particular use to us in the forthcoming discussion.

Number Fraction of Sequences of A Units

The number fraction of sequences of A units of length n, $N_A(n)$, is defined as the number of such sequences divided by the number of all possible sequences of length n:

$$N_A(n) = \frac{P_{n+2}\{BA_nB\}}{\sum_1^\infty P_{n+2}\{BA_nB\}} \qquad (5.42)$$

Obviously a sequence of n A units must be preceded and succeeded by a B unit. Since,

$$\sum_1^\infty P_{n+2}\{BA_nB\} = P_3\{B\underline{AB}\} + P_4\{BA\underline{AB}\} + P_5\{BAA\underline{AB}\} \dots\dots\text{etc.}$$

all the sequences end in AB. We therefore have:

$$\sum_1^\infty P_{n+2}\{BA_nB\} = P_2\{AB\} \qquad (5.43)$$

then:

$$N_A(n) = \frac{P_{n+2}\{BA_nB\}}{P_2\{AB\}} \qquad (5.44)$$

For example, the number fraction of say a sequence of three A's, i.e. AAA, is equal to the number fraction of BAAAB sequences divided by the number fraction of AB diads.

$$N_A(3) = \frac{P_5\{BAAAB\}}{P_2\{AB\}} \qquad (5.45)$$

Number Average Length of A or B Runs

The number average length of A or B runs, respectively, denoted by the symbols $\bar{l}_A$ and $\bar{l}_B$, have the definition:

$$\bar{l}_A = \frac{\sum_1^\infty n\, N_A(n)}{\sum_1^\infty N_A(n)} \qquad (5.46)$$

Now:

$$\sum_1^\infty N_A(n) = 1$$

and substituting equation 5.44:

$$\bar{l}_A = \frac{\sum\limits_{1}^{\infty} n\, P_{n+2}\{BA_nB\}}{P_2\{AB\}} \tag{5.47}$$

and since:

$$\sum_{1}^{\infty} n\, P_{n+2}\{BA_nB\} = (1 \cdot P_3\{B\underline{A}B\}) + (2 \cdot P_4\{B\underline{AA}B\}) + (3 \cdot P_5\{B\underline{AAA}B\}) + \text{...etc.}$$

which is equivalent to:

$$\sum_{1}^{\infty} n\, P_{n+2}\{BA_nB\} = P_1\{A\} \tag{5.48}$$

then:

$$\bar{l}_A = \frac{P_1\{A\}}{P_2\{AB\}} \quad \text{and, by analogy,} \quad \bar{l}_B = \frac{P_1\{B\}}{P_2\{BA\}} \tag{5.49}$$

Remember $P_2\{AB\} = P_2\{BA\}$, and $P_1\{A\} = 1 - P_1\{B\}$ so the number average length of A or B runs may be readily calculated if one has molar composition and diad sequence data. We will be using $\bar{l}_A$ and $\bar{l}_B$ information later in our discussions of copolymer microstructure.

The Run Fraction or Number

Although we will not use it much in this book, the run fraction or number, R, is included because it has been commonly used in the nmr literature. It is defined as the fraction of A and B sequences (runs) occurring in a polymer chain.

Consider, for example, the portion of a polymer chain depicted below:

$$\underline{A}\ \underline{B}\ \underline{AA}\ \underline{B}\ \underline{A}\ \underline{B}\ \underline{AAA}\ \underline{BB}\ \underline{A}\ \underline{BBBB}\ \underline{AA}\ \underline{B}$$

We will assume that the chain is sufficiently long so that end effects can be ignored. The chain above contains 20 units arranged in 12 alternating runs (underlined). The run number is thus $12 / 20 = 0.6$. Note that every run of A units is terminated by an A–B link. Similarly, B runs are terminated by B–A links.

Thus:

$$R = \text{Fraction of } (AB + BA) \text{ links}$$
$$= P_2\{AB\} + P_2\{BA\} = 2P_2\{AB\} \tag{5.50}$$

And since (equation 5.24):

$$P_1\{B\} = P_2\{BA\} + P_2\{BB\}$$

we arrive at the rather useful relationships:

$$P_2\{AA\} = P_1\{A\} - \frac{R}{2} \tag{5.51}$$

$$P_2\{BB\} = P_1\{B\} - \frac{R}{2} \tag{5.52}$$

A Measure of the Departure from Randomness

A convenient measure of the deviation from random statistics may be obtained from a parameter denoted by the symbol, χ, and defined as:

$$\chi = \frac{P_2\{AB\}}{P_1\{A\}\, P_1\{B\}} \tag{5.53}$$

and we know that for a completely random process:

$$P_2\{AB\} = P_1\{A\} \times P_1\{B\}$$

Accordingly, random (Bernoullian) statistics are obeyed when the number fraction of AB (or BA) diads equals the mole fraction of A's multiplied by the mole fraction of B's in the copolymer, or when $\chi = 1$.

If the number fraction of AB diads exceeds $P_1\{A\} \times P_1\{B\}$ in the copolymer, then we have an alternating tendency. In other words, the number of AB diads is greater than that calculated for a truly random case. For the limiting case of a completely alternating copolymer, $P_1\{A\} = P_1\{B\} = 0.5$, and as there are no AA or BB sequences, $P_2\{AB\} = P_2\{BA\} = 0.5$. Thus the value of χ is 2.

On the other hand, if the number fraction of AB diads is less than that calculated from $P_1\{A\} \times P_1\{B\}$ in the copolymer, then we have a "blocking" tendency. If you think about it the limit in this case is two homopolymers of A and B where there are no AB links and the value of χ is 0. For real block copolymers the value of χ will approach 0, the greater the molecular weight of the blocks, the fewer AB links, and the closer χ will get to 0.

In summary:

$\chi = 1$	Completely random copolymer
$\chi > 1$	Copolymer with an alternating tendency
$\chi = 2$	Completely alternating copolymer
$\chi < 1$	Copolymer with a "blocky" tendency
$\chi \approx 0$	Block copolymer

The Terminal Model

We will now revisit the copolymer equation (terminal model) that we discussed earlier in this chapter having obtained some necessary background in probability theory. When the rate of addition of monomer to a growing chain depends upon the nature of the terminal end group, there are two independent conditional probabilities. These two conditional probabilities can be derived in terms of the kinetics of polymerization using the following approach:

Terminal Group	Added Group	Rate	Final Product
---A*	A	$k_{AA}\,[A^*]\,[A]$	---AA*
---B*	A	$k_{BA}\,[B^*]\,[A]$	---BA*
---A*	B	$k_{AB}\,[A^*]\,[B]$	---AB*
---B*	B	$k_{BB}\,[B^*]\,[B]$	---BB*

In terms of conditional probabilities we can write, for example:

$$P\{A/A\} = \frac{\text{Rate of reaction producing AA*}}{\text{Sum of all reactions involving A*}} \qquad (5.54)$$

Therefore:

$$P\{A/A\} = \frac{k_{AA}[A^*][A]}{k_{AA}[A^*][A] + k_{AB}[A^*][B]} \qquad (5.55)$$

and using the same definition of reactivity ratios (equation 5.6) and dividing through by $[A^*][B]$:

$$P\{A/A\} = \frac{\dfrac{k_{AA}[A]}{k_{AB}[B]}}{\dfrac{k_{AA}[A]}{k_{AB}[B]} + 1} = \frac{r_A x}{1 + r_A x} \qquad (5.56)$$

where x is once again the monomer feed ratio = $[A]/[B]$. Similarly:

$$P\{B/B\} = \frac{r_B}{r_B + x} \qquad (5.57)$$

There are two further independent conditional probabilities that we shall require, $P\{A/B\}$ and $P\{B/A\}$. These are readily obtained from:

$$P\{A/B\} = 1 - P\{B/B\} = \frac{x}{r_B + x} \qquad (5.58)$$

$$P\{B/A\} = 1 - P\{A/A\} = \frac{1}{r_A x + 1} \qquad (5.59)$$

Recall that, *in general*, (equation 5.25):

$$P_2\{AB\} = P_2\{BA\} \qquad (5.60)$$

For the *terminal model* we can write:

$$P_1\{A\} P\{B/A\} = P_1\{B\} P\{A/B\} \qquad (5.61)$$

Rearranging and substituting for equations 5.58 and 5.59 we obtain:

$$y = \frac{P_1\{A\}}{P_1\{B\}} = \frac{P\{A/B\}}{P\{B/A\}} = \frac{1 + r_A x}{1 + \dfrac{r_B}{x}} \qquad (5.62)$$

This is the now familiar instantaneous copolymer composition first derived by Mayo and Lewis.

Similarly, the run number, the number average length of A or B runs, the deviation from randomness, the number and weight fractions of A or B sequences, can be expressed in terms of conditional probabilities or reactivity ratios. Recall:

$$R = 2P_2\{BA\} \tag{5.63}$$

As first order Markovian statistics are applicable to the *terminal model*, this is equivalent to:

$$R = 2P_1\{B\}P\{A/B\} = \frac{2P\{B/A\}}{P\{A/B\} + P\{B/A\}} P\{A/B\} \tag{5.64}$$

We'll leave the student to prove this (hint: start with equation 5.25). Substituting equations 5.58 and 5.59 we get:

$$R = \frac{2}{r_A x + 2 + \dfrac{r_B}{x}} \tag{5.65}$$

Similarly for χ:

$$\chi = P\{A/B\} + P\{B/A\} = \frac{r_A x + 2 + \dfrac{r_B}{x}}{r_A x + 1 + r_A r_B + \dfrac{r_B}{x}} \tag{5.66}$$

and for $\bar{l}_A$ and $\bar{l}_B$:

$$\bar{l}_A = \frac{1}{P\{B/A\}} = 1 + r_A x \tag{5.67}$$

$$\bar{l}_B = \frac{1}{P\{A/B\}} = 1 + \frac{r_B}{x} \tag{5.68}$$

and for $N_A(n)$ and $N_B(n)$:

$$N_A(n) = \left(\frac{r_A x}{1 + r_A x}\right)^{n-1} \left(1 - \frac{r_A x}{1 + r_A x}\right) \tag{5.69}$$

$$N_B(n) = \left(\frac{\dfrac{r_B}{x}}{1 + \dfrac{r_B}{x}}\right)^{n-1} \left(1 - \frac{\dfrac{r_B}{x}}{1 + \dfrac{r_B}{x}}\right) \tag{5.70}$$

and finally for the weight fraction of A (or B) sequences, $w_A(n)$ and $w_B(n)$:

$$w_A(n) = n \left(\frac{r_A x}{1 + r_A x}\right)^{n-1} \left(1 - \frac{r_A x}{1 + r_A x}\right)^2 \tag{5.71}$$

$$w_B(n) = n \left(\frac{\dfrac{r_B}{x}}{1 + \dfrac{r_B}{x}}\right)^{n-1} \left(1 - \frac{\dfrac{r_B}{x}}{1 + \dfrac{r_B}{x}}\right)^2 \tag{5.72}$$

Having derived all of these equations for the terminal model, let us now apply them to the problem of compositional drift.

Copolymer Composition as a Function of Conversion

For a batch system (i.e., one where we are not continuously introducing monomer) we can calculate the composition of the copolymer we obtain as a function of conversion, by which we mean the fraction of *total* monomer that has been used up. If we start with monomer compositions $[A]_0$ and $[B]_0$ and at some point the monomer composition remaining (i.e., unpolymerized) is $[A]$ and $[B]$, then we define the conversion as:

$$\text{Conversion} = \frac{\left([A]_0 + [B]_0\right) - \left([A] + [B]\right)}{\left([A]_0 + [B]_0\right)} \tag{5.73}$$

Obviously, we cannot simply use equation 5.62 and assume that $[A]$ and $[B]$ are constants over the entire course of the polymerization. Calculations must be performed using some variation or form of the following steps:

a) Choose a short period of time over which $[A]$ and $[B]$ are assumed to be constant.

b) Starting with the initial known concentrations, $[A]_0$ and $[B]_0$ at the very beginning of the reaction calculate how much monomer of each type is incorporated into the polymer in this chosen time period.

c) Calculate the amount of each type of monomer that remains after this period of time.

d) Repeat these steps for each subsequent time interval starting with the newly calculated monomer concentrations.

There are two points concerning these calculations that need to be mentioned. First, it is usually not convenient to pick a "time period" as such. Instead an interval in which a given amount of total monomer is used up (i.e., the amount of A plus the amount of B) is defined. Obviously, the errors in assuming constant monomer concentrations get smaller as we choose shorter and shorter intervals. This can make for a lot of calculations if we wish to know how the copolymer composition varies over the entire course of a polymerization. Fortunately, this is the type of problem that is ideally suited to solution by construction of a simple computer program. We have developed such a program[*] to calculate not only the composition of copolymers but also the microstructure as a function of conversion based upon the equations presented in the previous section above. The input necessary is the initial monomer concentrations $[A]$ and $[B]$ and the reactivity ratios r_A and r_B. A further input, denoted STEP, is the interval between successive calculations. The computer calculates and prints out the instantaneous copolymer composition, χ, $\bar{l}_A$, $\bar{l}_B$, $N_A(n)$, $N_B(n)$, $W_A(n)$, and $W_B(n)$ for n = 1 to 10 at the given interval. It then calculates the new monomer concentrations and repeats the process for the next interval etc. An interval of 0.01 gives one hundred points for complete reaction.

---

[*] M. M. Coleman and W. D. Varnell, *J.Chem. Ed.* **59**, 847, (1982).

To illustrate the type of information gained from these computer calculations several examples are presented. All reactivity ratios were taken from the published data listed in the polymer handbook and represent typical values. No attempt was made to ensure the validity of any specific set of reactivity ratios.

Case I. Vinylidene Chloride (VDC) and Vinyl Chloride (VC)

For this calculation equimolar initial concentration of the monomers was chosen (i.e., [VDC] = [A] = 0.5, [VC] = [B] = 0.5), and the reactivity ratios employed were r_{VDC} = 4.0 and r_{VC} = 0.2. The compositional variation as a function of conversion is illustrated in figure 5.5. The copolymer composition is initially rich in VDC, but this decreases as the polymerization proceeds. Polymer chains containing equimolar concentrations of VDC and VC are formed at approximately 63% conversion, after which chains rich in VC are formed. This graph convincingly demonstrates the large variation of copolymer composition with degree of conversion and illustrates that a wide compositional distribution of polymer chains will be present for a copolymer synthesized to moderately high conversions. In other words, if the polymerization was taken to completion, (i.e., 100% conversion), the *average* composition of the copolymer is still 50:50, but it is made up of a wide distribution of chains of different composition.

Table 5.1 shows a summary of the results obtained at six specific degrees of conversion (in our calculations, one hundred such intervals are determined). The number average lengths of a given sequence $\bar{l}_A$ or $\bar{l}_B$ vary considerably. Initially $\bar{l}_A$ the number average length of VDC sequences is 5.0 while the corresponding number average length of VC sequences, $\bar{l}_B$ is only 1.2. Throughout the polymerization, $\bar{l}_A$ decreases while $\bar{l}_B$ increases. At 80% conversion, polymer chains are being formed with $\bar{l}_A$ values of 1.1 and $\bar{l}_B$ of 7.5. The calculated values of χ, which are close to unity, indicate that the polymerization of VDC and VC is nearly random.

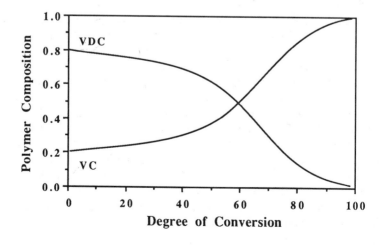

Figure 5.5 *Compositional variation for vinylidine chloride/vinyl chloride copolymers.*

***Table 5.1** Copolymerization of VDC/VC.*

	Monomer Concentration	% Conversion	Polymer Composition	$\bar{I}_A$, $\bar{I}_B$ values	χ value
VDC =	0.50	Initial	0.81	5.0	1.03
VC =	0.50		0.19	1.2	
VDC =	0.42	10	0.79	4.5	1.04
VC =	0.48		0.21	1.2	
VDC =	0.34	20	0.76	4.0	1.04
VC =	0.46		0.24	1.3	
VDC =	0.20	40	0.68	3.0	1.05
VC =	0.40		0.32	1.4	
VDC =	0.08	60	0.51	1.9	1.06
VC =	0.32		0.49	1.9	
VDC =	0.01	80	0.13	1.1	1.02
VC =	0.19		0.87	7.5	

***Table 5.2** Number Fraction of $(VDC)_n$ and $(VC)_n$ Sequences.*

Polymer Composition	n = 1	2	3	4	5	6	7	8	9	10
VDC = 0.81	0.20	0.16	0.13	0.10	0.08	0.07	0.05	0.04	0.03	0.03
VC = 0.19	0.83	0.14	0.02	0.00	0.00	0.00	0.00	0.00	0.00	0.00
VDC = 0.76	0.25	0.19	0.14	0.11	0.08	0.06	0.04	0.03	0.03	0.02
VC = 0.24	0.79	0.17	0.04	0.01	0.00	0.00	0.00	0.00	0.00	0.00
VDC = 0.51	0.52	0.25	0.12	0.06	0.03	0.01	0.01	0.00	0.00	0.00
VC = 0.49	0.54	0.25	0.12	0.05	0.02	0.01	0.01	0.00	0.00	0.00
VDC = 0.13	0.89	0.10	0.01	0.00	0.00	0.00	0.00	0.00	0.00	0.00
VC = 0.87	0.13	0.12	0.10	0.09	0.08	0.07	0.06	0.05	0.04	0.04

The table header for Table 5.2 reads: Number Fraction of $(A)_n$ and $(B)_n$ Sequences.

From the calculated number of A and B sequences shown in table 5.2, it is apparent that initially the vast majority of VC units exist as isolated units while there is a large distribution of sequence lengths of VDC units. At 60% conversion, the polymer composition is nearly equimolar and the distribution of the number fraction of VDC and VC is almost identical, with approximately 50% as single units (i.e., A or B), 25% as dimers (i.e., AA or BB) and 12% as trimers

Table 5.3 *Weight Fraction of (VDC)$_n$ and (VC)$_n$ Sequences.*

Polymer Composition	n =	Weight Fraction of (A)$_n$ and (B)$_n$ Sequences								
	1	2	3	4	5	6	7	8	9	10
VDC = 0.81	0.04	0.06	0.08	0.08	0.08	0.08	0.07	0.07	0.06	0.05
VC = 0.19	0.69	0.23	0.06	0.01	0.00	0.00	0.00	0.00	0.00	0.00
VDC = 0.76	0.06	0.09	0.11	0.11	0.10	0.09	0.08	0.07	0.06	0.05
VC = 0.24	0.62	0.26	0.08	0.02	0.01	0.00	0.00	0.00	0.00	0.00
VDC = 0.51	0.27	0.26	0.19	0.12	0.07	0.04	0.02	0.01	0.01	0.00
VC = 0.49	0.29	0.27	0.19	0.11	0.07	0.04	0.02	0.01	0.01	0.00
VDC = 0.13	0.79	0.17	0.03	0.00	0.00	0.00	0.00	0.00	0.00	0.00
VC = 0.87	0.02	0.03	0.04	0.05	0.05	0.05	0.05	0.05	0.05	0.05

(i.e., AAA or BBB sequences). At 80% conversion polymer chains are being formed with number fractions of A and B sequences that are almost the converse of that calculated at the start of the polymerization.

Table 5.3 shows the corresponding weight fraction of A and B sequences. It is apparent from a comparison of tables 5.2 and 5.3 that the number and weight fraction distributions vary significantly. Consider, for example, the number and weight fraction of VDC sequences at the start of the polymerization reaction (initial). On a number basis, isolated VDC units are present in greater amounts than any other species (i.e., dimers, trimers, etc.). However, on a weight basis, sequences containing four and five VDC units predominate. This is analogous to the number and weight average molecular weight distributions found for linear polycondensation.

Case II. Styrene (St) and Maleic Anhydride (MAH)

The copolymerization of St and MAH is a classic case of a highly alternating copolymer. In this example initial concentrations of 0.75 M for St and 0.25 for MAH, and reactivity ratios of $r_{St} = 0.04$ and $r_{MAH} = 0.015$ were used. Figure 5.6 shows the compositional variation as a function of conversion. Although we have started with a 3:1 molar feed ratio of St to MAH, the composition of the polymer chains formed is approximately equimolar up to about 50% conversion at which time St-rich, and eventually almost pure polystyrene, chains are formed.

Table 5.4 shows a summary of the calculated results for several specific degrees of conversion. At the start of the polymerization the polymer composition is calculated to be 53% St and 47% MAH. The $\bar{l}_A$ and $\bar{l}_B$ values are both approximately unity, which together with the calculated value of 1.89 for χ demonstrates that the polymer chains formed have a predominantly alternating microstructure. As the polymerization proceeds toward approximately 50% conversion there is a tendency toward increasingly rich St polymer chains being

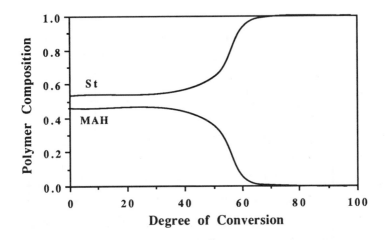

Figure 5.6 *Compositional variation for styrene/maleic anhydride copolymers.*

formed. However, the $\bar{l}_B$ value is invariably unity, which demonstrates that the MAH units are exclusively one unit in length. The $\bar{l}_A$ value, on the other hand, increases while χ decreases. The number fraction of A and B sequences is also informative (see table 5.5). Note that for MAH the fraction of isolated B groups is unity regardless of the extent of reaction. This illustrates that the MAH-radical cannot add to another MAH-monomer. Until about 50% conversion, the number fraction of isolated St sequences is also dominated by isolated units. In essence, this is an excellent example of a predominantly alternating copolymer.

Table 5.4 *Copolymerization of St/MAH.*

	Monomer Concentration	% Conversion	Polymer Composition	$\bar{l}_A$, $\bar{l}_B$ values	χ value
ST =	0.75	Initial	0.53	1.1	1.89
MAH =	0.25		0.47	1.0	
ST =	0.70	10	0.53	1.1	1.88
MAH =	0.20		0.47	1.0	
ST =	0.59	30	0.55	1.2	1.82
MAH =	0.11		0.45	1.0	
ST =	0.48	50	0.64	1.8	1.57
MAH =	0.03		0.36	1.0	
ST =	0.40	60	0.96	25	1.04
MAH =	0.001		0.04	1.0	

Table 5.5 *Number Fraction of (St)$_n$ and (MAH)$_n$ Sequences.*

Polymer Composition	n =	Number Fraction of (A)$_n$ and (B)$_n$ Sequences								
	1	2	3	4	5	6	7	8	9	10
St = 0.53	0.89	0.10	0.01	0.00	0.00	0.00	0.00	0.00	0.00	0.00
MAH = 0.47	1.00	0.00	0.00	0.00	0.00	0.00	0.00	0.00	0.00	0.00
St = 0.55	0.82	0.15	0.03	0.00	0.00	0.00	0.00	0.00	0.00	0.00
MAH = 0.45	1.00	0.00	0.00	0.00	0.00	0.00	0.00	0.00	0.00	0.00
St = 0.64	0.57	0.25	0.10	0.05	0.02	0.01	0.00	0.00	0.00	0.00
MAH = 0.36	1.00	0.00	0.00	0.00	0.00	0.00	0.00	0.00	0.00	0.00
St = 0.96	0.04	0.04	0.04	0.04	0.03	0.03	0.03	0.03	0.03	0.03
MAH = 0.04	1.00	0.00	0.00	0.00	0.00	0.00	0.00	0.00	0.00	0.00

Table 5.6 *Copolymerization of VDC/MA.*

	Monomer Concentration	% Conversion	Polymer Composition	$\bar{I}_A$, $\bar{I}_B$ values	χ value
VDC =	0.50	Initial	0.50	2.0	1.00
MA =	0.50		0.50	2.0	
VDC =	0.25	50	0.50	2.0	1.00
MA =	0.25		0.50	2.0	
VDC =	0.05	99	0.50	2.0	1.00
MA =	0.05		0.50	2.0	

Table 5.7 *Number Fraction of (VDC)$_n$ and (MA)$_n$ Sequences.*

Polymer Composition	n =	Number Fraction of (A)$_n$ and (B)$_n$ Sequences								
	1	2	3	4	5	6	7	8	9	10
VDC = 0.50	0.50	0.25	0.13	0.06	0.03	0.02	0.01	0.00	0.00	0.00
MA = 0.50	0.50	0.25	0.13	0.06	0.03	0.02	0.01	0.00	0.00	0.00

Case III. Vinylidine Chloride (VDC) and Methyl Acrylate (MA)

This is an interesting copolymer where the reactivity ratios for both monomers equal unity. For this example we chose equimolar monomer concentra-

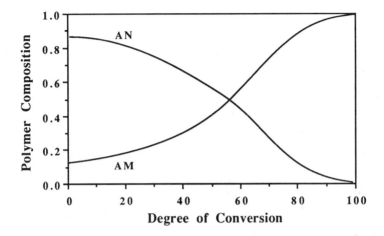

Figure 5.7 *Compositional variation for acrylonitrile/allyl methacrylate copolymers.*

tions, i.e., [VDC] = [A] = 0.5, [MA] = [B] = 0.5. Table 5.6 summarizes the computed data. The composition of the copolymer is also equimolar and does not vary throughout the polymerization reaction. Over the entire course of the polymerization the $\bar{l}_A$ and $\bar{l}_B$ values equal 2.0 and the χ value is unity, indicating a completely random polymerization process. The number fractions of A and B sequences are also identical (table 5.7) and reflect a random distribution of the monomers incorporated into the polymer chain.

Case IV. Acrylonitrile (AN) and Allyl Methacrylate (AM)

In this case, reactivity ratios chosen from the literature were r_{AN} = 9.55 and r_{AM} = 0.515 and represent a copolymerization of two monomers where the product $r_A r_B > 1$.

Table 5.8 *Copolymerization of AN/AM.*

	Monomer Concentration	% Conversion	Polymer Composition	$\bar{l}_A$, $\bar{l}_B$ values	χ value
AN =	0.50	Initial	0.87	10.6	0.76
AM =	0.50		0.13	1.5	
AN =	0.33	20	0.82	7.7	0.71
AM =	0.47		0.18	1.7	
AN =	0.09	55	0.50	3.2	0.62
AM =	0.37		0.50	3.2	
AN =	0.01	80	0.12	1.5	0.77
AM =	0.19		0.89	11.3	

Table 5.9 *Number Fraction of* $(AN)_n$ *and* $(AM)_n$ *Sequences.*

Polymer Composition	n = 1	2	3	4	5	6	7	8	9	10
	Number Fraction of $(A)_n$ and $(B)_n$ Sequences									
AN = 0.87	0.10	0.09	0.08	0.07	0.06	0.06	0.05	0.05	0.04	0.04
AM = 0.13	0.66	0.22	0.08	0.03	0.01	0.00	0.00	0.00	0.00	0.00
AN = 0.82	0.13	0.11	0.10	0.09	0.07	0.07	0.06	0.05	0.04	0.04
AM = 0.18	0.58	0.24	0.10	0.04	0.02	0.01	0.00	0.00	0.00	0.00
AN = 0.50	0.31	0.21	0.15	0.10	0.07	0.05	0.03	0.02	0.02	0.02
AM= 0.50	0.31	0.22	0.15	0.10	0.07	0.05	0.03	0.02	0.02	0.02
AN = 0.12	0.68	0.22	0.07	0.02	0.01	0.00	0.00	0.00	0.00	0.00
AM = 0.89	0.09	0.08	0.07	0.07	0.06	0.06	0.05	0.05	0.04	0.04

An equimolar monomer feed ratio was employed, i.e., [AN] = [A] = 0.5, [AM] = [B] = 0.5. Figure 5.7 illustrates the wide compositional variation of the copolymer during the polymerization reaction. Initially the copolymer chains are rich in AN but the amount of AM increases progressively as the reaction proceeds. At approximately 55% conversion, the polymer chains formed contain about the same molar concentration of AN and AM. Table 5.8 shows the computed results for polymer chains formed at four specific degrees of conversion. The χ values obtained are less than unity indicating a definite tendency toward a "blocky" microstructure. This is also reflected in the $\bar{l}_A$ and $\bar{l}_B$ values and the number fraction of $(A)_n$ and $(B)_n$ sequences (table 5.9).

Summary

It is convenient to compare χ, $\bar{l}_A$, $\bar{l}_B$, $N_A(n)$ and $N_B(n)$ for polymer chains that are formed from approximately equimolar concentrations of the monomers for the four different copolymers considered. This is displayed in table 5.10. For the VDC / MA copolymer $\bar{l}_A$ and $\bar{l}_B$ are equal to 2.0 and χ is unity, which is indicative of a completely random copolymer. Note also that the values of $N_A(n)$ and $N_B(n)$ are identical and the distribution of the sequence lengths is entirely consistent with random placement of the two monomers. The equimolar VDC / VC copolymer chain formed at approximately 60% conversion is also close to random. A χ value of 1.06 together with $\bar{l}_A = \bar{l}_B = 1.9$ suggests largely random placement of the two monomers with a very slight tendency toward alternation. The $N_A(n)$ and $N_B(n)$ values are close to those seen for the VDC/MA copolymer, which also reflects a predominantly random copolymer. In contrast, the St/MAH and AN/AM copolymers are significantly different. In the former copolymer the χ value is 1.89 and the $\bar{l}_A$ and $\bar{l}_B$ values both equal 1.1. This is indicative of a predominantly alternating copolymer. Furthermore, the $N_A(n)$ and $N_B(n)$ values

Table 5.10 *Comparison of Copolymer Microstructure.*

Polymer Composition	% Conversion	$\bar{l}_A, \bar{l}_B$ values	χ value	N(n) n = 1	N(n) n = 2	N(n) n = 3	N(n) n = 4	N(n) n = 5
VDC = 0.50	All	2.0	1.00	0.50	0.25	0.13	0.06	0.03
MA = 0.50		2.0		0.50	0.25	0.13	0.06	0.03
VDC = 0.50	61	1.9	1.06	0.53	0.25	0.12	0.06	0.03
VC = 0.50		1.9		0.53	0.25	0.12	0.06	0.03
St = 0.53	Initial	1.1	1.89	0.89	0.10	0.01	0.00	0.00
MAH = 0.47		1.0		1.00	0.00	0.00	0.00	0.00
AN = 0.50	55	3.2	0.62	0.31	0.21	0.15	0.10	0.07
AM = 0.50		3.2		0.31	0.22	0.15	0.10	0.07

are skewed toward those characteristic of isolated units, which also denote an alternating copolymer. In the case of the AN/AM copolymer the opposite effect is observed. The χ value of 0.62 together with $\bar{l}_A$ and $\bar{l}_B$ values of 3.2 suggests a copolymer with a "blocky" tendency. In addition, the distribution of $N_A(n)$ and $N_B(n)$ is skewed towards sequences of greater length, which again implies a "blocky" character.

The Penultimate Model

So far we have considered the terminal model, where the end group of the growing chain affects the addition of the incoming monomer unit, as described by the reactivity ratios. There are, however, well documented cases where the specific character of the last *two* monomer units of the growing chain affects the addition of the incoming monomer unit and this is called the *penultimate model*. We will briefly show how it is possible to differentiate between terminal (first order Markovian) and penultimate (second order Markovian) polymerization from a knowledge of the sequence distribution.

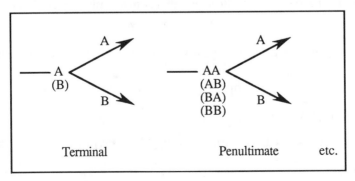

Conditional Probabilities for the Penultimate Model

When the nature of the penultimate unit has a significant effect upon the absolute rate constant in copolymerization, *four* independent conditional probabilities can be written with four monomer reactivity ratios, in a manner similar to the terminal model.

Penultimate Group	Added Group	Final Product
~~AA*	A	~~~AAA*
~~BA*	A	~~~BAA*
~~BB*	A	~~~BBA*
~~AB*	A	~~~ABA*
~~AA*	B	~~~AAB*
~~BA*	B	~~~BAB*
~~BB*	B	~~~BBB*
~~AB*	B	~~~ABB*

And:

$$P\{B/AA\} = \frac{1}{1 + r_A x} \qquad P\{A/AA\} = 1 - P\{B/AA\} \qquad (5.74)$$

$$P\{A/BA\} = \frac{r_A' x}{1 + r_A' x} \qquad P\{B/BA\} = 1 - P\{A/BA\} \qquad (5.75)$$

$$P\{B/AB\} = \frac{\frac{r_B'}{x}}{1 + \frac{r_B'}{x}} \qquad P\{A/AB\} = 1 - P\{B/AB\} \qquad (5.76)$$

$$P\{A/BB\} = \frac{1}{1 + \frac{r_B}{x}} \qquad P\{B/BB\} = 1 - P\{A/BB\} \qquad (5.77)$$

where:

$$r_A = \frac{k_{AAA}}{k_{AAB}} \quad r_A' = \frac{k_{BAA}}{k_{BAB}} \quad r_B = \frac{k_{BBB}}{k_{BBA}} \quad r_B' = \frac{k_{ABB}}{k_{ABA}} \qquad (5.78)$$

Again in a similar manner to the methodology used to derive the terminal copolymer equation, we start with a reversibility relationship. Thus *in general,*

$$P_3\{AAB\} = P_3\{BAA\} \qquad (5.79)$$

For the penultimate model we must employ 2nd order Markovian statistics, so we use:

$$P_2\{AA\}P\{B/AA\} = P_2\{BA\}P\{A/BA\} \qquad (5.80)$$

Recall that $P_2\{BA\} = P_2\{AB\}$ and therefore:

$$P_1\{A\} P\{A/A\} P\{B/AA\} = P_1\{A\} P\{B/A\} P\{A/BA\} \qquad (5.81)$$

And since:

$$P\{A/A\} = 1 - P\{B/A\} \qquad (5.82)$$

It follows that:

$$P\{B/A\} = \frac{P\{B/AA\}}{P\{B/AA\} + P\{A/BA\}} \qquad (5.83)$$

Similarly starting with $P_3\{BBA\} = P_3\{ABB\}$:

$$P\{A/B\} = \frac{P\{A/BB\}}{P\{A/BB\} + P\{B/AB\}} \qquad (5.84)$$

Recall from equation 5.62 that:

$$\frac{P_1\{A\}}{P_1\{B\}} = \frac{P\{A/B\}}{P\{B/A)\}} \qquad (5.85)$$

Substituting equations 5.83 and 5.84 leads to:

$$\frac{P_1\{A\}}{P_1\{B\}} = \frac{1 + \dfrac{P\{A/BA\}}{P\{B/AA\}}}{1 + \dfrac{P\{B/AB\}}{P\{A/BB\}}} \qquad (5.86)$$

Finally substituting for the reactivity ratios and molar feed ratio yields:

$$y = \frac{P_1\{A\}}{P_1\{B\}} = \frac{1 + \dfrac{r_A' x\,(1 + r_A x)}{(1 + r_A' x)}}{1 + \dfrac{\dfrac{r_B'}{x}\left(1 + \dfrac{r_B}{x}\right)}{\left(1 + \dfrac{r_B'}{x}\right)}} \qquad (5.87)$$

which is the copolymer composition equation for the penultimate model. While we would admit that this looks awful, the form of the equation is quite simple and easily derived using conditional probability.

Testing the Models

There have been a number of systems where penultimate effects have been established[*], including the free radical polymerization of MMA and 4-vinyl pyridine, styrene and fumaronitrile, etc. However, experimental accuracy is paramount and is the crucial factor in determining whether or not the effect is real. Very precise measurements of copolymer composition are required, especially at low and high monomer feed ratios, in order to distinguish between the terminal and penultimate models. It is convenient to write the copolymer composition equation in the general form:

---

[*] See, for example, G. Odian, *Principles of Polymerization*, 3rd Edition, Wiley, 1991.

$$\frac{P_1\{A\}}{P_1\{B\}} = \frac{1 + (r_A)x}{1 + \frac{(r_B)}{x}} \tag{5.88}$$

For the *terminal* model: $(r_A) = r_A$ and $(r_B) = r_B$, which are independent of x. while for the *penultimate* model, (r_A) and (r_B) are dependent upon x and given by:

$$(r_A) = r_A' \frac{1 + r_A x}{1 + r_A' x} \qquad (r_B) = r_B' \frac{1 + \frac{r_B}{x}}{1 + \frac{r_B'}{x}} \tag{5.89}$$

If we let $y = P\{A\}/P\{B\}$, it is theoretically possible to test whether or not the experimental curve of y versus x can be fitted adequately over the whole composition range by only two reactivity ratios, which implies a terminal model. If not, two further reactivity ratios, r_A' and r_B', can be estimated by an iterative process. Unfortunately, the greatest sensitivity occurs at high and low values of x, where the inherent errors are the greatest.

Another approach is to rearrange the penultimate copolymer equation into a Fineman-Ross type of equation:

$$\frac{y-1}{x} = (r_A) - (r_B)\frac{y}{x^2} \tag{5.90}$$

where a plot of $(y-1)/x$ versus y/x^2 linearizes the equation and permits a determination of the reactivity ratios. Again, the penultimate effect only produces a *slight deviation* from a straight line at the high and low values of x.

Finally, we can use the number average length of A or B runs. Recall that we derived the relationship between $\bar{l}_A$ or $\bar{l}_B$ and the reactivity ratios for the *terminal* model (equations 5.67 and 5.68):

$$\bar{l}_A = \frac{1}{P\{B/A\}} = 1 + r_A x \tag{5.91}$$

$$\bar{l}_B = \frac{1}{P\{A/B\}} = 1 + \frac{r_B}{x} \tag{5.92}$$

We will leave the student to derive the following for the *penultimate* model:

$$\bar{l}_A = 1 + \left(r_A' \frac{1 + r_A x}{1 + r_A' x} \right) x \tag{5.93}$$

and:

$$\bar{l}_B = 1 + \left(r_B' \frac{r_B + x}{r_B' + x} \right) \frac{1}{x} \tag{5.94}$$

If $\bar{l}_A$ is plotted against x (or $\bar{l}_B$ against 1/x) we should obtain a straight line with an intercept of unity and a slope of r_A (or r_B) for the terminal model. For the penultimate case, however, a concave or convex curve (also with an intercept of unity) should be observed—the shape of which depends upon whether r_A (or r_B) is greater or less than r_A' (or r_B').

It is possible to directly determine the four reactivity ratios by rearranging ("linearizing") the above equations:

$$\frac{\bar{I}_A - 2}{x} = r_A - \frac{1}{r_A'}\left(\frac{\bar{I}_A - 1}{x^2}\right) \tag{5.95}$$

and:

$$x\,(\bar{I}_B - 2) = r_B - \frac{1}{r_B}\,(\bar{I}_B - 1)\,x^2 \tag{5.96}$$

Always remember, however, that errors can lead to false conclusions and testing for the penultimate effect requires rigorous experimental accuracy.

Isomerism in Polymers—A Special Case of Copolymerization?

As we pointed out in chapter 1, it is perfectly legitimate to consider structural, sequence and stereo isomerism as special cases of copolymerization. The symbols A and B could equally be used to denote, respectively, *cis*-1,4- and *trans*-1,4- placements of the monomers in the chain of polybutadiene, or "normal" and "backwards" placements of the monomers in the chain of poly(vinylidine fluoride), or the insertion of a *meso* or *racemic* placement in the chain of poly(methyl methacrylate). Thus it would be logical to develop the relevant statistical relationships (with appropriate changes in nomenclature!) to describe the microstructure of this class of "copolymers" in this chapter. But we think that most students will have had enough of probability theory by now and need a break. To paraphrase General Douglas MacArthur, "we will return", after we savored the delights of spectroscopy, in the next chapter.

E. STUDY QUESTIONS

1. Starting from the steady state assumption:

$$k_{12}[M_1^{\bullet}][M_2] = k_{21}[M_2^{\bullet}][M_1]$$

Derive the copolymer equation for the terminal model.

2. The following data was obtained from a series of free radical copolymerization experiments employing the monomers styrene (M_1) and diethylfumarate (M_2). The copolymers were isolated at low conversions and the copolymer composition determined (m_1 and m_2 for styrene and diethylfumarate, respectively).

M_1 (weight %)	m_1 (weight %)
13.1	30.5
28.7	38.6
37.7	40.5
47.6	45.5
58.5	48.6
70.8	54.0

Determine the reactivity ratios r_1 and r_2 by:

(a) The method of Mayo and Lewis.
(b) The method of Fineman and Ross.
(c) The method of Kelen and Tüdõs.
(d) Comment on the accuracy of the results obtained from the three different methods.
(e) What do the reactivity ratios tell you about the microstructure of the resultant copolymers ?

3. Answer the following questions which pertain to the section on general statistical relationships:

(a) Assuming the presence of only two monomers A and B, express the triad probability $P_3\{ABB\}$ in terms of the sum of two tetrad probabilities.
(b) Starting from $P_1\{A\}$ show that $P_2\{AB\} = P_2\{BA\}$.
(c) Assuming only two monomers A and B, give an example of the equality of a reversible sequence using a pentad sequence.
(d) Express the pentad $P_5\{BABAB\}$ in terms of simple (first order) Markovian statistics.

4. From nmr spectroscopy the number fraction of BAAAB pentad and AB diad sequences was determined to be 0.020 and 0.16, respectively. What is the number fraction $N_A(n)$ of sequences of three A's (i.e. AAA) ?

5. Given that $P\{A/B\} = x / (r_B + x)$ and $P\{B/A\} = 1 / (1 + r_A x)$; derive the Mayo-Lewis copolymerization equation starting from $P_2\{AB\} = P_2\{BA\}$.

6. Assuming that the terminal model applies, show that:

$$R = 2P_1\{B\}P\{A/B\} = \frac{2}{r_A x + 2 + \dfrac{r_B}{x}}$$

and:

$$\chi = P\{A/B\} + P\{B/A\} = \frac{r_A x + 2 + \dfrac{r_B}{x}}{r_A x + 1 + r_A r_B + \dfrac{r_B}{x}}$$

7. Using the Mayo-Lewis copolymerization equation for the terminal model, calculate the instantaneous copolymer composition for a monomer feed ratio of 40:60 parts by weight of methacrylic acid (MAA) and styrene (St) at 70°C. From the literature the appropriate reactivity ratios are 0.7 and 0.15 for MAA and St, respectively. In addition, calculate the run number, R, the deviation from randomness parameter, χ, and the number average lengths of MAA and St runs.

8. Write a computer program to calculate the copolymer composition of monomers A and B as a function of conversion. Assume an initial equimolar feed ratio and reactivity ratios of $r_A = 4.0$ and $r_B = 0.06$. In addition, calculate the values of $\bar{l}_A$, $\bar{l}_B$ and χ for each interval of the degree of conversion. Comment upon the results you obtain. You will be expected to hand in a listing of your program and the output. You will not receive extra credit for beautiful graphics and tables, but you may impress your professor !

9. The following data was calculated for a copolymer synthesized from A and B assuming the terminal model.

Copolymer composition : A = 0.87 B = 0.13.
Number Ave Length of Runs : $\bar{l}_A = 10.6$ $\bar{l}_B = 1.5$
Deviation from Randomness : $\chi = 0.76$
Number Fraction of $(A)_n$ Sequences : $(A)_1 = 0.10$ $(A)_2 = 0.09$ $(A)_3 = 0.08$
Number Fraction of $(B)_n$ Sequences : $(B)_1 = 0.66$ $(B)_2 = 0.22$ $(B)_3 = 0.08$

What do these data tell you ?

10. $\bar{l}_A$ is defined as the number average length of A runs.

(a) Derive equations relating $\bar{l}_A$ to the reactivity ratio(s) for the terminal and penultimate models.

(b) The following experimental data was obtained from copolymerization studies at very low degrees of conversion involving monomers A and B:

x (= A/B mole ratio)	$\bar{l}_A$
2.8	3.80
2.2	3.34
1.8	3.05
1.4	2.74
1.0	2.40
0.6	2.00
0.4	1.76
0.2	1.44

Does the data fit the terminal or penultimate model? Estimate the values of r_A and r_A', if applicable.

F. SUGGESTIONS FOR FURTHER READING

(1)   J. L. Koenig, *Chemical Microstructure of Polymer Chains,*
        J. Wiley & Sons, New York, 1982.

(2)   G. Odian, *Principles of Polymerization,*
        3rd Edition, J. Wiley & Sons, 1991.

CHAPTER 6

Spectroscopy and the Characterization of Chain Structure

"I'm picking up good vibrations,
She's giving me excitations"
—The Beach Boys

A. INTRODUCTION

Spectroscopy, in all its forms, is a powerful tool for the study of structure. Atomic spectroscopy, for example, probes electronic transitions and provides direct, compelling evidence of the quantization of energy. Our concern here is molecular spectroscopy, which encompasses not only electronic transitions, but also those involving molecular vibrations, the reorientation of nuclei in magnetic fields, and other phenomena. In polymer science, these techniques provide important, if not indispensable, methods for identification, characterizing microstructure, orientation, intermolecular interactions and other wonders too numerous to mention. Clearly, this is a specialization that includes an extensive body of knowledge, so all we can describe in this text is an outline of the basics and an indication of how some of these tools are used in the most important tasks of polymer characterization. We shall focus our attention on what we believe are the two most significant techniques, infrared and nuclear magnetic resonance (nmr) spectroscopy, but do so in rather different ways. The results of nmr analysis can be related directly to sequence distributions and the use of probability theory discussed in the last chapter, so our discussion of this technique will involve some of these specifics. Our discussion of infrared spectroscopy will be more general, because it does not provide this level of detailed analysis. It will involve some mention of conformational order, crystallinity and intermolecular interactions, topics that are introduced in the following chapter. Students that are not familiar with some of these fundamentals may wish to jump ahead and read Chapter 7 first. We will commence this discussion with a review of the general fundamentals of spectroscopy, where we will also briefly mention other spectroscopic methods.

B. FUNDAMENTALS OF SPECTROSCOPY

Spectroscopy is the study of the interaction of light (in the general sense of electro-magnetic radiation) with matter. When a beam of light is focused on a sample a number of things can happen. It can be reflected, or if the sample is

145

transparent to the frequency or frequencies of the incident light, simply transmitted with no change in energy. Some of the light may also be absorbed or scattered. We will discuss the elements of light scattering later on in Chapter 10, as the scattering of electromagnetic radiation is a powerful tool for characterizing polymer structure at the level of overall chain conformation and sample morphology. In conventional light scattering the scattered radiation that is measured has the same frequency or energy as the incident beam, although a small portion of the light can also exchange energy with the sample and scatter at a different frequency. This latter effect provides the basis for *Raman spectroscopy*.

Our interest is the light that is absorbed by the sample, however, as this is the basis for *absorption spectroscopy**. The light is only absorbed if its energy and hence frequency correspond to the energy difference between two quantum levels in the sample. This is described by the Bohr frequency condition:

$$\Delta E = E_2 - E_1 = h\upsilon \tag{6.1}$$

where h is Planck's constant υ and is the *frequency* of the light in cycles per second (Hertz, Hz). This relationship, and many spectroscopic measurements are often expressed in terms of wavelength, λ (in units of length, e.g., cm, Å):

$$\lambda = \frac{\tilde{c}}{\upsilon} \tag{6.2}$$

where $\tilde{c}$ is the velocity of light, or (in infrared spectroscopy) in terms of wavenumber, $\tilde{\upsilon}$ (in units of inverse length, i.e., cm^{-1}), defined as:

$$\tilde{\upsilon} = \frac{\upsilon}{\tilde{c}} = \frac{1}{\lambda} \tag{6.3}$$

Molecular Processes and the Absorption of Radiation

There are various types of energy changes that are excited in molecules as a result of an interaction with light and the transitions that occur depend upon the energy or wavelength of the incident radiation. The electromagnetic spectrum is a continuum, but humankind has given various parts of it names, or associated parts of it with various applications, as illustrated in figure 6.1. Starting at the high energy or short wavelength region of the spectrum, we find that γ-rays cause transitions between energy states within nuclei (the basis for Mössbauer spectroscopy), while absorptions of x-rays involve the inner shell electrons of the atoms of a sample**. Visible and ultra-violet (uv) light also excite electronic transitions, and provide the basis for uv-visible spectroscopy, the less strongly bound electrons found in the π orbitals of conjugated system absorbing in the visible range of light, for example, while higher energy (frequency) uv light is required to excite more strongly bound electrons.

---

* There is such a thing as *emission spectroscopy*, also, but that will not concern us here.
** Radiation in this energy range can also beat the living hell out of a polymer, breaking bonds, knocking out atoms and causing other, usually undesirable, effects. However, this can be useful if applied cleverly, as in the use of γ-radiation to cross-link polyethylene and increase its environmental stability.

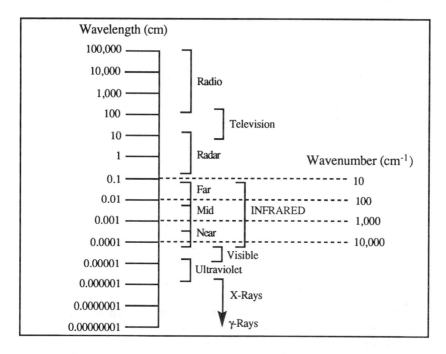

Figure 6.1 *Frequency ranges for different types of spectroscopy.*

As frequency is decreased (or wavelength increased) the energy of the radiation is no longer sufficient to excite electrons. Molecules have vibrational energy levels, however, so that specific types of vibrational motion of the chemical bonds of a molecule, called modes, are excited by infrared radiation, usually in the wavelength range of 0.1 to 0.00025 cm, or as it is more commonly expressed, 10 to 4000 cm^{-1} (wavenumbers). Continuing into the microwave and radiofrequency range, there are transitions associated with the rotational energy levels of small molecules (not polymers!). But, *in the absence of an external field*, there are no other absorptions associated with molecular processes in this frequency range*. Be that as it may, if we now trundle up a bloody great magnet and apply a large magnetic field to the sample, we would now, in principle, find a wealth of absorptions in this radiowave region of the spectrum. This is because specific nuclei (protons, deuterons, ^{13}C, ^{15}N, ^{19}F, etc.) have magnetic dipole moments by virtue of their "spin" (but, ^{12}C and ^{16}O, for example, have zero spin). The energy of the dipolar nucleus depends on whether or not it is aligned with an external magnetic field. If radiowaves having a frequency corresponding to the energy difference between the aligned and unaligned states (i.e., $\Delta E = h\upsilon$) now impinge on the sample, there's strong coupling or resonance

---

* However, the lagging of molecular dipole moments behind the oscillations of the electric field of light leads to some dissipation of energy and this is the basis for dielectric relaxation experiments.

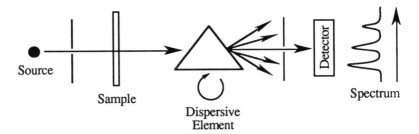

Figure 6.2 *Schematic diagram of a spectrometer.*

between the radiation and the nuclear spins, and absorption occurs, as the spin "flips". This is nuclear magnetic resonance or nmr[*].

Spectrometers

The technicalities of spectroscopic instrumentation and the fundamentals of the spectroscopists' art belong in a book devoted specifically to the subject, but it is important to have a feel for the nature of the experiment, whose basic elements are illustrated schematically in figure 6.2. The first thing you need is a source of radiation in the frequency range you're interested in. Assuming you can obtain such a source (and a lot of applied physics has gone into developing good sources and detectors) you need to arrange the optics (for infrared or uv-visible spectroscopy) so that the light is focused onto the sample. Usually, the light or radiation consists of a range of frequencies, only some of which are absorbed by the sample. Now it is necessary to determine how much of which frequencies have been absorbed, so you need to separate the light according to wavelength and measure the intensity of light at that wavelength. We use a picture of a prism which focuses light through a slit onto a detector to represent this in figure 6.2, so that as the prism is rotated different frequencies are detected. Such devices were indeed used in the first infrared instruments, but were subsequently

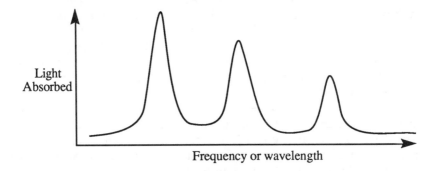

Figure 6.3 *Schematic diagram of a spectrum.*

---

[*] A close cousin, electron spin resonance, esr, occurs in the higher frequency microwave region and is a useful tool for studying unpaired electrons (free radicals).

replaced by gratings and more recently by interferometers (which don't require slits at all).

If you have a source that emits radiation of a given frequency and if this frequency can be changed so as to "sweep" through a given range, such as a tunable laser, then you don't need a dispersion device (i.e., the prism in figure 6.2) at all. This is essentially the way nmr experiments are performed.

Intensities and Band Shapes

The spectrum of a material is a plot of the intensity of light transmitted through the sample, or absorbed by it, as a function of frequency or wavelength. We have drawn a schematic representation of an absorption spectrum in figure 6.3 and there are two things you should notice. First, the *bands* have a finite width. We'll get back to this in a while. Second, the intensities of the bands are all different. We'll discuss this initially.

The intensity of a *spectral line* or *band* depends on a number of things. First, it must obviously depend upon the number of molecules or functional groups or nuclei giving rise to that particular absorption. Accordingly, it must depend upon the concentration of that species and the total amount (thickness) of the sample in the spectrophotometer beam. It must also depend upon the chemical character of the group or nature of the species. In infrared spectroscopy, for example, we will see that the vibrations of certain types of bonds inherently absorb more radiation than others. Finally, the amount of absorption must also depend upon the intensity of the incident beam. The relationship between the absorption (defined as the logarithm of the ratio of the intensity of transmitted to incident light), and these various factors is given by the Beer-Lambert law:

$$A = abc \qquad (6.4)$$

where A is the intensity (in terms of the amount of radiation absorbed, not transmitted) of the band of interest; integrated areas or peak heights can be used, but the units change accordingly; a is the absorptivity or extinction coefficient which depends on the nature of the absorbing species, b is the sample thickness, and c is the concentration of the component of interest. For multicomponent systems, the absorption of any particular band can be written:

$$A = \sum_{i=1}^{n} a_i b_i c_i \qquad (6.5)$$

assuming there is no interaction between the i components.

Note that for various reasons this law does not hold for very thick or concentrated samples (in infrared and uv-visible spectroscopy) and sample preparation and technique are crucial to good quantitative work.

Turning our attention to bandwidth, there are numerous reasons why bands are not infinitely narrow. There are instrumental effects, factors associated with the finite lifetimes of excited states, and so on. The spectra of polymers are usually considerably broader than their low molecular weight counterparts, however, and the key factors are interactions between the components of the system coupled to the fact that even ordered polymers are only semi-crystalline (see Chapter 7). In the amorphous state the elements of the chains are in slightly different local environments, giving rise to slightly different frequencies of absorption and hence broader bands.

C. BASIC INFRARED SPECTROSCOPY

Infrared spectroscopy is an old and familar friend to the organic chemist. It provided numerous insights in the early days of polymer science (we mentioned the detection of short chain branches in polyethylene in Chapter 2, for example), and for many years was the workhorse of all polymer synthesis laboratories, providing a routine tool for the identification of materials and the characterization of microstructure. But all good things must end and for many of these applications infrared spectroscopy has been superseded by the greater detail offered by nmr spectroscopy. Nevertheless, this is not to say that scientists have hurled their infrared spectrometers out of their laboratory windows. They are still ideally suited for specific tasks and provide certain types of information (including measurements of orientation and strong specific interactions), that cannot be obtained by nmr, so infrared spectroscopists still find things to do (which is just as well, because otherwise the authors of this text would be looking for work!) and the technique remains a fundamental characterization tool. The way we will approach it here is to describe the characteristics of the infrared spectra of various polymers, to give you a feel for the type of information that can be gained from infrared studies. We start by considering some of the principles of infrared absorption.

Conditions for Infrared Absorption

We mentioned in our discussion of basic spectroscopy that a fundamental and inherent condition for infrared absorption is that the frequency of the absorbed radiation must correspond to the frequency of what is called a normal mode of vibration and hence, a transition between vibrational energy levels. We will consider molecular vibrations in a little more detail shortly, but this condition is not by itself sufficient, and an additional fundamental requirement is that there must be some mode of interaction between the impinging radiation and the molecule. Even if infrared radiation with the same frequency as a fundamental normal vibration is incident on the sample, it will only be absorbed under certain conditions. The rules determining optical activity are known as *selection rules*, which have their origin in quantum mechanics. It is easier to obtain a physical picture of these interactions by considering the classical interpretation, however, and we will not discuss the equivalent quantum mechanical description in this text.

Infrared absorption is simply described by classical electromagnetic theory: an oscillating dipole is an emitter or absorber of radiation. Consequently, the periodic variation of the dipole moment of a vibrating molecule results in the absorption or emission of radiation of the same frequency as that of the oscillation of the dipole moment. The requirement of a change in dipole moment with molecular vibration is fundamental. Certain normal modes do not result in such a change, as illustrated by the two in-plane stretching vibrations of CO_2 depicted below:

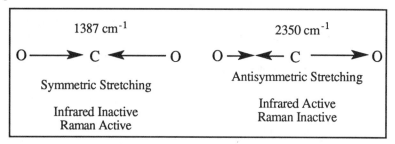

The net dipole moment of the symmetrical unperturbed molecule is zero. In the totally symmetric vibration, the two oxygen atoms move in phase sucessively from and towards the carbon atom. The symmetry of the molecule is maintained in this vibration and there is no net change in dipole moment, hence no interaction with infrared radiation occurs for this motion. Conversely, in the antisymmetric stretching vibration, the symmetry of the molecule is perturbed and there is a change in the net dipole moment with consequent infrared absorption.

The selection rules for the symmetric and antisymmetric stretching modes of CO_2 were determined just by inspection. For this simple molecule, it is easy to see that the symmetric stretch does not result in a change in dipole moment. For more complex molecules, the number and activity of the normal modes can actually be predicted from symmetry considerations alone using the methods of group theory, a tool which is also important in other characterization techniques (such as x-ray diffraction). A discussion of this requires a more specialized treatment, however, and the important point you should grasp here is that infrared absorption involves a change in dipole moment with molecular vibration[*]. Now let us look at the nature of molecular vibrations in a little more detail.

Normal Modes of Vibration

We presented above a schematic picture of two *normal modes of vibration* of the CO_2 molecule, but said virtually nothing about the nature of a normal mode or how we determine what it is. In fact, if you made a model of the CO_2 molecule by suspending in the air three balls of appropriate weight attached by springs, you could apparently excite all sorts of different vibrations by walking up to it and giving it a good whack with different degrees of force from different angles.

---

[*] Although we have only made a passing reference to Raman scattering, it is interesting to note that it also involves molecular vibrations, but the selection rules are different from infrared. Specifically, for a vibrational mode to be Raman active there must be a change of polarizability with molecular vibration. Of particular interest is the rule of mutual exclusion, which states that if the molecule has a center of symmetry, vibrational modes active in the infrared will be Raman inactive and vice versa. We will see an example of this later.

It turns out, however, that all the motions that you excite by whacking can be broken down into the sum (or difference) of just a few types of motion, called the normal modes of vibration. Note that these do not include translational or rotational movement of the molecule as a whole, but just motions of the balls (or atoms) relative to one another. The number of normal modes available to a molecule depends upon the number of atoms it contains and hence its *degree of freedom*. The motion of any atom in a molecule, be it the result of an internal vibration or rotations and translations, can be resolved into components parallel to the x, y and z axis of a Cartesian system and the atom is described as having three degrees of freedom. A system of N nuclei therefore has 3N degrees of freedom. However, for a non-linear molecule six of these degrees of freedom correspond to translations and rotations of the molecule as a whole and therefore have zero vibrational frequency, so there are 3N - 6 vibrational degrees of freedom or normal modes of vibration. For strictly linear molecules, such as carbon dioxide, rotation about the molecular axis does not change the position of the atoms and only two degrees of freedom are required to describe any rotation. Consequently, linear molecules have 3N - 5 normal vibrations. For a theoretically infinite polymer chain, only the three translations and one rotation have zero frequency and there are 3N - 4 degrees of vibrational freedom. Each normal mode consists of vibrations, although not necessarily significant displacements, of all the atoms in the system. As an example, consider the displacement of the atoms, shown diagrammatically in figure 6.4, which represent the normal modes (and hence absorptions) occurring at 1163, 1074, 1031 and 925 cm^{-1} in the infrared spectrum of *trans*-1,4-poly(2,3-dimethylbutadiene). The displacements of the atoms in the course of the vibration are proportional to the length of the arrows, which are greatly exaggerated in this figure.

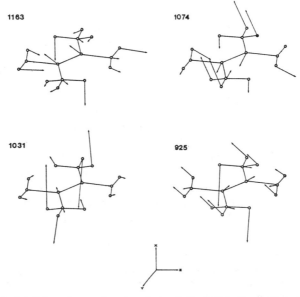

Figure 6.4 *Selected normal modes of* trans-1,4-poly(2,3-dimethylbutadiene).

A Simple Model

The normal modes of a polymer can appear to be complicated and you may be wondering how we determine the form of the vibration. As it turns out, the vibrations can be calculated in a conceptually simple manner using the methods of classical mechanics. The nuclei are considered to be point masses and the forces acting between them springs that obey Hooke's Law. The motion of each atom is assumed to be simple harmonic, meaning that the force on the atoms is proportional to their displacement from an equilibrium position. In a normal vibration, each particle carries out a simple harmonic motion of the same frequency and, in general, these oscillations are in phase; however, the amplitude may be different from atom to atom, as can be seen from figure 6.4. These normal modes of vibration are excited upon infrared absorption (or Raman scattering). Naturally, different types of vibrations will have different energies and so absorb or inelastically scatter radiation at different frequencies.

In the early days of infrared spectroscopy, before all this was worked out, it was empirically determined that certain functional groups absorb infrared radiation at characteristic frequencies (the basis of the so-called *group frequency approach* widely used for identification of unknown samples). Since normal vibrations involve displacements of all atoms it might be expected that the constitution of the rest of the molecule might have a more profound effect than the relatively small frequency shifts often observed for these groups. However, the intensity of an infrared band depends upon the extent of the displacement of atoms in a particular vibration, as larger displacements would lead to a bigger change in dipole moment. All the atoms may be vibrating with the same frequency, but the largest displacements from an equilibrium position can be localized in a small group of atoms.

The details of normal mode calculations are beyond the scope of this book[*], but it is important to obtain a feel for the factors that affect the frequencies of a band, so we will consider a simple example of two point masses, m_1 and m_2, connected by a Hookean spring as shown diagrammatically below.

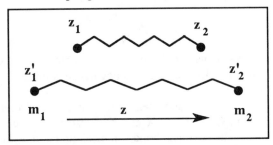

For this one-dimensional model, the z axis is coincident with the molecular axis and movements of the atoms in the x and y directions are not allowed. Let us define the displacement of the two atoms relative to each other by $z = [(z'_1 -$

---

[*] Those with truly masochistic tendencies might wish to consult another of our books for further details; P. C. Painter, M. M. Coleman and J. L. Koenig, *The Theory of Vibrational Spectroscopy and Its Application to Polymeric Materials*, J.Wiley & Sons, New York 1982.

z'_2) - (z_1 - z_2)] where z_1 - z_2 is the equilibrium separation between m_1 and m_2 and (z'_1 - z'_2) is the distance after a finite extension or compression. Assuming the spring obeys Hooke's Law, the exerted force is given by -fz where f is the force constant or ("stiffness") of the spring. It is also convenient at this stage to define m_r, the reduced mass, equal to $(m_1 m_2)/(m_1 + m_2)$.

Since force is defined as mass times acceleration, for a conservative field (i.e., no frictional forces) the following equation holds:

$$- fz = m_r \frac{d^2 z}{dt^2} \qquad (6.6)$$

You could obtain this equation by considering the individual displacements of the atoms, but when you combine everything together you end up with this expression in terms of the reduced mass. Rearranging:

$$m_r \frac{d^2 z}{dt^2} + fz = 0 \qquad (6.7)$$

This is a 2nd order differential equation which has a periodic solution of the form:

$$z = A \cos(2\pi \upsilon t + \varepsilon) \qquad (6.8)$$

where A is an amplitude, ε is a phase angle and υ is the frequency of vibration. Differentiating twice with respect to time leads to:

$$\frac{d^2 z}{dt^2} = - 4\pi^2 \upsilon^2 z \qquad (6.9)$$

Substituting we obtain:

$$(- 4\pi^2 \upsilon^2 m_r + f) z = 0 \qquad (6.10)$$

Assuming $z \neq 0$ we obtain what should be the familiar equation for harmonic motion:

$$\upsilon = \frac{1}{2\pi} \sqrt{\frac{f}{m_r}} \qquad (6.11)$$

Although the model considered is extremely simple, it provides us with the important result that the vibrational frequency depends inversely on mass and directly on the force constant.

To see how this helps us, consider as an example the isolated stretching vibration of the bond C-X, where X can be substituents such as hydrogen, chlorine, or oxygen. The chemical bond can be assumed to be the focus of the forces acting between the atoms, i.e., the "spring". Accordingly, as the mass of the substituent is increased there should be a decrease in frequency (or wavenumber). In fact, the C-H stretching modes absorb near 2900 cm[-1] while C-Cl stretching vibrations occur near 600 cm[-1]. Conversely, increasing the force constant between atoms, say by formation of a double bond, increases the frequency. The C-O stretching vibrations are found near 1100 cm[-1] while C=O frequencies are characteristically observed at about 1700 cm[-1]. For more complex vibrating systems, the vibrational frequencies naturally depend upon the type of motion and the geometry, in addition to the mass of the atoms and the forces acting between them. This brings us to polymers.

D. CHARACTERIZATION OF POLYMERS BY INFRARED SPECTROSCOPY

Infrared spectroscopy can be applied to the characterization of polymeric materials at various levels of sophistication. As most commonly used, it is now a rapid and easy method for the qualitative identification of major components through the use of group frequencies and distinctive patterns in the "fingerprint" region of the spectrum. At one time this was an art, because the voluminous number of standard spectra available in the literature demanded an excellent memory and the dedication of a Sherlock Holmes. Today, gross identification of polymeric materials has become a rather trivial task using modern computer assisted instruments. Libraries of standard spectra, stored in the computer memory and used in conjunction with search routines, have reduced the problem to that of applying a "black box."

At the next level of sophistication, infrared spectroscopy may be employed to characterize the structure of polymeric materials. By this we mean not only the overall chemical composition of the polymer chain but also how individual units are distributed. Thus, it is feasible to obtain information concerning the nature and concentration of structural and conformational units present in a particular sample. This, in turn, leads to a consideration of the spectroscopic features observed when a polymer orients or crystallizes. Symmetry plays a major role in the observed spectra of ordered systems, so that the spectrum is sensitive to the conformation of polymer chains. Infrared spectroscopy may also be employed to study the changes occurring upon chemical modification, degradation and oxidation of polymers. Finally, at the highest level of sophistication, infrared spectroscopy is used in conjunction with computer calculations of normal modes in order to understand the fundamental features of the vibrational motion of polymers. As mentioned above our approach here is more of a survey, so we will start with a simple polymer system.

Amorphous Polymers with Weak Intermolecular Interactions

What we are considering here are non-crystalline polymeric materials in which van der Waals forces predominate*. Examples would include atactic polystyrene, styrene-co-butadiene and ethylene-co-propylene copolymers etc. Let us commence by considering the spectrum, shown in figure 6.5, of a film of an amorphous polymer glass, atactic polystyrene (*a*-PS). One is initially struck by the large number of infrared bands present and the apparent complexity of the spectrum. Upon further thought, however, one can also take the opposite view

---

* We assume at this point that you have a basic knowledge of the various types of inter-molecular interactions. If you don't, or have forgotten, jump ahead to the beginning of the next chapter where we review this material.

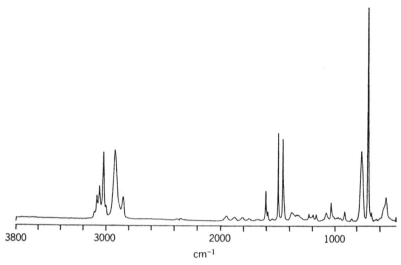

Figure 6.5 *The infrared spectrum of atatic polystyrene.*

and consider the spectrum to be remarkably simple, given that we are dealing with an amorphous material where the average molecular weight might be in excess of 10^5 g/mole and where there exist broad distributions of molecular weight, conformational states and sequences of stereoisomers. There is an element of truth in both viewpoints.

The polymer chain of a-PS is composed of chemical repeat units (-CH$_2$-CHC$_6$H$_5$-) that contain 16 atoms (see table 1.2). The chemical repeat unit is relatively large (compared to that of polyethylene, for example—see later) and, to a first approximation, we can assume that a given repeat unit does not know, in infrared terms, that the adjacent unit exists. Accordingly, the gross features of the spectrum will reflect those of a low molecular weight analogue of the polymer repeat unit. Thus, we might anticipate 3N - 4 = 44 fundamental vibrations. As there is no translational symmetry between the units of the chain or inherent symmetry in the structure of the repeat unit, all these normal modes are predicted to be infrared active. There are also bands due to overtones and combinations of the fundamental modes, but we neglect these here. In addition, we will assume that the a-PS sample is chemically "pure" and that there are no additional bands in the spectrum attributable to oxidation, degradation, incorporation of impurities or the presence of chemically distinct end groups etc.

Let us return to the spectrum of a-PS and ask the question, "What can we immediately gain from just a cursory glance?" First of all, from group frequency correlations we can determine that the sample contains aliphatic and aromatic groups from the bands observed in the 2800 to 3200 cm^{-1} region of the spectrum[*]. Secondly, we can initially eliminate such groups as hydroxyls,

---

[*] A word of caution: the C-H stretching region is often complicated by something called Fermi resonance, which arises from the interaction of a fundamental and an overtone occurring at about the same frequency.

amines, amides, nitriles, carbonyls etc., which all have distinctive group frequencies. We must not be dogmatic, however, because as we have mentioned, symmetry may dictate that in ordered materials particular normal modes are inactive in the infrared. Third, the presence of a group of distinctive and relatively sharp bands that are characteristic of monosubstituted aromatic rings readily leads one to the conclusion that the spectrum resembles that of a styrenic polymer. It is also readily apparent that the spectrum of *a*-PS is characterized by infrared bands that vary quite substantially in breadth. In simple terms, the relatively narrow bands can be attributed to localized normal modes that are conformationally insensitive (i.e., are unaffected by the shape of the chain). These would include such modes as the C–H stretching and the aromatic ring breathing vibrations. Be that as it may, we must not oversimplify the situation, as normal coordinate calculations indicate a degree of mixing of specific ring modes with backbone vibrations. Nonetheless, the above is a reasonable "rule of thumb" and the corollary is that the relatively broad bands in the spectrum are conformationally sensitive. In the amorphous state the polymer resembles a "bowl of spaghetti". Polymer chains essentially obey random flight statistics (see next chapter) and there is a myriad of conformations taken up by the individual units in the chain. In infrared spectroscopic terms, bands that contain significant contributions from vibrations that involve the polymer backbone will reflect the distribution of conformational states by broadening, although in certain cases, bands characteristic of certain types of local conformations may be resolved.

Amorphous Polymers with Strong Intermolecular Interactions

In this case, we consider amorphous polymers in which there are hydrogen bonds as these are the most commonly found strong intermolecular interactions in polymers. Examples include atactic poly(4-vinyl phenol) (PVPh), epoxy resins, poly(vinyl alcohol), amorphous polyamides and polyurethanes.

PVPh is structurally similar to *a*-PS, except that there is a hydroxyl group present in the *para* position of the aromatic ring.

Poly(4-vinyl phenol)

Infrared spectra of PVPh, recorded as a function of temperature (A = 30°C, B = 50°C, C = 100°C, D = 150°C, E = 200°C, F = 250°C), are displayed in figure 6.6. The figure is split into two frequency ranges, 3800–2800 and 2000–450 cm^{-1}. For clarity of presentation, the higher frequency region has been expanded on the absorbance scale by a factor of about four times that of the lower frequency region. In common with the spectrum of *a*-PS, there are infrared bands that are relatively narrow, attributable to normal vibrations that are conformationally insensitive, and broader bands that are sensitive to the distribution of conformations. Using the group frequency approach, one may confidently assign the bands occurring at 825, 1100, 1170, 1445 and 1595/1610 cm^{-1}, which are essentially temperature independent, to the aromatic ring. In contrast, the relatively broad bands in the 1200-1400 cm^{-1} region of the spectrum change markedly as a function of temperature. These bands can be assigned to mixed vibrations containing contributions from O–H deformation and C–O stretching vibrations. The changes observed with temperature may be attributed to two main factors: conformational sensitivity and intermolecular interactions.

Perhaps the most interesting region of the spectrum of PVPh is that occurring between 3100 and 3600 cm^{-1}, where the O–H stretching frequency appears. This region is characterized at room temperature by a very broad band centered at 3360 cm^{-1}, assigned to a broad distribution of hydrogen bonded hydroxyl groups (hydrogen bonding between chains and functional groups of the same type is often called "self-association") and a much narrower band at 3525 cm^{-1} which is attributed to "free" non-hydrogen bonded hydroxyl groups. This latter band appears to increase at the expense of that attributed to the hydrogen bonded groups as the temperature is increased. While this is intuitively pleasing, as it is consistent with our preconceived notions of temperature effects upon equilibrium, it is not as dramatic as it might appear at first glance from figure 6.6.

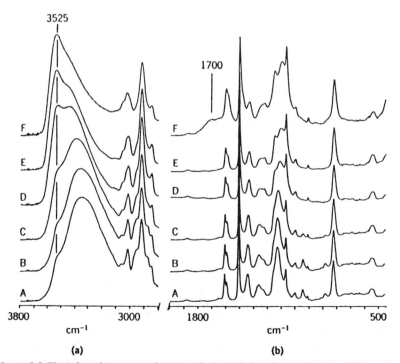

Figure 6.6 The infrared spectrum of atatic poly(4-vinyl phenol) as a function of temperature:
(A) 30 ℃; (B) 50 ℃; (C) 100 ℃; (D) 150 ℃; (E) 200 ℃ and (F) 250 ℃.

As the temperature and hence molecular motion increase the average strength of the hydrogen bonds decreases, which is reflected by a shift to higher frequency. This causes greater overlap with the "free" band at 3525 cm^{-1}. In addition, we now know that the absorption coefficient is a strong function of the strength of the hydrogen bond. Accordingly, as the strength of the hydrogen bond diminishes, the absorption coefficient decreases, leading to a concomitant decrease in area of the hydrogen bonded band. Although this may appear complicated, there is a wealth of information available from infrared temperature studies of what we call strongly self-associated polymers that provides direct evidence of the number and strength of the hydrogen bonds present. This is very useful in a number of areas (e.g., mixing polymers to form blends).

Ordered Polymers with Weak Intermolecular Interactions

The effect of order or "crystallinity" on the infrared spectrum of a polymer is one of the most interesting, and at the same time, most generally misunderstood topics in polymer vibrational spectroscopy. In the vast majority of cases infrared spectroscopy *cannot* be used as an absolute method to determine three dimensional crystallinity. More on this later. We will initially consider the general features that are seen in the spectra of ordered polymers that do not "self-associate" (i.e. do not have strong intermolecular interactions such as hydrogen bonds) and which contain large chemical repeating units, such as isotactic polystyrene (*i*-PS), *trans*-1,4-polyisoprene and *trans*-1,4-poly(2,3-dimethyl butadiene). Then we will turn our attention to polyethylene; an ordered polymer containing a relatively small chemical repeating unit.

Crystallinity vs. Preferred Conformation: Isotactic Polystyrene

Let us proceed with an examination of the infrared spectrum of *i*-PS, as illustrated in figure 6.7. In contrast to *a*-PS, where the chemical repeat units are arranged in a random sequence of stereochemical placements, in pure *i*-PS every adjacent unit is isotactic to the first. This structural regularity results in the possibility of crystalline order and *i*-PS prefers to fold into a regular shape called a 3_1 helical chain conformation (more on this in Chapter 7) which has a crystalline melting point (T_m) of 230°C. If one rapidly quenches a sample of *i*-PS from above the T_m to room temperature, however, crystallization is prevented and the material becomes an amorphous glass. We might therefore anticipate that the spectrum of quenched *i*-PS should closely resemble that of *a*-PS. And it certainly does. A comparison of the spectra presented in figures 6.5 and 6.7(B) leaves little doubt that one would be hard pressed to differentiate between them. In fact, there are some very subtle differences in a few infrared bands which may be ascribed to weak vibrational coupling between different stereochemical repeating units, but they are truly subtle, and to all intents and purposes, can be ignored. If we now anneal the quenched *i*-PS, above the glass transition temperature (T_g) but below the T_m, crystallization occurs and the sample transforms into a semi-crystalline material. The spectrum of the annealed (semi-crystalline) *i*-PS is shown in figure 6.7(A). This spectrum may be considered to

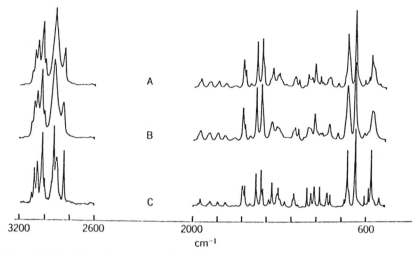

Figure 6.7 *The infrared spectrum of isotactic polystyrene. (A) Annealed, (B) quenched from the melt, and (C) difference spectrum (A - B).*

be composed of two major "non-interacting" components; amorphous and "crystalline". Since the intermolecular interactions are relatively weak, these two components do not recognize, in infrared spectroscopic terms, the existence of one another in the semi-crystalline material. Accordingly, this is an ideal situation for what has become known as difference spectroscopy, as the two components can be separated without significant distortion to the band shapes.

The *difference spectrum* is obtained by subtracting the spectrum of the quenched (amorphous) *i*-PS sample from that of the annealed (semi-crystalline), and is also shown in figure 6.7(C), so that the amorphous bands are removed (this takes a degree of spectroscopic judgement). This spectrum, and others like it, are commonly referred to as the "crystalline" infrared spectrum. This is somewhat misleading. To be pedantic, in the case of a weakly self-associated polymer where there is little, if any, intermolecular vibrational coupling between the units in the polymer chain, the sharp bands seen in the "crystalline" difference spectrum actually reflect an individual repeat unit in "preferred" conformations. In other words, this spectrum has little to do with three dimensional order as such, its just that in crystalline domains structural units are in the preferred conformation. This is the reason why infrared spectroscopy is not truly a direct measure of three dimensional crystallinity in such cases (although some useful empirical correlations have been established, because extended ordered sequences, for all intents and purposes, will only be found in crystalline domains).

Polymorphism: Trans-1,4-Polyisoprene

We have seen that in the case of a weakly interacting polymer containing a relatively large, structurally regular repeat unit, crystallinity results in the presence of sharp bands in the infrared spectrum which can be attributed to a narrow distribution of preferred conformations. If we therefore observe significant differences in the infrared spectra of two polymorphic* forms of such a polymer, it suggests that there must also be significant differences in the chain conformation of the polymorphs. *Trans*-1,4-polyisoprene (TPI) is an excellent example of a polymer that exists in polymorphic forms which are easy to prepare. Difference spectra corresponding to the α and β crystalline forms are shown in figure 6.8. There are obvious and distinct differences between the two spectra (examine, for example, the region between 750 and 900 cm^{-1}). These are spectra that can be directly assigned to two dissimilar preferred polymer chain conformations. The β form contains one monomer per translational repeat unit with torsional angles of 180° between adjacent CH_2 groups and +105° between the CH_2 and -C=C groups**. In contrast, the α form contains two monomers per translational repeat unit and the preferred conformation is reported to be *trans*-CTS-*trans*-CT$\overline{\text{S}}$. To reiterate, the main point is that the infrared spectrum of TPI is sensitive not to crystalline form *per se*, but to the different preferred chain conformations present in the two forms. With very few exceptions, infrared spectroscopy is not sensitive to polymorphism if there is not a parallel change in polymer chain conformation.

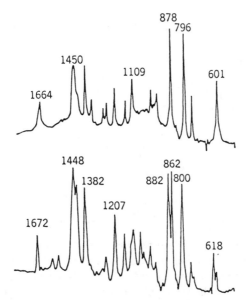

Figure 6.8 *The infrared spectrum of the α (bottom) and β (top) crystalline forms of* trans-1,4- polyisoprene.

---

* Polymorphism, in this context, is when a polymer can have more than one different crystalline form.

** Just as a chemical repeat unit can be taken as the basic element of the chain in terms of chemical structure, the transitional repeat unit describes the chain in terms of translational symmetry. A translational repeat unit often, but not always, contains more than one chemical repeat unit.

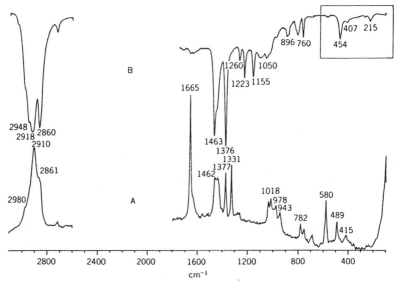

Figure 6.9 *The infrared (top) and Raman (bottom) spectra of* trans-1,4-poly-(2,3-dimethyl butadiene).

The Effect of Symmetry: Poly(2,3-dimethylbutadiene)

Although we are not going to discuss Raman spectroscopy in this book, it is interesting to consider the infrared and Raman spectrum of a highly crystalline sample of *trans*-1,4-poly(2,3-dimethylbutadiene) (TPDMB). The infrared spectrum is shown in transmission in figure 6.9(B)*.

$$\left[\begin{array}{c} CH_3 \qquad CH_2 \\ \quad C=C \\ CH_2 \qquad CH_3 \end{array}\right]_n$$

trans-1,4-Poly(2,3-dimethylbutadiene)

At first glance, the spectrum resembles that of a saturated hydrocarbon. There is certainly no convincing evidence for a band in the 1660-1670 cm⁻¹ region of the spectrum, where C=C double bonds absorb. On the other hand, the Raman spectrum of the same polymer [figure 6.9(A)] is dominated by a line at 1665 cm⁻¹. We are seeing an excellent example of "mutual exclusion" where bands observed in the infrared spectrum are not observed in the Raman spectrum, and *vice versa* (this is also what occurs in CO_2 considered earlier). The preferred chain conformation of TPDMB is similar to that of the β form of TPI. The additional methyl group on the C=C double bond, however, leads to a center of inversion or symmetry. This dictates what is called mutual exclusion.

---

* Transmission spectra are just absorption spectra plotted upside down on a log scale. Nearly all infrared spectra were reported this way until about 20 years ago.

Bands that appear in the infrared are absent in the Raman, and *vice versa*. A combination of infrared and Raman studies is necessary to obtain information on the conformation of this type of polymer chain. Suffice it to say, if one suspects high symmetry and only infrared data are available, beware of absent bands arising from the dictates of selection rules.

An Exception: Polyethylene

It is perhaps ironic that one of the simplest polymers in terms of the structure of the chemical repeat unit, polyethylene (PE), is at the same time, one of the most complex and interesting from an infrared spectroscopic point of view. It should come as no surprise that PE and the low molecular weight paraffin analogues have been the subject of voluminous vibrational spectroscopic studies. Here, however, all we wish to do is highlight the differences between the spectral features observed in PE compared to those polymers discussed above, i.e., *i*-PS, TPI and TPDMB.

PE, or as some people call it, polymethylene, contains a simple and relatively small CH_2 chemical repeat. Unlike the polymers mentioned above, there exists significant vibrational coupling between the units in the polymer chain. Accordingly, the infrared spectrum of PE is considerably different from that of a low molecular weight analogue. This is in marked contrast to those polymers that contain large chemical repeat units. If you study polymer vibrational spectroscopy in more detail, you will find that this means that there are a number of normal vibrations in the spectrum of paraffins that are sensitive to the number and conformation of CH_2 groups in a sequence. For example, the frequency of the CH_2 rocking mode in compounds represented by $C(X)–(CH_2)_n–C(X)$ occurs at 815, 752, 733, 726 and 722 cm^{-1}, respectively, as n progresses from 1 to 5. Above n = 5 the difference in frequency becomes insignificant. Nonetheless, this information is particularly useful in the characterization of polymers containing sequences of CH_2 units and has been successfully employed in the analysis of the sequence length distribution in ethylene/propylene copolymers.

PE, especially the more structurally pure linear polymer, crystallizes readily, rapidly and extensively. In the preferred chain conformation (planar zig-zag, see next chapter) there are two chemical repeats per translational repeat unit. The symmetry of the planar zig-zag polymer chain contains a center of inversion. As we have mentioned above, this implies mutual exclusion of infrared bands and Raman lines and there are only five fundamental vibrations predicted to be active in the infrared spectrum. (A total of 3(6) - 4 = 14 normal modes are predicted, 8 are Raman active and 1 is inactive in both the infrared and Raman.) One glance at an infrared spectrum of a highly crystalline spectrum of PE (figure 6.10) will convince the reader that there are considerably more than five bands present! And we don't even show the region below 600 cm^{-1}.

Even if we allow for the presence of overtone and combination bands and bands characteristic of amorphous conformations, end groups, impurities, oxidation and degradation etc., we still have to account for more than five obvious infrared bands. Under normal conditions PE crystallizes in an orthorhombic unit cell containing two chains. There are now two ethylene units

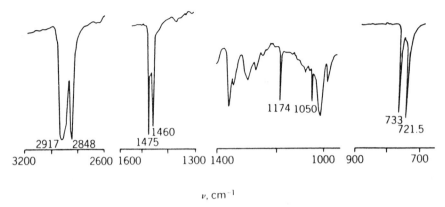

Figure 6.10 *The infrared spectrum of polyethylene.*

that have to be considered in the calculation of the number of active infrared bands. In brief, 12 infrared bands are predicted because the infrared spectrum depends upon the number of units in the crystal's translational repeat unit. In most polymers, these units do not "see" each other spectroscopically, and their modes are superimposed (i.e., at the same frequency). However, as PE has a small repeat unit that packs very efficiently in a crystalline lattice, a given ethylene unit of one chain "knows" that it is in close proximity to a similar group on an adjacent chain (there is an interaction between them). In essence, we have a doubling of the number of the normal modes as one interacts with and perturbs the other (actually, it is a bit more complicated than this, but this is enough for now).

Symmetry again rears its head and will dictate which of these normal modes is active in the infrared. In any event, what we observe is the presence of additional bands and the phenomenon of what is termed *crystal field splitting*. The classic 733/721 and 1460/1475 cm^{-1} doublets in PE are examples of crystal field splitting. It is important to recognize the distinction between crystal field splitting, which is greater than a one dimensional effect and categorically requires two or three dimensional order, and the bands due to a single chain preferred conformation, which simply depend upon interactions between adjacent groups in the one dimension along the chain.

Strongly Interacting Ordered Polymers

The outstanding examples in this category are the structurally regular polyamides (nylons) and polyurethanes. Extensive interchain hydrogen bonding exists in these materials and the ability to form ordered ("crystalline") structures is generally accepted as being responsible for the useful physical properties of these materials. Spectroscopically, they are fascinating. No longer can we simply assume, as in the majority of cases for polymers that do not self-associate, that the development of order is reflected primarily in the infrared spectrum by changes in the distribution of conformations. With strong intermolecular interactions, large perturbations of certain vibrational modes of a single chain are found and we must consider both conformational and interactional effects. At this time, there remains much to be wrested from the infrared spectra of nylons and polyurethanes.

Hydrogen Bonding: Nylon 11

Poly(aminoundecanoic acid), or more commonly, nylon 11, has a regular structure with a chemical repeat of $-(CH_2)_{10}-NH-CO-$.

Poly(aminoundecanoic acid) - nylon 11

Extensive intermolecular hydrogen bonding occurs at ambient temperatures between the N-H group of one amide unit and the C=O of another. The polymer has a T_g of 45°C and a T_m of 196°C. We will briefly discuss two regions of the infrared spectrum of nylon 11; between 1600 and 1700 cm^{-1}, where the so-called Amide I mode absorbs, and between 3150 and 3500 cm^{-1}, where we find the N–H stretching vibration.

The Amide I mode is a mixed mode comprised of contributions from the C=O stretching, the C–N stretching and C–C–N deformation vibrations. For our purposes, however, we will not make serious errors if we consider the Amide I mode to be equivalent to the carbonyl stretching vibration. It is sensitive to local order, but this conformational sensitivity does not arise through mechanical coupling to the main chain. Instead it is due to differences in the pattern of hydrogen bonds that determine the relative arrangements of C=O groups and the degree of dipole-dipole interactions. In other words, in highly ordered domains, where the C=O groups are spatially arranged in a specific array, a given C=O group "knows" that others exist within a sphere of influence through dipole-dipole interactions. The upshot is that the frequency of the Amide I mode for ordered structures is significantly different from that of disordered amorphous structures. Infrared spectra of nylon 11, recorded as a function of temperature, are displayed in the range between 1600-1800 cm^{-1} in figure 6.11. The Amide I mode at 30°C is characterized by a relatively sharp band, skewed to the high frequency side, and centered at 1638 cm^{-1}. Significant changes in the spectra are observed with temperature. Actually, there are three spectral contributions to this region which may be resolved by a curve fitting. The first is a narrow band (width at half-height, $w_{1/2} = 18$ cm^{-1}) which systematically decreases in intensity and increases in frequency from 1636 to 1641 cm^{-1} when the temperature is increased to 190°C. This is assigned to hydrogen bonded carbonyl groups in ordered domains. Above the T_m (196°C), this band can no longer be seen. The second is a much broader band ($w_{1/2} = 38 \pm 4$ cm^{-1}) which systematically increases in intensity and frequency (from 1645 to 1654 cm^{-1}) as the temperature

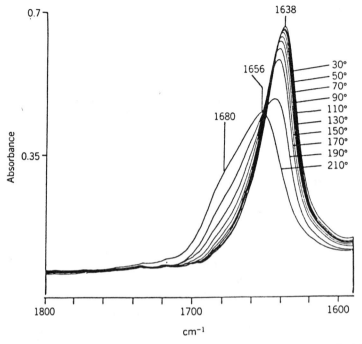

Figure 6.11 *The infrared spectrum of nylon 11 in the Amide I region as a function of temperature.*

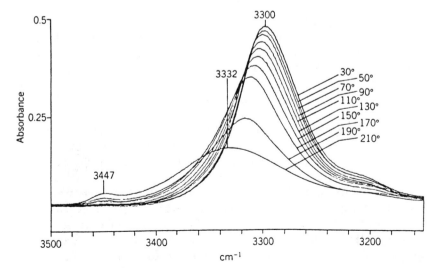

Figure 6.12 *The infrared spectrum of nylon 11 in the N–H stretching region as a function of temperature.*

is increased from 30 to 220°C. This band is attributed to hydrogen bonded carbonyl groups in disordered amorphous structures. Finally, the third band which occurs between 1679 and 1683 cm^{-1} ($w_{1/2} = 26 \pm 2$ cm^{-1}), is due to the "free" (non-hydrogen bonded) carbonyl groups. It can be seen that there is a wealth of information to be gained from studies of the Amide I mode of polyamides.

The N–H stretching mode is a conformationally insensitive, essentially isolated, vibration. It is, however, extraordinarily sensitive to hydrogen bonding. Figure 6.12 shows the infrared spectral region from 3150 to 3500 cm^{-1} of the nylon 11 sample recorded as a function of temperature. The spectra are characterized by a very weak band occurring at 3447 cm^{-1}, attributed to "free" N–H groups, and a relatively broad dominant band which varies in frequency from 3300 to 3332 cm^{-1} as the temperature is raised from 30 to 210°C. This latter band is assigned to hydrogen bonded N–H groups. As the temperature is raised the band broadens and shifts to higher frequency, which is consistent with a reduction of the average strength of the hydrogen bonds and a broadening of the distribution. There is also a marked reduction in absolute intensity (area) of the hydrogen bonded N–H band with increasing temperature which cannot be solely ascribed to the transformation of hydrogen bonded N–H groups to "free" groups. In fact, the loss of area is, in large part, a direct consequence of a reduction of the value of the absorptivity coefficient with the reduction in strength of the hydrogen bond. Unlike the Amide I, the hydrogen bonded N–H stretching vibration does not appear to be composed of two obvious contributions attributable to ordered and disordered hydrogen bonded amide groups. Rather, the N–H band envelope reflects the distribution of the strengths of the hydrogen bonded N–H groups regardless of whether they are in ordered or disordered domains. It is interesting to note, however, that the breadth of the hydrogen bonded N–H band correlates nicely with the degree of order in the material. One might ask the question, "Why do we see distinct bands associated with ordered and disordered structures in the Amide I region of the spectrum but not in the N–H stretching region?" There is, after all, a one to one correspondence between hydrogen bonded N–H and C=O groups. The answer lies in the differing sensitivities of the two normal vibrations.

Shades of Grey: Polymers with "Moderate" Intermolecular Interactions

Between those polymers that strongly self-associate and those that do not are a large number of polymers that may be categorized as moderately self-associated. These would include polymers containing functional groups that can interact through weak hydrogen bonds and/or dipolar interactions. Poly(vinyl chloride) (PVC) and simple aliphatic or aromatic polyesters are typical examples. It is important to recognize that the pigeonholes we have chosen to categorize the major features of the infrared spectra of polymers are somewhat arbitrary and are only useful as "rules of thumb." Many of the effects we have described so far can be assumed to be present in the spectra of moderately self-associated polymers; it is just a question of which dominate and which can be ignored. For example, the carbonyl stretching vibration of poly(ε-caprolactone) (PCL), a polymer containing the chemical repeat unit -$(CH_2)_5$-COO-, is sensitive to order in much the same way as the nylons mentioned above.

Copolymers and Structural Irregularities

"True" copolymers are derived from two or more co-monomers, as in the case of styrene-co-butadiene (SBR), ethylene-co-propylene (EP) or ethylene-co-methacrylic acid (EMAA) copolymers. However, as we mentioned in chapter 1, there are many polymers synthesized from a single monomer that might better be characterized as copolymers. For example, the homopolyester PCL, mentioned above, could be described as an alternating copolymer of pentamethylene units and ester moieties. In addition, in the free radical polymerization of chloroprene the predominant structural repeat unit is head-to-tail *trans*-1,4, but there are significant concentrations of *cis*-1,4; 1,2; 3,4 and head-to-head *trans*-1,4 units present. Thus, polychloroprene (TPC) can also be considered a rather complex copolymer. In any event, as infrared spectroscopists we should ask the question "What differences should we anticipate when we compare the spectra of homo- and copolymers?"

If we have two or more relatively large, essentially uncoupled repeat units present in the polymer chain, the sequence distribution will make little difference to the infrared spectroscopic features that are observed. In other words, it would be difficult to tell the difference between a random and a block SBR copolymer.

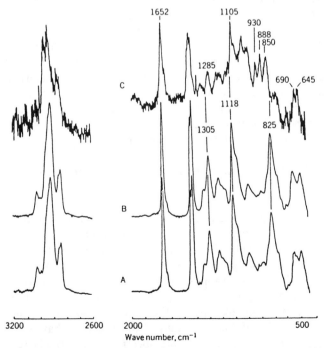

Figure 6.13 The infrared spectrum of polychloroprene polymerized at (A) -40°C and (B) -20°C. (C) Difference spectrum (B - A).

The spectrum is dominated by the characteristic bands of the individual chemical repeat units. A similar argument can be made for TPC. Figure 6.13 shows the difference spectrum representative of the structural "irregularities" *cis*-1,4; 1,2 and 3,4 units present in polychloroprene after the spectral contribution from the predominant *trans*-1,4 structural units have been subtracted out of the spectrum. Characteristic bands are readily identified and the reason the subtraction works so well is that the individual structural units act essentially independently. Obviously, for polymers and copolymers such as this, infrared spectroscopy can provide a useful measure of overall composition, but it is not ordinarily sensitive to sequence distributions. At the other extreme, we have copolymers synthesized from relatively small monomers where one might expect significant vibrational coupling between the structural units in the polymer chain, as in ethylene-co-propylene copolymers (EP). The infrared spectra of EP copolymers are complex and contain information concerning sequence distribution and degree of order.

Let us now mix things up a bit and consider the infrared spectrum of a random copolymer of ethylene and methacrylic acid.

Ethylene-co-methacrylic acid

From what we have seen so far we might expect a rather complex spectrum composed of features characteristic of an unassociated aliphatic chain containing a relatively small chemical repeat unit (like PE), together with features attributable to a strongly hydrogen bonded polymer. This is not a bad approximation. Sequence and ordering effects pertaining to the ethylene portion of the copolymer are indeed observed. The spectroscopic contribution from the methacrylic acid (MAA) portion is also very interesting. The carbonyl stretching frequency of a random EMAA copolymer containing 4 mole % MAA is observed at 1700 cm[-1] [see Figure 6.14(C)]. This frequency corresponds to the formation of an intermolecular hydrogen bonded dimer. There is scant, if any, evidence for "free" MAA units at ambient temperatures, which absorb at a much higher frequency of 1750 cm[-1]. So even though we have a random copolymer containing only 4 mole % of MAA units, these units efficiently find one another. This illustrates the strong driving force for association that is a property of such strongly interacting groups. It also nicely leads us into our next topic.

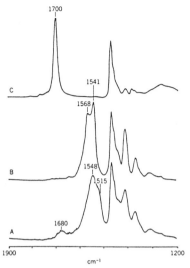

Figure 6.14 *The infrared spectrum of ethylene-co-methacrylic acid containing 4 mole %
MAA. (A) Calcium ionomer; (B) sodium ionomer, and (C) nonionized acid copolymer.*

Very Strong Intermolecular Interactions: Ionomers

Ionization of some or all of the acid groups present in the EMAA copolymers
leads to a class of materials referred to as ionomers. Dramatic effects are
observed in the infrared spectra upon ionization, the most important being the
reduction or loss of the band at 1700 cm^{-1}, attributed to hydrogen bonded acid
dimers, and the appearance of an intense band at about 1550 cm^{-1} which is
characteristic of carboxylate groups. This is in itself interesting and useful, but
there are subtle effects occurring in the carboxylate stretching region that reveal
information concerning the local structure of ionic domains so-called (multiplets).
In figure 6.14 the spectra denoted A and B are the completely ionized calcium
and sodium salts, respectively, and there are obvious differences We will not
dwell on the interpretation of the spectral differences, but rather just state that we
can infer the local structure of the ionic domains from a consideration of the
known coordinating tendencies of different cations together with a symmetry
analysis of models of most probable structures.

Oxidation, Degradation etc.

The "rules of thumb" implied above also pertain to the spectroscopic changes
seen upon the chemical reaction of polymers. In other words, the larger the unit
representing chemical modification and the smaller the effects of vibrational
coupling, the greater the confidence in assuming that the spectroscopic changes
simply reflect isolated structural defects. Accordingly, the vast majority of
studies involving chemical modification rely upon the simple group frequency
approach.

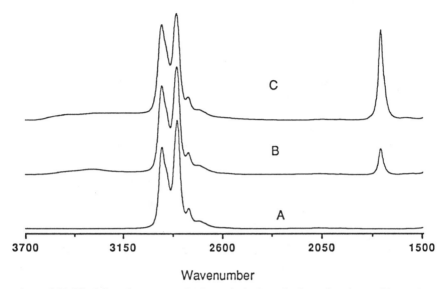

Wavenumber

Figure 6.15 *The infrared spectrum of polytetrahydrofuran in the carbonyl stretching region recorded at 150°C as a function of time; (A) 0, (B) 1.5, and (C) 3.5 h.*

Fortunately, many reactions result in the formation of carbonyl containing moeities and the vibrational modes of carbonyl groups, as we have seen, are some of the most intense and useful in infrared spectroscopy. One example should suffice. Figure 6.15 shows infrared spectra of a film of pure polytetrahydrofuran, $(-CH_2-CH_2-CH_2-CH_2-O-)_n$, recorded at 150°C as a function of time. The oxidation of the polymer may be readily followed by monitoring the carbonyl band at 1737 cm^{-1}.

Multicomponent Systems: Polymer Mixtures and Complexes

Infrared spectroscopy is one of many techniques that have been applied to unravel the complexities of the interactions occurring in polymer mixtures. At the extremes, polymer mixtures may be categorized as miscible (single phase) or immiscible (multiphased). More on this in Chapter 9. For now, however, you should recognize that in multiphased systems the two phases do not necessarily consist of just the pure components of the mixture. For a grossly phase separated system, where the two polymers exist in essentially separate and distinct phases, one can assume that in infrared spectroscopic terms, one polymer does not recognize the existence of the other, and *vice versa*. Thus, the spectrum of the blend would reflect the simple addition of the spectra of the two individual components. It is the case of miscible, or immiscible polymer blends where the separate phases are both mixtures of the two polymers, that is much more interesting. To simplify matters, it is known that the nature, relative strength and number of intermolecular interactions occurring between the polymeric components of the blend play a key role in determining miscibility.   As we have

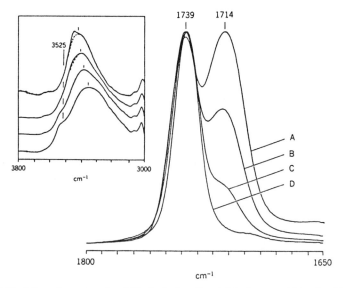

Figure 6.16 *Infrared spectra of poly(4-vinyl phenol)-poly(vinyl acetate) blends; (A) 80:20, (B) 50:50, (C) 20:80, and (D) 0:100.*

seen, infrared spectroscopy is sensitive to strong intermolecular interactions especially hydrogen bonding. An excellent example is shown in figure 6.16. Here we see the spectra in the carbonyl stretching region of poly(vinyl acetate) (PVAc) and miscible blends of PVAc and PVPh. The band at 1739 cm^{-1} is attributed to the PVAc carbonyl groups that are "free" (non-hydrogen bonded) while that at 1714 cm^{-1} is representative of PVAc acetoxy carbonyl groups hydrogen bonded to the phenolic hydroxyl group of PVPh, as depicted below:

Phenolic hydroxyl - acetoxy carbonyl
hydrogen bond

Not only do the spectra indicate mixing at the molecular level, but it is possible to obtain a quantitative measure of the number and relative strength of the intermolecular interaction as a function of temperature. These parameters can then be used to model phase behavior*.

---

* See, if you must, a book entitled *Specific Interactions and the Miscibility of Polymer Blends*, by M. M. Coleman, J. F. Graf and P. C. Painter. Same authors, same publisher, same lousy jokes; you get the picture!

E. BASIC NMR SPECTROSCOPY

Nmr spectroscopy occupies a special place in the arsenal of experimental methods that have been employed to study polymeric materials. During the past three decades there has been a veritable plethora of diverse (and very clever) experiments reported in the literature. Instruments have improved enormously and the emergence of solid state nmr, two dimensional (2D) techniques, nmr imaging etc., suggests that the best is yet to come. The subject has become increasingly mathematical, however, and is loaded with jargon. Phrases and acronyms like, "spin-spin relaxation", "COSY", "cross polarization", "dipolar decoupling", "CRAMPS", "magic angle spinning", and the like, can conjure fear into the hearts of students approaching this subject for the first time. Fortunately, in this introductory text we do not have to be concerned with such advanced applications of nmr spectroscopy. Rather, we will emphasize the straightforward applications of the method to the chemical microstructure of polymer chains and copolymer compositional analysis. This is a vast subject in its own right and we can only present what we consider to be representative and informative exam-ples, which hopefully will whet your appetite and send you scurrying to the li-brary for more. But first we present a brief summary of the basic fundamentals of the technique, paying particular attention to those aspects that can be used to obtain information about polymer composition and microstructure.

What Is NMR?

Nuclei of certain isotopes possess what can be thought of as a mechanical spin, or *angular momentum*, which is described by the *nuclear spin*, or *spin number*, I. A given nucleus has a specific spin number, i.e, $I = 0$, 1/2, 1, 3/2, ... etc., which is related to the mass number and the atomic number, as shown in table 6.1. The spinning nucleus gives rise to a magnetic field and the nucleus may be viewed as a small magnet of magnetic moment, μ. Significantly, the most common isotopes of carbon, ^{12}C, and oxygen, ^{16}O, are nonmagnetic ($I = 0$) and do not exhibit nmr spectra. For our purposes here, we will only be considering the ^{1}H, ^{13}C, and ^{19}F nuclei, which, fortuitously, because it makes things simple, all have spin numbers of 1/2. If we introduce a magnetic nucleus into a uniform external magnetic field, it assumes a discrete (i.e., quantized) set of $(2I + 1)$ of orientations. Thus, ^{1}H, ^{13}C, and ^{19}F nuclei ($I = 1/2$) will assume only one of two possible orientations that correspond to energy levels of $\pm \mu H_0$ in an applied magnetic field (H_0 is the strength of the external magnetic field). This is depicted in figure 6.17. The low-energy orientation corresponds to that

Table 6.1 *Spin Numbers of Isotopes.*

Mass Number	Atomic Number	Spin Number, I
odd	even or odd	1/2, 3/2, 5/2, ...
even	even	0
even	odd	1, 2, 3, ...

Table 6.2 Characteristics of Certain Isotopes.

Isotope	Abundance (%)	nmr Frequency [a] (MHz)	Relative Sensitivity [b]	Spin Number I
^{1}H	99.98	42.6	1.000	1/2
$^{2}H(D)$	0.016	6.5	0.0096	1
^{13}C	1.11	10.7	0.0159	1/2
^{14}N	99.64	3.01	0.0010	1
^{15}N	0.37	4.3	0.0010	1/2
^{19}F	100	40.01	0.834	1/2

[a] In 10 kG field.   [b] For equal numbers of nuclei at constant H_0.

state in which the nuclear magnetic moment is aligned parallel to the external magnetic field, and the high-energy orientation corresponds to that state in which the nuclear magnetic moment is aligned antiparallel (opposed) to this field. The transition of a nucleus from one possible orientation to the other is a result of the absorption or emission of a discrete amount of energy, such that $E = h\upsilon = 2\mu H_0$, where υ is the frequency of the electromagnetic radiation that is absorbed or emitted. For ^{1}H in a magnetic field of 14,000 gauss, the frequency of such energy is in the radio-frequency region—about 60 megacycles per second (60 MHz). A comparison of the nmr frequencies of nuclei that are of major interest for polymer spectroscopy is given in table 6.2.

The Chemical Shift

If the resonance frequencies for all nuclei of the same type in a molecule were identical, only one peak would be observed, and protons of a methyl group, for example, would resonate at the same frequency as those in an aromatic ring. This is not so, however, and subtle differences in nmr frequencies are observed. These are a consequence of the different molecular environments of the nuclei. Surrounding electrons shield the nuclei to different extents, depending upon

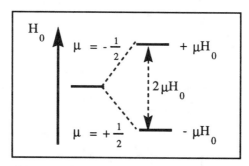

Figure 6.17 The two orientations of a ^{1}H nucleus.

chemical structure, and the effective magnetic field felt by a nucleus is not identical to that of the applied field. To be able to resolve these subtle effects requires an nmr spectrometer that consists of a very strong magnet that produces a stable and homogeneous field, a radio-frequency oscillator (i.e., the source of radiation—see figure 6.2), a radio-frequency receiver (the detector) and a device that can vary the magnetic field or frequency over a relatively narrow range (analogous to the prism or grating in figure 6.2). This separation of resonance frequencies of nuclei in different structural environments from some arbitrarily chosen standard [e.g., tetramethylsilane (TMS)] is termed the *chemical shift*.

The extent of electronic shielding is directly proportional to the strength of the applied field, so that chemical shift values are proportional to field strength (or equivalently, the oscillator frequency). In order to express chemical shifts in a form independent of the applied field (or oscillator frequency), a chemical shift parameter, δ', has been introduced, defined as $\delta' = (H_r - H_s)/H_r$, where H_s and H_r are the field strengths corresponding to resonance for a particular nucleus in the sample (H_s) and a reference (H_r). If TMS is used as the internal standard, the common equation for the chemical shift, δ (in units of parts per million, ppm), is obtained:

$$\delta \text{ (ppm)} = \frac{(\upsilon_{TMS} - \upsilon_s) \cdot 10^6}{\text{spectrometer frequency (cps)}} \qquad (6.12)$$

where $(\upsilon_{TMS} - \upsilon_s)$ is the difference in absorption frequencies of the sample and the reference in cps.

Typically, for ^1H nuclei a range of 10 ppm covers most organic molecules. The corresponding range for ^{13}C nuclei is much greater, ≈ 600 ppm (see below). We should also mention another parameter commonly encountered in the ^1H nmr spectroscopic literature, τ, which is defined as $10 - \delta$.

It is now appropriate to look at a very simple nmr spectrum that illustrates the principles outlined above. Figure 6.18 shows a low resolution ^1H nmr spectrum of ethanol. There are three absorption peaks that have an area ratio of 1:2:3, corresponding to the protons in $-OH$, $-CH_2-$, and $-CH_3$ groups, respectively.

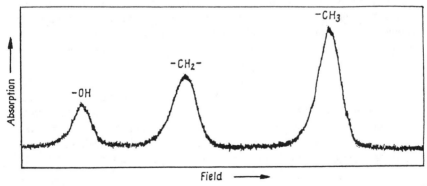

Figure 6.18 Low resolution ^1H nmr spectrum of ethanol (reproduced with permission from L. M. Jackman and S. Sternhell, Nuclear Magnetic Resonance Spectroscopy in Organic Chemistry, Pergamon Press, 1969).

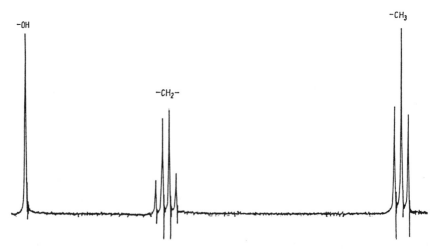

Figure 6.19 High resolution ^1H nmr spectrum of ethanol (reproduced with permission from L. M. Jackman and S. Sternhell, Nuclear Magnetic Resonance Spectroscopy in Organic Chemistry, *Pergamon Press, 1969).*

This nicely illustrates first, the different chemical shifts for protons in the three different groups and second, that a quantitative measure of the number of protons in each group is readily obtained[*]. All well and good, but there is even more useful information to be gained if the nmr spectrum is recorded at a higher resolution, which leads us to the fascinating subject of spin-spin interactions.

Spin-Spin Interactions

If the nmr spectrum of ethyl alcohol is obtained at a higher resolution (figure 6.19), then the peaks due to the methylene and methyl protons appear as multiplets, but the total relative area of each group is again $\approx$ 1:2:3, corresponding to the –OH, –CH_2, and –CH_3 groups, respectively. The *methyl* absorption is split into a *triplet* (relative areas $\approx$ 1:2:1), and the *methylene* absorption is split into a *quartet* (relative areas $\approx$ 1:3:3:1). These splitting patterns are caused by the magnetic field of the protons of one group being influenced by the spin arrangements of the protons on the adjacent group. The observed multiplicity of a given group of equivalent protons depends on the number of protons on adjacent atoms and is equal to n + 1, where n is the number of protons on adjacent atoms. Thus the two CH_2 protons in the ethyl group split the CH_3 protons into a triplet, the three CH_3 protons of the ethyl group split the CH_2 protons into a quartet, etc. In simple cases of interacting nuclei, the relative intensities of a multiplet are symmetric about the mid-point and are approximately numerically proportional to the coefficients seen in Pascal's triangle (i.e., for a doublet, 1:1; a triplet, 1:2:1, a quartet, 1:3:3:1 etc.)

---

[*] Note that unlike infrared spectroscopy, where the absorption coefficients for the bands are all different, in ^1H nmr the intensities of the bands give a direct measure of the number of nuclei that they represent. Things are not quite so easy in ^{13}C nmr (see later).

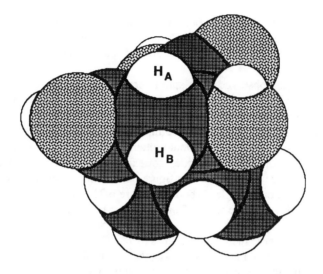

Figure 6.20 *Space filling model of (±)-2-isopropylmalic acid.*

Interactions between magnetically equivalent protons in the same group (for example, the two protons in the CH$_2$ group) are not observed in simple systems like ethanol. You should also be aware that the spacing (in cps) of the three components of the methyl group triplet is equal to the spacing of the four components of the methylene group quartet and is *independent* of the strength of the applied field. This spacing is called the *spin-spin coupling constant*, and is denoted by the symbol *J*. The simple rules for determining the multiplicities for spin-spin interactions of adjacent groups hold only for cases in which the separation of resonance lines of interacting groups (the symbol $\Delta\upsilon$) is much larger than the coupling constant J of the groups ($\Delta\upsilon \gg J$). In systems of interact-

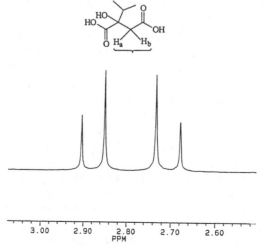

Figure 6.21 *^1H nmr spectrum of (±)-2-isopropylmalic acid.*

ing nuclei in which the coupling constant is of the same order of magnitude as the separation of resonance lines ($\Delta\upsilon \cong J$), the simple multiplicity rules no longer hold; more lines appear and simple patterns of spacings and intensities are no longer found.

As we have just observed, spin-spin interactions between protons within a group of magnetically equivalent protons (for example, the two CH_2 group protons in ethanol) are not ordinarily observed. However, there are cases (which are particularly relevant to our forthcoming discussions of polymer tacticity) where two magnetically non-equivalent protons on the same carbon can give rise to a maximum of four lines. Consider, for example, ($\pm$)-2-isopropylmalic acid illustrated as a space filling model in figure 6.20. This molecule has two non-equivalent protons on a single carbon. In figure 6.21 a scale expanded ^1H nmr spectrum is displayed in the region from 2.5 to 3.5 ppm. The lines arising from the two protons, H_a and H_b, are shown directly below the methylene group itself. You will observe there are four lines there. The two central lines are more intense than the satellite lines. In this case, the chemical shift difference is of the same order of magnitude as the coupling constants. This system is designated an *AB pattern,* in that there are two protons involved in a slightly different chemical shift. Of course, if the two methylene protons were equivalent, a single line would be seen. It is situations similar to these we shall look for in tacticity measurements on polymers.

^{13}C NMR

There are some features of ^1H and ^{13}C nmr spectroscopy that are common, but it is their dissimilarities that are more interesting. The most abundant isotope of carbon, ^{12}C, has no nuclear spin ($I = 0$). Thus it cannot be observed in nmr experiments. On the other hand, ^{13}C has the same nuclear spin ($I = 1/2$) as ^1H, but unlike ^1H, which has a natural abundance of > 99.9%, ^{13}C is only present to the extent of 1.1% (table 6.2). One major consequence of this low natural abundance is that ^{13}C-^{13}C spin-spin coupling interactions are unlikely in unenriched compounds (assuming random placements of the ^{13}C nuclei, the probability of two adjacent ^{13}C nuclei is ≈ 0.0001). Incidentally, if ^{13}C did exist in greater natural abundance, the early ^1H nmr spectroscopists would have had serious problems in interpreting ^1H nmr spectra, because of the complication of ^{13}C-^1H spin-spin coupling. The effect of this coupling in ^{13}C nmr spectra may now be readily eliminated by a technique called proton decoupling.

The inherently low natural abundance of ^{13}C obviously lowers the effective sensitivity of ^{13}C *vis-a-vis* ^1H nmr experiments. To compound the problem further, ^{13}C nuclei only produce 1/64 of the signal that ^1H nuclei yield on excitation. Thus the relative sensitivity of a ^{13}C nmr experiment is some 6000 times less than that of a ^1H nmr experiment. While this appears to be a daunting practical limitation, instrument manufacturers have effectively solved the problem and obtaining high quality ^{13}C nmr spectra of organic materials is now routine (see, for example, figure 6.22 which shows the proton decoupled ^{13}C nmr spectrum of 2-ethoxyethanol obtained in our laboratory.) Perhaps the most

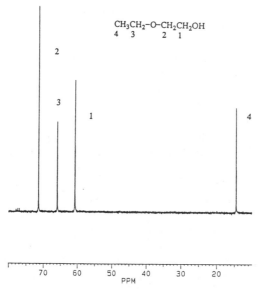

Figure 6.22 *^{13}C nmr spectrum of 2-ethoxyethanol.*

striking difference between ^{13}C and ^{1}H nmr spectroscopy, however, is the enhanced effective resolution possible in ^{13}C nmr. ^{13}C resonances of organic compounds are found over an enormous chemical shift range of 600 ppm. This is some 50 times greater than that for ^{1}H nuclei. Frequently, one can identify individual resonances for each carbon in a compound, as in the spectrum of 2-ethoxyethanol which shows well resolved nmr lines attributable to each individual carbon. Note, however, that the lines are not of equal intensity, even though they are each assigned to a single carbon.

It is not possible to discuss here the many subtleties of ^{13}C nmr, but suffice it to say that quantitative analysis is much more difficult than in ^{1}H nmr. One cannot just simply relate the relative intensities to the number of equivalent carbons in the molecule. Additional information pertaining to relaxation mechanisms and the so-called nuclear Overhauser effect is required. The interested reader is refered to the texts on the subject mentioned at the end of this chapter.

^{13}C nmr spectroscopy has the following additional advantages over ^{1}H nmr for the analysis of organic polymers; (1) the direct observation of molecular backbones; (2) the direct observation of carbon-containing functional groups that have no attached protons (e.g., carbonyls, nitriles) and (3) the direct observation of carbon reaction sites of interest. However, ^{1}H nmr spectroscopy has not passed into oblivion, but in fact has a number of competing advantages relative to ^{13}C nmr including: (1) the ease of quantitative analysis; (2) the rapidity of analysis time; (3) the enhanced sensitivity; (4) the direct observation of OH and NH groups (undetectable by ^{13}C nmr) and (5) the separation of olefinic and aromatic protons, which appear in different regions of the ^{1}H nmr spectrum while olefinic and aromatic carbons overlap one another in the ^{13}C nmr spectrum.

F. CHARACTERIZATION OF POLYMERS BY NMR SPECTROSCOPY

In common with infrared spectroscopy, nmr spectroscopy may be applied to the characterization of polymeric materials at various levels of complexity. Modern chemists involved in polymer synthesis would be lost without ready access to nmr spectrometers. They serve as routine analytical techniques for the screening, identification and analysis of monomers, (co)polymers and reaction products. At the next level of sophistication, nmr spectroscopy has been extensively employed to characterize the microstructure of polymeric materials. Here nmr excels and the technique has dominated the field of polymer sequence distribution in all its myriad forms, including those involving stereoisomerism, sequence isomerism, structural isomerism and copolymer composition. Finally, at the highest level of sophistication, nmr spectroscopy is used to probe the dynamics of polymer systems through the study of relaxation phenomena, but this latter topic must be left for more specialized texts.

Our approach here will be to start by presenting a few representative examples of the simple use of nmr spectroscopy in polymer analysis. We will then turn our attention to the meat of this section and describe the application of nmr spectroscopy to the characterization of polymer sequence distributions. In Chapter 1, and again at the end of Chapter 5, we mentioned that isomerism in homopolymers can be viewed as a special case of copolymerization. We will develop this theme, spending some time on the classic ^{1}H nmr spectroscopic studies of the tacticity of poly(methyl methacrylate), and show how probability theory can been used to describe the sequence distribution in terms of the different tactic placements in the polymer chain. Finally, we will close by describing a few examples of analogous studies pertaining to sequence and structural isomerism.

Copolymer Analysis

As we mentioned above, nmr spectroscopy is one of the major instrumental methods used routinely to identify and analyze the composition of copolymers. Three representative examples, two employing ^{1}H and one using ^{13}C nmr spectroscopy, taken from our own recent studies, will be used to illustrate the utility of the nmr technique.

Methyl Methacrylate-co-Hexyl Methacrylate Copolymers

Let us start with a very straightforward example. We recently synthesized a series of methyl methacrylate-co-hexyl methacrylate copolymers (MMA-co-HMA) for the purpose of testing a model that we have developed to predict the phase behavior of polymer blends. ^{1}H nmr was employed to determine the copolymer composition*. Figure 6.23 shows a typical spectrum of a MMA-co-HMA copolymer. The rather complicated collection of nmr lines occurring

---

* M. M. Coleman, Y. Xu, S. R. Macio and P. C. Painter, *Macromolecules*, **26**, 3457 (1993).

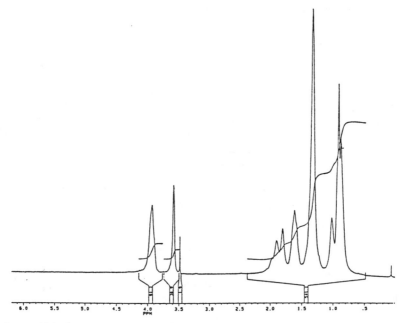

Figure 6.23 *¹H nmr spectrum of a copolymer of methyl methacrylate and hexyl methacrylate.*

between 0.5 and 2.5 ppm are assigned to the alkyl methylene and methyl protons of the copolymer. Fortunately, for the purposes of measuring copolymer composition, we can ignore these nmr lines and focus our attention solely on the two isolated nmr lines occurring at ≈ 3.6 and 3.9 ppm. These are assigned to the three protons of the –OCH₃ methoxy substituent of MMA and the two alkoxy protons of the –OCH₂– methylene group, respectively, as depicted above. Thus, if we divide the relative areas of these two nmr lines by 3 and 2, respectively, quantitative compositional analysis becomes remarkably straightforward, i.e.:

$$\% \text{ MMA} = \frac{A_{3.6\,ppm}/3}{A_{3.6\,ppm}/3 + A_{3.9\,ppm}/2} \times 100 \qquad (6.13)$$

Styrene-co-Vinyl Phenol Copolymers

Another set of copolymers that we have synthesized, also for our ongoing research on polymer blends, are those containing styrene and 4-vinyl phenol. Because side reactions involving the phenolic hydroxyl group occur during the direct polymerization of 4-vinyl phenol (VPh), protected monomers are commonly employed. In our studies we used VPh protected with the *t*-butyl-dimethylsilyl group (t-BSOS) to prepare copolymers of styrene and t-BSOS. These were subsequently hydrolyzed to produce styrene-co-vinyl phenol (STVPh) copolymers, as summarized below[*].

[*]H nmr spectroscopy was used to identify the copolymers and follow the deprotection step. Figure 6.24 shows a comparison of the [*]H nmr spectra of a parent styrene-co-t-BSOS copolymer (denoted A) and the product after the desilylation process, STVPh (B). The nmr lines at $\delta = 0.95$ and 0.16 ppm in spectrum (A) are assigned to the methyl and t-butyl substituents in the t-butyldimethylsilyl monomer, and their absence in the nmr spectrum of the deprotected copolymer (B) clearly indicates the elimination of the t-butyl-dimethylsilyl group.

---

[*] Y. Xu, J. F. Graf, P. C. Painter and M. M. Coleman, *Polymer*, **32**, 3103 (1991).

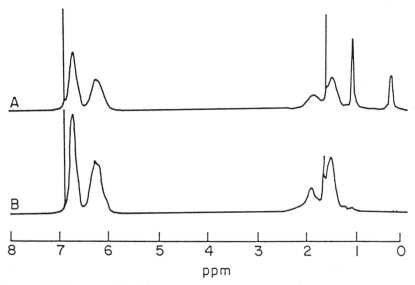

Figure 6.24 *¹H nmr spectrum of (A) a styrene-co-t-BSOS copolymer and (B) the corresponding styrene-co-vinyl phenol copolymer after deprotection.*

The copolymer composition of the various styrene-co-t-BSOS copolymers synthesized was also determined from ¹H nmr spectra similar to that shown in figure 6.24(A). It is a little more complicated than the methyl methacrylate-co-hexyl methacrylate copolymers that we just discussed, but the principle is the same, and you just have to keep track of the number and type of protons that contribute to a given nmr line or group of lines. As we inferred above, the nmr line at 0.16 ppm corresponds to the 6 protons present in the two methyl substituents of the t-BSOS. Similarly, the nmr line at 0.95 ppm corresponds to the 9 protons present in the three methyl substituents of the t-butyl group in t-BSOS. Thus the normalized area per proton corresponding to the t-BSOS chemical repeat may be determined from either:

$$A^{H}_{t\text{-}BSOS} = \frac{\text{Total area of } 0.16 \text{ ppm line}}{6}$$

$$= \frac{\text{Total area of } 0.95 \text{ ppm line}}{9} \tag{6.14}$$

Now we need the normalized area per proton corresponding to the styrene chemical repeat. But there no isolated nmr line in the spectrum that is solely characteristic of styrene repeat unit. We should not despair, however, because we can obtain the needed information in an indirect manner. The relatively broad lines appearing between 6.2 and 7.2 ppm in spectrum A correspond to the aromatic protons that occur in both St and t-BSOS. So if we measure this total area (which reflects the contributions from the 5 aromatic protons of the styrene repeat and the 4 aromatic protons of the t-BSOS repeat), subtract out the contribution from the t-BSOS repeat (i.e., 4 times the normalized area per proton calculated from equation 6.14) and then divide by 5 (the number of aromatic protons in the styrene repeat), we will have the normalized area per proton corresponding to the St chemical repeat. In summary:

$$A_{St}^{H} = \frac{(\text{Total area of region from 6.2 to 7.2 ppm}) - 4\,A_{t\text{-BSOS}}^{H}}{5} \qquad (6.15)$$

Thus, the % styrene in the copolymer is simply given by:

$$\% \text{ Styrene} = \frac{A_{St}^{H}}{A_{St}^{H} + A_{t\text{-BSOS}}^{H}} \times 100 \qquad (6.16)$$

Ethyl Methacrylate-co-4-Vinyl Phenol Copolymers

A final example in this section is taken from our work on copolymers of methacrylates and vinyl phenol which were synthesized using similar chemistry to that shown above for the styrene-co-vinyl phenol copolymers. Figure 6.25 shows a typical ^{13}C nmr spectra of an ethyl methacrylate (EMA) copolymer containing 52 mole % EMA before (top) and after (bottom) deprotection. The absence of the nmr peaks at around 0 ppm (the two methyl carbons attached to silicon), 19 ppm (the tertiary carbon of the t-butyl group), and 26 ppm (the three methyl carbons on the t-butyl group) after desilylation, clearly indicates the absence of any residual t-butyldimethylsilyl. While ^{13}C nmr can be used to obtain quantitative data concerning the copolymer composition, after due attention is given to relaxation times and the nuclear Overhauser effect, it is easier to use ^{1}H nmr. The aromatic protons of the phenol ring and the methylene group that is adjacent to the oxygen of the ester side chain (i.e., the $-OCH_2-$ between 3.2 and 4.2 ppm) may be employed as a quantitative analytical probe[*].

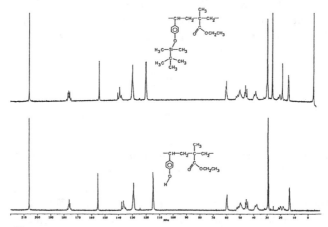

Figure 6.25 ^{13}C nmr spectrum of ethyl methacrylate-co-t-BSOS (top) and ethyl methacrylate-co-VPh copolymers.

---

[*] Y. Xu, P. C. Painter and M. M. Coleman, *Polymer*, **34**, 3010 (1993).

The Observation of Tacticity: Poly(methyl methacrylate) (PMMA)

The observation and measurement of tacticity in polymer systems is one of the areas of polymer characterization where nmr spectroscopy reigns supreme. Nowadays, ^{13}C nmr spectroscopy is used extensively to study stereoisomerism in polymers and the detailed information that can be obtained, in terms of the number and size of the different sequences of tactic placements, is truly incredible. However, we are getting ahead of ourselves. Before we look at some examples of ^{13}C nmr spectra, we will initially discuss ^{1}H nmr studies of poly(methyl methacrylate), because this is where it all started and there is insight to be gained.

Recall that we mentioned above that two magnetically non-equivalent protons, whether they be on the same carbon or on an adjacent carbon, can give rise to a maximum of four lines. If the two protons involved have only slightly different chemical shifts and the chemical shift difference is of the same order of magnitude as the coupling constants, we observe a complex AB pattern. Were the two protons magnetically equivalent, of course, a single line would be observed. It is these types of situations that we shall look for in the ^{1}H nmr spectra of tactic polymers.

Common or garden PMMA, produced by free radical polymerization, is atactic (*a*-PMMA), but the essentially pure isotactic (*iso*-PMMA) or syndiotactic (*syn*-PMMA) forms of PMMA can be synthesized.

Poly(methyl methacrylate)

Schematic representations of these different chain structures are presented in figure 6.26. Let us first consider the top structure in the figure, labelled *isotactic*, where all groups of the same type are on the same side of the chain backbone. The large spheres represent ester groups; the smaller dark spheres opposite represent methyl groups. The other two small spheres attached to the backbone represent the hydrogen atoms of methylene units. Thus, this structure depicts pure *iso*-PMMA. Note that the *two methylene protons* on the carbon backbone are *not magnetically equivalent*. The upper methylene protons, all along the chain, are always in the environment of the two methyl groups. Conversely, the lower protons on the same carbon are in all cases flanked by two ester groups. Accordingly, the magnetic environments of the two methylene protons are different and we would expect a small difference in chemical shift, which should give rise to a four line AB pattern, as shown schematically in figure 6.27.

The middle structure in figure 6.26 shows an alternating structure, in that the methyl groups alternate up and down as you move along the carbon chain. This

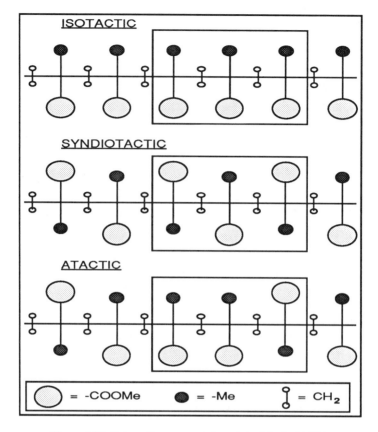

Figure 6.26 *Schematic representation of tacticity in PMMA.*

is the perfect *syndiotactic* structure corresponding to pure *syn*-PMMA. If we now consider the methylene protons in this structure, we observe that they *are magnetically equivalent*. Each methylene proton has the same environment; flanked on one side by a methyl group and on the other side by an ester group. The same is true for the other proton attached to that same carbon. Since they are magnetically equivalent, there will be no spin-spin interaction and these protons should give rise to a single resonance line. The location of this resonance line, that is its chemical shift, must be exactly in the center of the resonance line pattern observed for the isotactic case, as shown in figure 6.27.

The lower structure depicted in figure 6.26 represents the *atactic* polymer. As we proceed along the carbon backbone and look at the environment of the methylene protons, in some cases we will find they are equivalent (which would give rise to a single line), and in other cases we will find they are non-equivalent (giving rise to the four line AB pattern). Hence, the methylene protons of an atactic polymer will give rise to an nmr spectrum which will resemble a combination of the isotactic and the syndiotactic spectra (figure 6.27). This leads to an important conclusion; if we observe a spectrum having both the singlet and the AB pattern in the methylene proton region of the nmr spectrum, we cannot

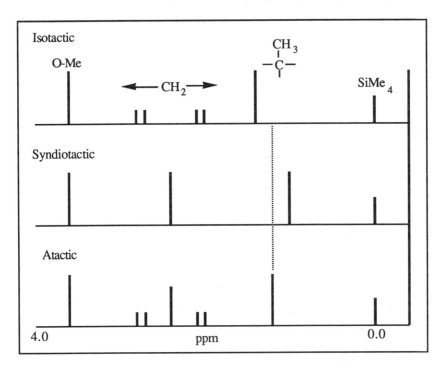

Figure 6.27 *Schematic representation of the ^1H nmr spectrum of PMMA triads.*

differentiate between a *mixture* of isotactic and syndiotactic polymers or a single atactic polymer. This is simply due to the fact that the methylene protons only "see" the influence of diad structures.

The picture is improved, however, if we consider the backbone methyl groups in these three structures. Here we see the *triad* information as highlighted by the small boxes in figure 6.26. The differences in magnetic environment of the methyl groups caused by the relative positions of the ester groups on each *adjacent* monomeric unit lead to subtle differences in the chemical shift of the methyl protons. Thus, the methyl line arising from an isotactic triad appears at one specific location; the methyl line for the syndiotactic triad appears at a different location, and the methyl line attributed to the atactic (heterotactic) triad, (where there is one meso and one racemic placement), appears at a third location between the isotactic and the syndiotactic methyl triad resonance lines. This is also summarized schematically in figure 6.27. An important point to emphasize is that triad (or higher sequence) data is required to adequately describe stereoregularity.

Proton nmr spectra recorded on a 60 MHz instrument of two different PMMA samples were first reported by Bovey and Tiers and are reproduced in figure 6.28. The top spectrum (a) is that of a predominantly *syn*-PMMA, while that of the bottom (b) is predominantly *iso*-PMMA. Note that the essential features of the experimental nmr spectra are the same as those depicted schematically in figure 6.27. There are, however, signs of additional splittings

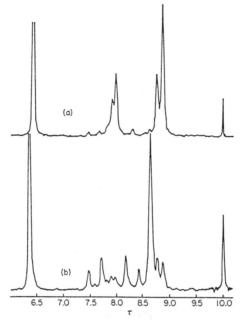

Figure 6.28 *60 MHz ^{1}H nmr spectra of PMMA. Reproduced with permission from F. A. Bovey*, High Resolution NMR of Macromolecules, *Academic Press (1972)*.

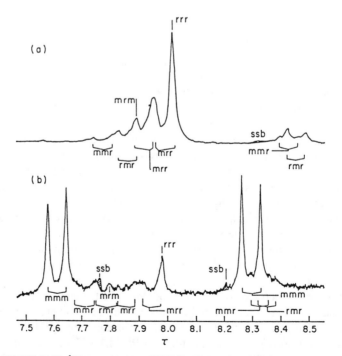

Figure 6.29 *220 M Hz ^{1}H nmr spectra of PMMA. Reproduced with permission from F. A. Bovey*, High Resolution NMR of Macromolecules, *Academic Press (1972)*.

that would indicate sensitivity to higher order sequences. In fact, with the advent of more powerful spectrometers larger sequences have been observed and measured using ^{1}H nmr. Figure 6.29 shows one example of a 220 MHz ^{1}H nmr spectrum of the β-methylene proton region of predominantly *syn*-PMMA (a) and *iso*-PMMA (b), where tetrad data is resolved (incidentally, pentad data is resolved in the α-methyl proton region). The tetrads are denoted as mmm, mmr, mrm, etc. in figure 6.29, and to appreciate the significance of this nomenclature we must again savor the delights of probability theory.

A Return to Probability Theory—Stereoisomerism

The reader may recall that in Chapter 1 under our broad definition of the word "copolymer", we argued that the polymerization of vinyl polymers could be viewed conceptually as a special case of copolymerization—that of the placement of two different "monomers" with opposite steric configurations into the chain. Whereas a copolymer may be described as a mixture of A and B sequences, a stereoregular homopolymer may be viewed as a mixture of $\underline{d}$ and $\underline{l}$ sequences linked together. Two adjacent monomer units in the chain are called meso diads when they have the same configuration (dd or ll) and racemic diads when the configurations are opposite (dl or ld). This was illustrated in Chapter 1 (figure 1.5). The degree of tacticity refers to the fraction of the appropriate tactic bonds present in a polymer and the fraction of each of these types is determined by the probabilities of forming the respective configurations.

As we have indicated, high resolution nmr is particularly suitable for the study of stereoisomerism in polymers and direct experimental measurements of configurational sequences in vinyl polymers are routinely made. Unfortunately, the prevailing nomenclatures employed to describe configurational sequences, which were developed from an nmr perspective, are significantly different from that described in Chapter 5 for copolymers. We will concentrate on the formalism attributed to Bovey[*].

The Generation of Configurational Sequences

In table 6.3, symbols representing diad, triad and tetrad sequences are summarized. The meso diad is designated m and the racemic diad, r. This system of nomenclature can be extended to sequences of any length. Thus, an *isotactic* triad is mm, a *heterotactic* triad mr, and a *syndiotactic* triad rr. Let us initially assume that the probability of generating a meso sequence, when a new monomer unit is formed at the end of a growing chain, can be described by a single parameter, P_m. In these terms, the generation of the chain obeys *Bernoullian* statistics. Conceptually, this is same as reaching into a large jar of balls marked "m" or "r" and withdrawing a ball at random. The proportion of "m" balls in the jar is P_m. The probability of forming a racemic sequence, r, is thus $(1 - P_m)$. Table 6.3 lists the Bernoullian probabilities for the various triads and tetrads and a plot of the triad relationships is shown in figure 6.30. Note that

---

[*] F. A. Bovey, *High Resolution NMR of Macromolecules*, Academic Press (1972).

Table 6.3 Configurational Sequences.

Type	Designation	Projection	Bernoullian Probability
Diad	meso, m		P_m
	racemic, r		$(1 - P_m)$
Triad	isotactic, mm		P_m^2
	heterotactic, mr		$2 P_m (1 - P_m)$
	syndiotactic, rr		$(1 - P_m)^2$
Tetrad	mmm		P_m^3
	mmr		$2 P_m^2 (1 - P_m)$
	rmr		$P_m (1 - P_m)^2$
	mrm		$P_m^2 (1 - P_m)$
	rrm		$2 P_m (1 - P_m)^2$
	rrr		$(1 - P_m)^3$

the proportion of mr units rises to a maximum at $P_m = 0.5$, corresponding to a completely random propagation, where the proportions mm:mr:rr will be 1:2:1. It is important to recognize, however, that P_m may take values other than 0.5.

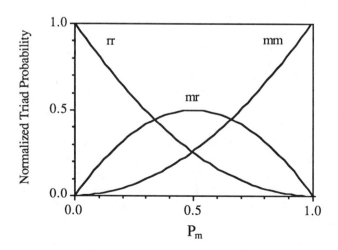

Figure 6.30 Bernoullian Triad Probabilities.

For any given polymer, if propagation obeys Bernoullian statistics, the mm, mr, and rr sequence frequencies, which may be determined from nmr data, would lie on a single vertical line in figure 6.30, corresponding to a single value of P_m. If this is not the case, then the configurational sequence of the polymer deviates from Bernoullian and might obey higher order Markovian statistics. Methyl methacrylate polymers produced by free radical initiators usually follow Bernoullian statistics, within experimental error, while those produced by anionic initiators do not. Tetrad and higher sequences can also be calculated and plotted in the same manner.

Some Necessary Nomenclature

In the nmr literature there are some shorthand notations that we must understand. First, the symbolism $P_n\{ \ \}$ that was used in Chapter 5 for copolymer sequences (e.g., $P_3\{ABA\}$), denoting the probability of a n-ad*, is dropped. Hence:

(m) is equivalent to $P_2\{m\}$, the number fraction of m *diads*

(mr) is equivalent to $P_3\{mr\}$, the number fraction of mr *triads*

Second, there is a distinction made by the type of brackets between observable, *distinguishable* (n-ads) and *indistinguishable* [n-ads]. It has been our experience that a significant fraction of students has difficulty with this concept and it is worthwhile to try to make this clear. Consider the schematic

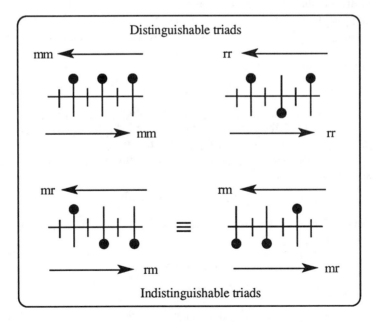

Figure 6.31 Schematic "ball and stick" diagram depicting triad sequences.

---

* An n-ad is a sequence of n species in the progression: diad, triad, tetrad, pentad, ... n-ad.

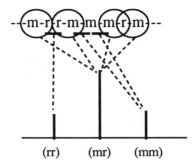

Figure 6.32 *Schematic diagram illustrating indistinguishable sequences.*

diagram shown in figure 6.31. At the top are depicted the mm and rr triads. It does not matter from which direction of the chain we read the tactic placements, the result is the same. "Flipping" either chain horizontally makes no difference to designation of the sequence. On the contrary, we see that if we read the chain sequence depicted at the bottom left hand side of the diagram, from left to right, we determine an rm sequence, but if we read it from the other direction, right to left, we obtain an mr sequence. Alternatively, if we compare the original chain to that "flipped" horizontally, as depicted in the lower right hand side of figure 6.31, reading both from left to right, we obtain an rm and an mr sequence, respectively. This is important, because it means that in the nmr experiment, we cannot differentiate between an mr or an rm sequence. In other words, the mr and rm sequences are indistinguishable and occur at the same chemical shift in nmr spectra, as we have tried to depict in the schematic diagram shown in figure 6.32.

Thus we see for example:

$$(mr) = [mr] + [rm] \qquad (6.17)$$

where (mr) symbolizes the number fraction of distinguishable triads, which is the sum of [mr] and [rm], the number fractions of the indistinguishable triads. With a bit of luck, you may recall from Chapter 5, that reversibility relationships are general (e.g., $P_2\{AB\}=P_2\{BA\}$), and therefore [mr] must equal [rm]. Accordingly:

$$(mr) = 2[mr] = 2[rm] \qquad (6.18)$$

The relative proportion of each sequence length is subject to the following general relationships:

$$(m) + (r) = 1$$

$$(mm) + (mr) + (rr) = 1$$

$$\sum (\text{all tetrads}) = 1 \quad \text{etc.} \qquad (6.19)$$

As in the case of the general probability relationships developed for copolymerization, reversibility relationships will be particularly useful:

$$[mr] = [rm]$$

$$[mmr] = [rmm]$$

$$[mrr] = [rrm] \tag{6.20}$$

Additionally, lower order sequences can be expressed as sums of the two appropriate higher order sequences, since any sequence has a successor or predecessor which can only be either an m or an r. Thus:

$$[mr] = [mmr] + [rmr]$$

$$[rmr] = [rrmr] + [mrmr] \tag{6.21}$$

Other relationships between distinguishable (observable in the nmr spectra) and higher order sequences, are necessary, such as diad-triad[*] :

$$(m) = (mm) + \frac{1}{2}(mr)$$

$$(r) = (rr) + \frac{1}{2}(rm) \tag{6.22}$$

Similarly for triad-tetrad:

$$(mm) = (mmm) + \frac{1}{2}(mmr)$$

$$(mr) = (mmr) + 2(rmr) = (mrr) + 2(mrm)$$

$$(rr) = (rrr) + \frac{1}{2}(mrr) \tag{6.23}$$

It is important that you understand the above relationships, because they have a direct bearing on the characterization and measurement of tacticity in polymers. Let us say, for example, that the schematic diagram shown in figure 6.32 represents the data obtained from a PMMA sample and you were asked to determine the % isotacticity (m) from this data. First, you must be convinced that you are simply observing triad data. Assuming this to be the case, you would now measure the areas of the three bands, $A_{(rr)}$, $A_{(mr)}$ and $A_{(mm)}$. The number fractions, (rr), (mr) and (mm), are simply calculated by dividing each individal area by the total area $A_T = A_{(rr)} + A_{(mr)} + A_{(mm)}$. The % isotacticity is *not* $A_{(mm)}/A_T$, but:

$$\% \text{ isotacticity} = (m) = \frac{A_{(mm)} + 1/2\,A_{(mr)}}{A_T} \tag{6.24}$$

---

[*] Note that [m] = [mm] + [mr], and (mr) = [mr] + [rm], therefore (m) = (mm) + 1/2(mr). A useful rule of thumb that can be employed for these relationships is as follows: if the sequence is different when it is reversed then 1/2 precedes the distinguishable sequences 1/2(n-ad) = [n-ad]. On the other hand, if the sequence is the same on reversal then (n-ad) = [n-ad].

The Terminal or Simple Markov Model

The terminal model for stereospecific homopolymerization involves only two parameters and follows simple Markovian statistics. They are:

$$P(r/m) = 1 - P(m/m) = u$$

$$P(m/r) = 1 - P(r/r) = w \tag{6.25}$$

These two parameters may be related to the measurable n-ads by the following type of derivation. We again start with a general reversibility relationship:

$$[mr] = [rm] \tag{6.26}$$

which may be written as:

$$(m) \, P(r/m) = (r) \, P(m/r) \tag{6.27}$$

now:

$$(r) = 1 - (m) \tag{6.28}$$

thus:

$$(m) \, P(r/m) = \big(1 - (m)\big) \, P(m/r) \tag{6.29}$$

and:

$$(m) = \frac{P(m/r)}{P(r/m) + P(m/r)} = \frac{w}{u + w} \tag{6.30}$$

Similarly:

$$(r) = \frac{u}{u + w} \tag{6.31}$$

In an identical fashion for the tetrads, starting with:

$$[mmr] = [rmm] \tag{6.32}$$

then:

$$(mm) \, P(r/m) = (r) \, P(m/r) \, P(m/m) \tag{6.33}$$

Hence for (mm) and for the other two triads we obtain:

$$(mm) = \frac{w(1 - u)}{u + w} \quad (mr) = \frac{2uw}{u + w} \quad (rr) = \frac{u(1 - w)}{u + w} \tag{6.34}$$

Using the same methodology, tetrads are given by:

$$(mrr) = \frac{2uw(1 - w)}{u + w} = \frac{(mr)\,(rr)}{(r)} \tag{6.35}$$

$$(rmr) = \frac{u^2 w}{u + w} = \frac{(mr)^2}{4\,(m)} \tag{6.36}$$

$$(rrr) = \frac{u(1 - w)^2}{u + w} = \frac{(rr)^2}{(r)} \tag{6.37}$$

Testing of Models

If *triad* data can be determined by experiment, it becomes possible to decide whether or not the mechanism governing the propagation of sequence type is consistent with *Bernoullian* statistics.

$$P(r/m) = u = \frac{\frac{1}{2}(mr)}{(m)} \qquad P(m/r) = w = \frac{\frac{1}{2}(rm)}{(r)}$$

$$P(m/m) = 1 - u = \frac{(mm)}{(m)} \qquad P(r/r) = 1 - w = \frac{(rr)}{(r)} \qquad (6.38)$$

Now:

$$\text{If:} \quad P(r/m) = P(r/r) = 1 - P_m$$

$$\text{and:} \quad P(m/r) = P(m/m) = P_m \qquad (6.39)$$

then the *propagation follows Bernoullian statistics.*

Note that if $P(r/m) \neq P(r/r)$ and/or $P(m/r) \neq P(m/m)$ one can only say that Bernoullian statistics are not obeyed. It is possible that propagation may follow the terminal or simple Markovian statistics, but it is also possible that it has a higher order.

If *tetrad* data is available it becomes possible to decide whether or not the mechanism governing the propagation of sequence type is consistent with *simple Markovian* statistics; ie. a terminal model.

$$P(m/mm) = \frac{(mmm)}{(mm)} \qquad P(m/rm) = \frac{(mmr)}{(mr)}$$

$$P(m/mr) = \frac{2(mrm)}{(mr)} \qquad P(m/rr) = \frac{(mrr)}{2(rr)} \qquad (6.40)$$

Now:

$$\text{If:} \quad P(m/mm) = P(m/rm) = P(m/m)$$

$$\text{and:} \quad P(m/mr) = P(m/rr) = P(m/r)$$

$$P(r/rr) = P(r/mr) = P(r/r)$$

$$P(r/rm) = P(r/mm) = P(r/m) \qquad (6.41)$$

then the *propagation follows First Order Markovian statistics.*

The Observation of Tacticity by ^{13}C NMR

PMMA was one of the more successful examples of the application of ^1H nmr to the characterization of stereoregularity, but the technique was not found to be universally applicable to all stereoregular polymers. Enter ^{13}C nmr spectroscopy, which has proved to be enormously sensitive to the sequence

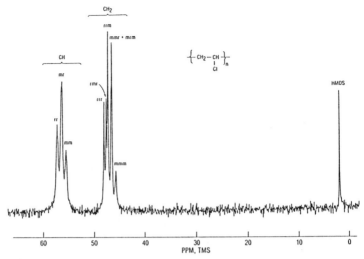

Figure 6.33 *Proton noise decoupled* ^{13}C *nmr spectrum of a free radically initiated poly(vinyl chloride). Reproduced with permission from J. C. Randall,* Polymer Sequence Determination, *Academic Press, 1977.*

distribution of stereoregular polymers in general. A couple of examples should suffice. Figures 6.33 and 6.34 show proton decoupled ^{13}C nmr spectra of poly(vinyl chloride) (PVC) and poly(vinyl acetate) (PVAc), respectively. At 25.2 MHz, the methine and methylene carbons of PVC are beautifully resolved into triads (rr, mr and mm) and tetrads (rrr, rmr, rrm, mmm and mmr + mrm). Similar resolution is seen in the PVAc spectrum. For the interested reader, a quick review of the polymer science literature over the past decade will reveal many such examples with even greater resolution.

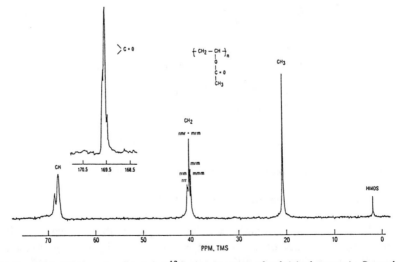

Figure 6.34 *Proton noise decoupled* ^{13}C *nmr spectrum of poly(vinyl acetate). Reproduced with permission from J. C. Randall,* Polymer Sequence Determination, *Academic Press, 1977.*

Sequence Isomerization

Nmr spectroscopy can provide clear evidence of sequence isomerization in polymers. The first example we will examine is a classic, poly(vinylidine fluoride) (PVDF), and allows us to introduce ^{19}F nmr spectroscopy. As noted in table 6.2, and in common with ^1H and ^{13}C nuclei, the ^{19}F nucleus has a spin number of 1/2, and has an abundance of 100%. Accordingly, ^{19}F nmr spectroscopy is a method commonly used to study fluoropolymers. Figure 6.35 shows the ^{19}F nmr spectrum of PVDF and it can be seen that there are four lines at 91.6 (denoted A), 94.8 (B), 113.6 (C) and 115.5 (D) ppm. The line at 91.6 ppm is dominant and the other three lines appear to have approximately the same relative intensities. Vinylidine fluoride ($CH_2=CF_2$) contains no asymmetric carbon atom, which eliminates the possibility of stereoisomerism in the polymer. Thus the only sound explanation for the three small peaks shown in figure 6.35 is the presence of head-to-head and tail-to-tail sequences, as indicated in the schematic diagram shown below.

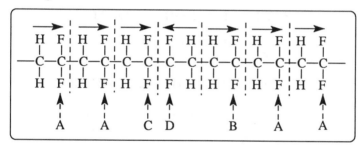

The arrows marked A, B, C and D correspond to the ^{19}F nmr peaks designated in figure 6.35. But what about the relative intensities and how would you determine the number fraction of VDF monomers that were incorporated backwards into the polymer chain from such data? For this we must revisit probability theory.

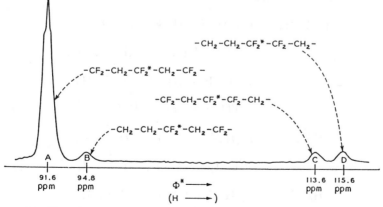

Figure 6.35 *^{19}F nmr spectrum of poly(vinylidine fluoride) at 56.4 MHz. Reproduced with permission from C. W. Wilson III and E. R. Santee, Jr., J. Polym. Sci. Part C, 8, 97 (1965).*

In the general case of a vinyl monomer $CH_2=CXY$, the head of the unit is defined as the CXY end while the tail is defined as the CH_2 end. A unit may add onto the chain in two ways and for convenience we will define the probability of "normal" addition as $P_1\{TH\}$ and "backward" addition as $P_1\{HT\}$:

$$R^* + CH_2=CXY \quad \begin{array}{c} \nearrow \quad R-CH_2-CXY^* \; = P_1\{TH\} \\ \\ \searrow \quad R-CXY-CH_2^* \; = P_1\{HT\} \end{array}$$

$$P_1\{TH\} + P_1\{HT\} \; = \; 1 \qquad (6.42)$$

Diads are related through:

$$\begin{aligned} P_1\{TH\} &= P_2\{TH\text{-}TH\} + P_2\{TH\text{-}HT\} \\ &= P_2\{TH\text{-}TH\} + P_2\{HT\text{-}TH\} \end{aligned} \qquad (6.43)$$

and:

$$\begin{aligned} P_1\{HT\} &= P_2\{HT\text{-}HT\} + P_2\{HT\text{-}TH\} \\ &= P_2\{TH\text{-}HT\} + P_2\{HT\text{-}HT\} \end{aligned} \qquad (6.44)$$

Hence:

$$P_2\{HT\text{-}TH\} \; = \; P_2\{TH\text{-}HT\} \qquad (6.45)$$

This means, of course, that the amount of head-to-head diads *must* always equal that of the tail-to-tail diads and this should not be a surprise at this stage of the game. In common with the general relationships described previously for copolymerization and stereoisomerism, from the higher order sequences we can always calculate lower orders.

$$\begin{aligned} P_2\{TH\text{-}TH\} &= P_3\{TH\text{-}TH\text{-}TH\} + P_3\{TH\text{-}TH\text{-}HT\} \\ &= P_3\{TH\text{-}TH\text{-}TH\} + P_3\{HT\text{-}TH\text{-}TH\} \end{aligned} \qquad (6.46)$$

$$\begin{aligned} P_2\{TH\text{-}HT\} &= P_3\{TH\text{-}HT\text{-}TH\} + P_3\{TH\text{-}HT\text{-}HT\} \\ &= P_3\{TH\text{-}TH\text{-}HT\} + P_3\{HT\text{-}TH\text{-}HT\} \end{aligned} \qquad (6.47)$$

$$\begin{aligned} P_2\{HT\text{-}TH\} &= P_3\{HT\text{-}TH\text{-}TH\} + P_3\{HT\text{-}TH\text{-}HT\} \\ &= P_3\{TH\text{-}HT\text{-}TH\} + P_3\{HT\text{-}HT\text{-}TH\} \end{aligned} \qquad (6.48)$$

$$\begin{aligned} P_2\{HT\text{-}HT\} &= P_3\{HT\text{-}HT\text{-}TH\} + P_3\{HT\text{-}HT\text{-}HT\} \\ &= P_3\{TH\text{-}HT\text{-}HT\} + P_3\{HT\text{-}HT\text{-}HT\} \end{aligned} \qquad (6.49)$$

And the reversibility relationships are also valid:

$$P_3\{HT\text{-}HT\text{-}TH\} = P_3\{TH\text{-}HT\text{-}HT\}$$

$$P_3\{TH\text{-}TH\text{-}HT\} = P_3\{HT\text{-}TH\text{-}TH\} \qquad (6.50)$$

Often diad, triad or even higher sequence information is available from nmr spectroscopic studies and we must be careful to assign the nmr bands correctly if we are to determine the fraction of units that are incorporated "backwards" in the polymer chain. We need to calculate $P_1\{TH\}$ and/or $P_1\{HT\}$ from such data. There are 4 possible diads, 8 possible triads, 16 possible tetrads, etc., but only a limited number of them are distinguishable (similar in principle to that described for stereoisomerism). For example, nmr cannot distinguish between {TH-TH-TH} and {HT-HT-HT} sequences, as this is equivalent to reading the chain in one direction or the other. A summary of the observable diad and triad sequences together with their Bernoullian probabilities is given in table 6.4. The results are rather interesting and predict that we should see 3 nmr lines in the case of diad and 4 in the case of triad information (5 for tetrads, 6 for pentads etc.). Even more interesting, n of the observed nmr lines (assuming they are all resolved) will have the same intensity if n-ad information is available.

Table 6.4 Observable NMR Diad and Triad Sequences.

Type	nmr Observables	Indistinguishable sequences	Bernoullian Probability
Diads	(TH-TH)	$P_2\{TH\text{-}TH\} + P_2\{HT\text{-}HT\}$	$P_{TH}^2 + (1\text{-}P_{TH})^2$ $= 2P_{TH}^2 - 2P_{TH} + 1$
	(TH-HT)	$P_2\{TH\text{-}HT\}$	$P_{TH}(1\text{-}P_{TH})$
	(HT-TH)	$P_2\{HT\text{-}TH\}$	$P_{TH}(1\text{-}P_{TH})$
Triads	(TH-TH-TH)	$P_3\{TH\text{-}TH\text{-}TH\}$ $+ P_3\{HT\text{-}HT\text{-}HT\}$	$P_{TH}^3 + (1\text{-}P_{TH})^3$ $= 3P_{TH}^2 - 3P_{TH} + 1$
	(TH-TH-HT)	$P_3\{TH\text{-}TH\text{-}HT\}$ $+ P_3\{TH\text{-}HT\text{-}HT\}$	$P_{TH}^2(1 - P_{TH}) + P_{TH}(1 - P_{TH})^2$ $= P_{TH}(1\text{-}P_{TH})$
	(TH-HT-TH)	$P_3\{TH\text{-}HT\text{-}TH\}$ $+ P_3\{HT\text{-}TH\text{-}HT\}$	$= P_{TH}(1\text{-}P_{TH})$
	(HT-TH-TH)	$P_3\{HT\text{-}TH\text{-}TH\}$ $+ P_3\{HT\text{-}HT\text{-}TH\}$	$= P_{TH}(1\text{-}P_{TH})$

For example, assume that $P_{TH} = 0.9$; if we can observe triad information in the nmr spectrum, we would expect a dominant line with a normalized intensity of 0.73 and 3 other lines of equal normalized intensities of 0.09 each. The ^{19}F nmr spectrum of poly(vinylidine fluoride) shown in figure 6.35 is an excellent example, where 5-6% of the monomers were found to be reversed.

The Observation of Sequence Isomerism by ^{13}C NMR

Polychloroprene is a particular favorite of ours, because it is a polymer that exhibits almost all possible configurational, structural and sequence isomers (it also contains long chain branches which we will discuss in chapter 10). Ferguson[*] first studied the ^1H nmr of predominantly trans-1,4-polychloroprene. He identified the peaks attributed to olefinic protons of cis-1,4- and trans-1,4- units and further observed that the methylene resonance was split in a way which could only be accounted for as being due to substantial proportions of head-to-head and tail-to-tail sequences. Again, the ^{13}C nmr spectrum of polychloroprene contains a wealth of information. In some of our own work we have shown that not only is one able to differentiate between and quantitatively measure the trans-1,4-; cis-1,4- (denoted CS); 1,2-; 3,4- and isomerized-1,2 structural units (see table 1.1—page 13), but it is also feasible to assign nmr lines to sequence isomers of the trans-1,4 units (i.e., {TH} and {HT} placements). Figure 6.36 shows the olefinic region of the ^{13}C nmr spectrum of a polychloroprene sample[**] and a summary of the assignments of the nmr lines is given in table 6.5.

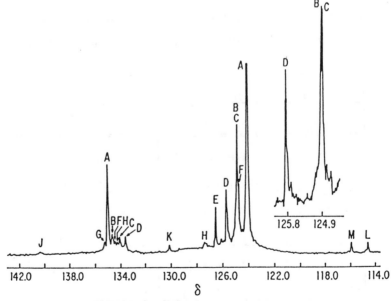

Figure 6.36 ^{13}C nmr spectrum of polychloroprene.

[*] R. Ferguson, J. Polym. Sci. A2, 4735 (1964).
[**] M. M. Coleman, D. L. Tabb and E. Brame, Jr., Rubber Chem Technol., 50(1), 49 (1977).

Table 6.5 *Assignment of the Major ^{13}C NMR Lines in the Olefinic Region of Polychloroprene.*

$-CH_2-*CCl=CH-CH_2-$	$-CH_2-CCl=*CH-CH_2-$	Designation	Assignment
134.9	124.1	A	{TH-TH-TH} + {HT-HT-HT}
134.6	124.9	B	{HT-TH-TH} + {HT-HT-TH}
133.9	124.9	C	{TH-TH-HT} + {TH-HT-HT}
133.5	125.8	D	{HT-TH-HT} + {TH-HT-TH}
134.1	126.6	E	{TH-CS-TH}
134.3	124.7	F	{TH-TH-CS}
135.1	124.1(?)	G	{CS-TH-TH}
-	127.6	H	{HT-CS-TH}

Structural Isomerization

The analysis of structural isomers in diene polymers is another feather in the cap of high resolution nmr. It is as effective for such measurements as infrared or Raman spectroscopy, and in many cases much more sensitive to sequence distribution. Although it is possible to detect structural isomerization in polyisoprenes (PI) and polybutadienes (PB) with ^{1}H nmr spectroscopy, there are only minimal chemical shift differences between the *cis* and *trans*-1,4 isomers. In PI, for example, the methyl groups of *trans*-1,4-isoprene units are only approximately 0.07-0.10 ppm more shielded than those of *cis*-1,4 units . For PB there is even less resolution. In common with the case of polychloroprene, the situation is vastly improved with ^{13}C nmr and proton decoupled spectra of the polybutadiene samples exhibit widely separated lines for the *cis*-1,4- and *trans*-1,4- isomers. Finally, just to illustrate the enormous sensitivity of ^{13}C nmr to sequences of different structural isomers, figure 6.37 shows the ^{13}C nmr spectrum of a polybutadiene containing 34% *trans*-1,4- (denoted t), 24% *cis*-1,4- (denoted c) and 42% 1,2- (denoted v). At least 18 peaks are observed that have be assigned to specific triad sequences (e.g., {c / v / t} etc.—see below).

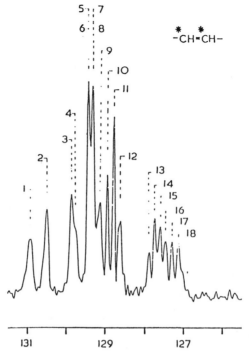

Figure 6.37 ^{13}C *nmr spectrum of a polybutadiene containing 34% trans-1,4-, 24% cis-1,4- and 42% 1,2- placements. Reproduced with permission from K-F Elgert, G. Quack and B. Stutzel,* Polymer, *16, 154, (1975).*

Copolymer Sequence Distributions

From our previous discussion of the statistics of copolymer sequence distributions in Chapter 5, and the examples of nmr spectra of polymers shown above, one might expect that the observation and assignment of sequences in the high resolution nmr spectra of copolymers would be reasonably straightforward. Indeed, if one is dealing with comonomers that do not contain asymmetric centers such as $CH_2=CX_2$, we find the number of *distinguishable* sequences to be manageable. For example if the two monomers are denoted A and B, then there are are three distinguishable diad sequences, AA, BB and AB ($\equiv$ BA), six distinguishable triad sequences, AAA, BBB, ABA, BAB, AAB ($\equiv$ BAA) and BBA ($\equiv$ ABB), etc. In general, for sequences of of length n, the number of distinguishable sequences N(n) is given by:

n =	2	3	4	5	6
N(n) =	3	6	10	20	36

However, if one factors in the additional complications of stereochemistry, which occurs if one uses comonomers that do contain asymmetric centers, such as $CH_2=CXY$, the problem rapidly becomes unmanageable. For example, figure 6.38 shows a schematic representation of the 6 distinguishable diad sequences.

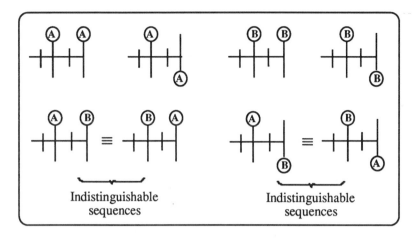

Figure 6.38 *Schematic representation of the number of distinguishable diads for two monomers of the type $CH_2=CXY$.*

In general, for the case of two $CH_2=CXY$ monomers the number of *distinguishable* sequences, N(n), of length n is given by:

n =	2	3	4	5	6
N(n) =	6	20	72	272	1056

Now add the possibility of sequence isomerization ("backwards addition") and the resulting high resolution nmr spectrum becomes a nightmare, as the sheer number of lines, many of which are unresolved and overlap, overwhelm our ability to analyze the data. In fact, it can sometimes be an advantage to limit the amount of information by degrading resolution or recording the nmr spectra on instruments with less powerful magnetic fields. (Also if you've thrown your infrared spectrometer out of the window, see page 142, this might prompt you to try to put it back together again!)

For our purposes in this book we will just show one example where nmr spectroscopy has been employed successfully to study copolymer sequences. Figure 6.39 shows the 60 MHz ^1H nmr spectrum recorded by Fischer and coworkers[*] in the range between 6–9 τ of a vinylidine chloride-co-isobutylene (VDC-co-IB) copolymer containing 70 mole % VDC. Also included in the figure are the spectra of pure PVDC [denoted (a)] and pure PIB (b) in the same region. Note that both VDC and IB fall under the category of monomers of the type $CH_2=CX_2$. The following tetrad structures (marked on the figure) have been assigned:

1 = VDC-VDC-VDC-VDC	2 = VDC-VDC-VDC-IB
3 = IB-VDC-VDC-IB	4 = VDC-VDC-IB-VDC
5 = IB-VDC-IB-VDC	6 = VDC-VDC-IB-IB
7 = IB-VDC-IB-IB	

[*] T. Fischer, J. B. Kinsinger and C. W. Wilson, III, *J. Polym. Sci. Polym. Letters Ed.*, **4**, 379 (1966).

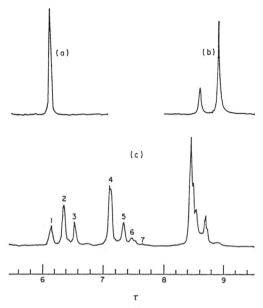

Figure 6.39 *^1H nmr spectra of (a) PVDC homopolymer, (b) PIB homopolymer and (c) a copolymer of vinylidine chloride and isobutylene. Reproduced with permission from F. A. Bovey,* High Resolution NMR of Macromolecules, Academic Press (1972).

G. STUDY QUESTIONS

1. Answer the following questions that pertain to the basics of infrared spectroscopy:

   (a) What is generally considered to be the frequency range (in cm or wavenumbers) of infrared radiation ?

   (b) What is the essential requirement (selection rule) for infrared adsorption ?

   (c) How many fundamental or normal modes of vibration are predicted for the repeat unit of poly(vinyl chloride) $[-CH_2-CHCl-]_n$?

   (d) Briefly, why does a "fingerprint" region exist in the infrared spectrum of organic molecules ?

   (e) Write down the basic harmonic oscillator equation that relates the frequency of vibration to the force constant and the reduced mass. Be sure to define all the symbols used.

2. Compare and contrast the infrared spectrum of atactic polystyrene with that of atactic poly(4-vinyl phenol). Be sure to include in your discussion the effect of temperature on the respective polymer spectra and explain the origin of the difference in the relative breadth of specific infrared bands.

3. (a) Two infrared bands attributed to N-H stretching modes are observed in the spectrum of nylon 11. Why are there two bands? What are they due to? Describe the character of these bands and how they vary with temperature.

(b) In the infrared spectrum of polyethylene the methylene rocking mode appears as a sharp doublet at 733/721 cm^{-1}. Why is this?

(c) The infrared spectrum of a highly crystalline sample of *trans*-1,4-poly(2,3-dichlorobutadiene), $[-CH_2-CCl=CCl-CH_2-]_n$, does not have a band in the 1640-1660 cm^{-1} region where the C=C stretching mode is expected to appear. Why is this?

4. (a) The infrared spectrum of a sample of *trans*-1,4-polyisoprene is recorded at an elevated temperature well above its crystalline point (i.e. in the molten state). The sample is then left in the spectrometer and permitted to slowly cool down to room temperature whereupon it crystallizes. A second infrared spectrum is now recorded of this crystallized sample. Describe and explain the changes that you would expect to observe between these two infrared spectra.

(b) If the same experiment as 4(a) was performed using *trans*-1,4-poly(2,3-dimethylbutadiene) instead of *trans*-1,4-polyisoprene, what additional changes might you anticipate between the two infrared spectra ?

5. The infrared spectra of two polymer blend films are recorded at room temperature. The first spectrum was that obtained from a 50:50 poly(vinyl phenol) (PVPh) blend with poly(ethyl methacrylate) and two carbonyl bands at $\approx$ 1730 and 1705 cm^{-1} are observed. In contrast, the second spectrum, obtained from a 50:50 PVPh blend with poly(n-hexyl methacrylate) reveals only one carbonyl band at 1730 cm^{-1}. How would you explain this and what is the origin of these two bands ?

6. Consider the proton nmr spectrum of a methyl methacrylate-co-hexyl methacrylate copolymer shown in figure 6.23. The areas of the lines at 3.6 and 3.9 ppm were determined to be 36.9 and 61.3, respectively. How much hexyl methacrylate was incorporated into the polymer ?

7. (a) Proton nmr spectroscopy has been sucessfully used to study tacticity in poly(methyl methacrylate) (PMMA). Draw and label a schematic diagram of the nmr spectrum you might expect to observe for the methylene and α-methyl protons of an atactic PMMA.

(b) Draw and label a schematic diagram of the proton nmr spectrum you might expect for the methylene and α-methyl protons for a copolymer consisting of blocks of pure syndiotactic followed by pure isotactic sequences of PMMA.

(c) You are given two powders. One is atactic PMMA and the other is a physical blend of pure isotactic PMMA and pure syndiotactic PMMA. Briefly, describe two methods which you could use to distinguish between these samples.

8. (a) From nmr studies of a polystyrene it was determined that (mm)=0.42 and (mr)=0.12. Calculate the percent isotacticity (m).

(b) The conditional probabilities P(r/m) and P(r/r) were determined from nmr data. If these two values are the same within error, is this consistent with the terminal model ?

9. The following normalized experimental results were obtained from nmr analysis of two poly(methyl methacrylates) (denoted I and II) that were synthesized under different polymerization conditions.

	I	II
(mm)	0.060	0.040
(mr)	0.340	0.175
(rr)	0.600	0.785
(rrr)	0.455	0.710
(mrr)	0.280	0.155

Determine whether these results are consistent with Bernoullian or Markovian statistics.

10. If 20% of the structural units in poly(vinylidene fluoride) (PVDF) are incorporated "backwards" into the chain and the ^{19}F NMR spectrum is sensitive to *triad* information, what would be the normalized intensity of the most intense line in the spectrum, assuming Bernoullian statistics?

H. SUGGESTIONS FOR FURTHER READING

(1)   P. C. Painter, M. M. Coleman and J. L. Koenig, *The Theory of Vibrational Spectroscopy and Its Application to Polymeric Materials*, John Wiley & Sons, New York, 1982.

(2)   H. W. Siesler and K. Holland-Moritz, *Infrared and Raman Spectroscopy of Polymers*, M. Dekker, New York, 1980.

(3)   J. L. Koenig, *Spectroscopy of Polymers,* American Chemical Society, Washington, 1992.

(4)   N. B. Colthup, L. H. Daly and S. E. Wiberley, *Introduction to Infrared and Raman Spectroscopy*, 3rd Edition, Academic Press, Boston, 1990.

(5)   F. A. Bovey, *High Resolution NMR of Macromolecules,* Academic Press, New York, 1972.

(6)   J. C. Randall, *Polymer Sequence Determination*, Academic Press, New York, 1977.

CHAPTER 7

Structure

> *"They move in the void and catching each other up jostle together,*
> *and some recoil in any direction that may chance, and others become*
> *entangled with one another in various degrees according to the symmetry*
> *of their shapes and sizes and positions and order, and they remain*
> *together and thus the coming into being of composite things is effected"*
> —Simplicius, 530 BC

A. INTRODUCTION

In this chapter we are going to discuss structure and to introduce the subject we will borrow some ideas from biochemists and molecular biologists. (We're not proud; if we find good ideas in other fields, we'll steal them!) In describing protein structure it is usual to consider four levels of organization, termed *primary*, *secondary*, *tertiary* and *quaternary* structure. Primary structure refers to the sequence of amino acids that makes up the polymer chain of a particular protein (or synthetic polypeptide). Secondary structure is the ordered conformation that the chains (or usually parts of chains) can twist itself into. The two most common are the α-helix and the β-sheet; the former looks like a coiled spring, while the latter looks like an extended zig-zag (you will see examples of these types of *conformations* later). Tertiary structure refers to how a single chain can be folded in on itself (globular proteins are usually tightly folded and look like a knotted up piece of string). Finally, quaternary structure refers to how different molecules can pack to form an organized unit.

In synthetic polymers we can make a similar classification. We refer to primary structure as microstructure, however, as we have discussed in the preceding chapters. We can also have ordered and disordered chain "shapes" (secondary structure), but this we simply refer to as chain conformation. We will see that chains with ordered conformations are organized into different types of larger scale structures (quaternary structure) when they crystallize*. This we call morphology. There is perhaps also a synthetic polymer equivalent of tertiary structure when polymers are in dilute solution, as the individual chains can be swollen or contracted and in special circumstances even collapsed in on themselves (these different shapes could just as easily be considered conformations, however). In this chapter we are going to be largely concerned with conformation and morphology (dilute solutions will be considered in Chapter 9). But instead of starting at the molecular scale, we are going to start

---

* Chains that have no regular shape can also form large scale ordered structures if they are in the form of block copolymers, but this is an advanced topic that we won't discuss.

with macroscopic considerations. What are the "states of matter" (i.e., gas, liquid, solid) available to a polymer? Macroscopic properties are, of course, ultimately determined by structure (or lack thereof) at the molecular level and we want to start by giving you a feel for the range of physical characteristics of polymeric materials. We will then review aspects of bonding and intermolecular interactions as a prelude to our discussion of conformation and morphology.

B. STATES OF MATTER AND BONDING IN POLYMER MATERIALS

The basic "states of matter" with which most people are familiar are gas, liquid and solid. In most treatments of the physics or physical chemistry of these states each is usually treated separately and then some attention is paid to the transitions between them. We shall start our discussions in this conventional manner, but we will quickly see that polymers are far too awkward to fit easily into such neat categories.

Covalent Bonds and the Nature of the Polymer Liquid and Solid State

For most students just starting their studies of materials the term "solid state" usually implies the regular arrangement of atoms or molecules in a crystalline lattice. The conversion of a solid to a liquid (melting) then involves the application of sufficient thermal energy that the forces that maintain the crystal are overcome and the atoms or molecules no longer have fixed, regular positions relative to one another, but constantly jiggle about at the urgings of thermal motion. Or course, even at very low temperatures there is motion in the crystalline state, but this is confined largely to small vibrational displacements about fixed positions. In contrast, in the liquid state the atoms or molecules have *translational motion relative to one another* and hence a liquid does not hold its shape, but takes up the shape of its container. Further heating can result in a another *phase transition*, this time from the liquid to the gaseous state, in a process we usually refer to as boiling. Just as in a liquid, the atoms or molecules of a gas are randomly arranged and constantly moving, but the densities of the two states are entirely different. In a liquid there is *random close packing* of the components, which for low molecular weight materials can be thought of in terms of the arrangements you would obtain by throwing a pot full of peas into a colander to drain them. There is no regular long-range order, but there is a fairly well-defined average number of neighbors with which all molecules are usually in contact (except those right at the surface). In a gas, however, there is a lot of space between the molecules and they only bump into one another occasionally.

The transformations between the crystalline solid, the liquid and the gaseous states are known as *first order transitions*. They are called this because they are accompanied by a discontinuity in thermodynamic quantities that are defined by the first derivative of the free energy* (e.g., volume = $(\partial G/\partial P)_T$, etc.). They also

---

* If you have forgotten what this is, we will remind you at the beginning of the next chapter.

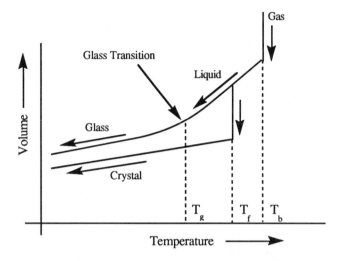

Figure 7.1 *Schematic diagram depicting two cooling paths of an assembly of atoms.*

involve a latent heat, but we will discuss this in more detail in Chapter 8. For now it is sufficient to consider general characterstics, as illustrated in figure 7.1.

We start at high temperatures, where the hypothetical material we have chosen is a gas. Upon cooling it forms a liquid at a sharply defined temperature (a "condensation" temperature close to the boiling point, T_b). This transition involves a very large change in volume, a discontinuity, at this temperature. Further cooling leads to a steady contraction in volume, but at the freezing point there is again a discontinuous sudden drop in volume as crystallization occurs. (For low molecular weight materials the freezing point and the melting point are usually almost identical temperatures. This will not be true in polymers, as we will see.)

In certain crystallizable materials fast cooling can take the material past the freezing point and the liquid apparently "solidifies" without crystallizing and without a discontinuity in quantities such as the volume. There is a change in the slope of the volume/temperature curve, however, as also illustrated in figure 7.1. At this point, known as the glass transition temperature, or T_g, there is no discontinuity in volume, but there are apparent discontinuities in quantities such as the specific heat, which are defined by the *second* derivative of the free energy (in the case of specific heat it is with respect to temperature). Hence, this could be called a second order transition (it is still open to question whether or not the T_g is related to a true second order transition). There are also certain phenomena that could be classified as third order transitions, and so on. As Goodstein[*] points out, however, this classification only works well for first order transitions. Furthermore, there is the question of whether the glassy material produced by fast cooling is a distinguishable state of matter, or just an extension

---

[*] D. L. Goodstein, *States of Matter*, Dover Publications, 1985.

of the liquid state to a region where translational motions are so slow (years and years) that in time frames of minutes and hours the material behaves like a solid. This is also a subject we will come back to discuss more completely later on.

For low molecular weight materials it is usually obvious that the glassy state is one of metastable equilibrium. In the situations described above we cooled the material so fast that crystals did not have time to form (as we will see, crystallization is governed by kinetic factors). Once the material is below the T_g the molecules lack sufficient mobility to organize themselves (in ordinary time spans). If such materials are heated above the T_g, crystallization can and does occur, however. In contrast, certain polymers can *never* crystallize, so perhaps here it would be reasonable to regard the glassy state as distinguishable from the liquid state. Nevertheless, the material is not at equilibrium and its properties depend upon thermal history and can change with time (ageing). The ability to crystallize, the location of the T_g (i.e., at high or low temperatures) and any other property must ultimately be related to the structure of the particular polymer chain we are considering. Clearly, the properties of a particular polymer will also depend upon the interactions of the polymer chain segments with one another and, as we will see later, chain flexibility. So, as mentioned in the introduction, our purpose in most of the rest of this chapter is to examine the types of bonds or interactions that are found in and between polymers, the range of shapes or *conformations* polymer chains can bend themselves into, and the degree of long range order (or disorder) found in a collection of such chains (morphology). Ultimately, we would like to know how all of these things determine properties.

We will start here by reemphasizing that the distinguishing characteristic of polymer materials is that they are made up of a large number of atoms that in nearly all cases are linked together by covalent bonds. The most common types of chain architecture or microstructures that can be formed have been described in Chapter 1. It is not within our brief to go into the nature of the covalent bond and we will assume that a student brings to the study of polymers some fundamental basic knowledge of physics and chemistry. This is not always a good assumption, but fortunately for us it is sufficient for most of our purposes to think of the covalent bond as a sharing of electrons between the atoms involved. The "sharing" is not always equal and sometimes seems like something out of the Communist Manifesto; from each according to its abilities, to each according to its needs; but in chemistry we talk far less grandly and define things like electronegativity, and so on. The important thing as far as we are concerned is that as a consequence of the distribution of electrons in its bonds certain molecules are non-polar, while others can exhibit degrees of polarity that influence intermolecular interactions. We will come back to this in the following section.

There is one immediate consequence of linking atoms into a long chain with strong bonds; we can take such a material and form a melt, but cannot "boil" it to form a gas of long polymer molecules*. In forming the vapor state from low

---

* The closest we can come to a gas-like state is to take a very dilute solution of a polymer, where all the chains are floating around as separate coils, exhibiting random Brownian motion. This isn't a real gas because of interactions with the solvent, but there are circumstances where pairwise interactions cancel out, and this leads to some interesting analogies.

molecular weight materials, such as water or carbon tetrachloride, we simply need to add sufficient energy to break the weak (relative to covalent bonds) interactions *between* molecules. The same types of weak forces also act between polymer molecules, but the effect is much greater because the polymer is so large. One way of looking at this is to consider a polymer to be a chain of *segments*, each roughly equal in size to a molecule such as benzene or ethylene (for example). The forces of attraction between chains will then be roughly equal to those between such small molecules, *multiplied by the number of segments in the chain*, which is often of the order of a thousand or more. Add to this the fact that molecules can get tangled up in one another (more on this later), then the energy required to vaporize a long polymer molecule is in nearly all cases larger than that required to break apart the covalent linkages holding it together. In practice, most polymers degrade before this point as a result of reactions, often involving atmospheric oxygen (oxidation). So, we don't have to consider the liquid to gas transition, but we do have some new things to worry about. We mentioned that if we take a polymer that is not capable of crystallizing, it can form a glass at low temperatures. If we now raise the temperature so that it is just above the T_g, then it can look like a solid but will deform irreversibly under a load (even a load that is just its own weight—like silly putty). At higher temperatures it will look like a liquid, but will still have some of the "elastic" properties that we associate with solids. Furthermore, if we now covalently bond these chains together with just a few cross-links to form a network, then we obtain a material that has property of rubber elasticity, being able to stretch seven or eight (or more) times its original length (think of a rubber band). This property is so unique that it is also tempting to regard rubbers as a distinguishable state of matter, but this is going too far and we should just think of these materials as solids with very unusual and valuable properties. In any event, if we now cool our cross-linked rubber we can again turn it into a conventional solid, a glass, by taking its temperature below its T_g.

In a similar manner certain crystallizable polymers can be obtained in a state that is outside, or perhaps between, the usual categories of solid and liquid. If the chain is particularly stiff, then instead of melting to form some type of coiled up shape, which most ordinary polymers do (see later), it can instead form a so-called *liquid crystalline* state, where there is a degree of order or alignment of chains in certain directions or planes, but not others. Certain types of low molecular weight materials also have this property and it is often accompanied by unique optical (and other) properties. Even when we consider polymers dissolved in low molecular weight solvents we can, in the right circumstances, form something that seems to be an unusual state of matter, in this case a *gel* (think of Jello).

By now we are sure that you just can't wait to learn why polymers are such neat materials and have all these intriguing in-between (liquid and solid) or apparently unique properties. But you will have to wait just a bit longer, because in order to understand what is going on we will need to consider how the chains interact with one another. We will assume that you should have learnt something about intermolecular interactions in basic courses before you decided to study polymers, but in case you fell asleep in all the lectures we will briefly describe the main types that we will have to take into account and then move on to a discussion of conformations.

Intermolecular Interactions

For "small," more-or-less spherical molecules (e.g., CCl_4), it is usual to consider interactions between molecules treated as a whole, but, as we mentioned above, for molecules that consist of a number of chemical units linked together in some fashion (e.g., the n-alkanes, any polymer) it is often far more useful to consider interactions between *segments*, sometimes defined in terms of identifiable chemical units, but which can also be defined in terms of some reference volume that may, for example, include parts of a polymer chemical repeat unit, or a number of such units.

The interaction energy between molecules or segments can be considered to consist of two components, arising from repulsive and attractive intermolecular forces, respectively. Repulsive forces become significant at short distances and it is usual to represent the repulsive potential by a term of the form $(\sigma/d)^{12}$, where d is the intermolecular distance. An exponential form $[A\exp(-Bd)]$ has also been used, but there appears to be no fundamental theoretical justification for either choice. Perhaps the most crucial point to keep in mind is that the repulsive potential gets very large, very quickly, at short distances, so that to a first approximation it is sometimes useful to treat the atoms that are linked together in a polymer molecule as hard spheres.

Turning now to attractive forces, we have somewhat arbitrarily categorized the most frequently encountered forces between molecules or polymer segments in table 7.1. We have employed the usual criterion of "interaction strength," where interactions between non-polar molecules are considered to be "weak" relative to those that are more polar. Such dispersion interactions have their origin in fluctuating charge distributions within molecules. The time average of these fluctuations is zero, but at some moment an instantaneous dipole can induce a corresponding dipole in a neighboring molecule or segment, so that there is a net attractive force between them. This perturbation of the electronic motion of one molecule by another can be related to its perturbation by light as a function of energy (frequency). This in turn can be expressed in terms of the variation of refractive index with frequency, or the *dispersion* of light (hence, the name dispersion forces). The attractive component of the potential energy, E, for spherical molecules that interact in this manner can (to a first approximation) be expressed in terms of the ionization energy of the molecules (I) and their polarizabilities (α), which for interactions between molecules 1 and 2 is given by:

$$E(d) = -\frac{3}{2} \frac{\alpha_1 \alpha_2}{d^6} \frac{I_1 I_2}{(I_1 + I_2)} \qquad (7.1)$$

This is a short range interaction, varying as the inverse sixth power of the distance between molecules. Accordingly, in some models and calculations it is often assumed that only nearest neighbor interactions need to be considered. Furthermore, equation 7.1 was originally formulated to describe the interactions

Table 7.1 Frequently Encountered Forces.

Type of Interaction	Characteristics	Approximate Strength
Dispersion forces Dipole / dipole (freely rotating)	Short range varies as $1/r^6$	≈ 0.2 to 2 kcal./ mole
Strong polar forces and hydrogen bonds	Complex form, but short range	≈ 1 to 10 kcal. / mole
Coulombic, as found in ionmers	Long range, varies as $1/r$	≈ 10 to 20 kcal. / mole (?)

between spherical, symmetrical molecules, so that its application to dispersion forces in asymmetric polyatomic molecules involves some problems.

In the most commonly employed treatments it is assumed that interactions of this type are pairwise additive (i.e., we only have to consider interactions between pairs of molecules or segments at a time, then add all of these up to get the total energy). The potential energy of the liquid can then be found by using, for example, the pair correlation function method described by Hildebrand and Scott[*]. The energy term is usually identified with the energy of vaporization, E^v, so that the energy of interaction between like molecules can be obtained from an experimentally determinable quantity. This energy of interaction can be expressed in terms of the *cohesive energy density*, C, of a substance, equal to $\Delta E^v/V$, where V is the molar volume, or alternatively in terms of a *solubility parameter*, δ, equal to $(\Delta E^v/V)^{0.5}$. This is an important concept and one we will return to later when we consider mixing (Chapter 9).

It should be kept in mind that a number of assumptions have already been piled up (spherical molecules, the additivity of pair interaction potentials, the neglect of higher order terms omitted from equation 7.1, etc.) and we haven't considered polar molecules yet. Nevertheless, the above description provides a useful basis for the definition of interaction parameters.

Polar molecules or polymers are those where the distribution of electrons in certain bonds is such that it can have a permanent dipole moment (e.g., polyacrylonitrile, PVC etc.):

---

[*] J. H. Hildebrand and R. L. Scott, *The Solubility of Nonelectrolytes, Third Edition*, American Chemical Society Monographs Series, 1950.

There is a certain force of attraction between such dipoles that will depend upon their distance apart and relative orientation. Furthermore, the presence of such dipoles can induce dipoles in other neighboring groups. Accordingly, for polar molecules or polymer segments we would need to consider the sum of these effects and also contributions from dispersion forces. For molecules where the dipoles are not too large we can assume that there is a "randomizing" effect due to thermal motion and obtain an equation for the total attractive potential energy which also has a d^{-6} dependence (i.e. is short range) and can be written;

$$E(d) = -\frac{C'_{12}}{d^6} \quad (7.2)$$

where:

$$C'_{12} = \frac{3}{2}\frac{\alpha_1\alpha_2 I_1 I_2}{(I_1 + I_2)} + (\alpha_1\mu_1^2 + \alpha_2\mu_2^2) + \frac{2\mu_1^2\mu_2^2}{3kT} \quad (7.3)$$

and C'_{12} can also be related to the cohesive energy density. The first term represents dispersion interactions, the second dipole/induced-dipole interactions and the last term represents dipole/dipole interactions where the orientations of the molecules are disrupted by thermal motion and an average has been obtained by applying Boltzmann statistics (hence, the d^{-6} dependence, rather than the d^{-3} dependence characteristic of interaction between two dipoles in a specific orientation relative to one another)[*]. These interactions, taken together, are often called van der Waals forces.

Using equation 7.3 it is possible to separate the cohesive energy and hence, solubility parameters into non-polar and (weak) polar contributions. We will not make this distinction. In any event, Hildebrand and Scott have pointed out that for molecules which have buried dipoles, the contribution of polar interactions is small. For molecules with strong dipole moments there is a significant effect, however, and such interactions must be accounted for in some manner.

Moving on to interactions that we describe as intermediate or strong, we need to consider:

a) strong dipoles
b) hydrogen bonds
c) ionic interaction in ionomers

For molecules that interact because of the presence of strong permanent dipoles, the attractive component of the potential energy is given by:

$$E(d) = -\frac{\mu_1^2\mu_2^2}{d^3} f(\theta,\phi) \quad (7.4)$$

where $f(\theta,\phi)$ is a geometric function that depends upon the relative orientation of the dipoles and is equal to 1 for dipoles arranged in a parallel fashion. For $-E(d) \gg kT$, dimers and other associated species will persist for "significant"

[*] For those who know a little more about these interactions, higher terms in London dispersion energy are neglected, as are multipole interactions, which have terms in d^{-8}, d^{-10} etc.

lifetimes. In liquids the relative orientation of the dipoles will fluctuate with thermal motion, but the degree of orientation will depend not only on the strength of the interaction relative to kT, but also upon the shape of a molecule and its rotational entropy. (We haven't talked about entropy yet, but we will, and at this point we will assume that you have a rough idea of what this means and discuss it more thoroughly in the next chapter.) Clearly, for polymers that have functional groups containing strong dipoles, this will depend upon chain flexibility, which will be a significant factor. A parameter g, defined originally by Kirkwood*, has been used to account for such hindered rotations. Although values of this parameter can be determined from dielectric measurements, it has not yet been employed in a model that provides useful interaction parameters, but it does provide a measure of the significance of polar interactions in a given material or mixture. For many molecules with strong dipoles the Kirkwood g factor has values close to unity, indicating an absence of significant degrees of association, but molecules that hydrogen bond are set apart as a class in which rotation is strongly hindered and we will now turn our attention to interactions of this type.

There is no simple, universally accepted definition of a hydrogen bond, but the description given by Pauling** comes close to capturing its essence;

> . . . under certain circumstances an atom of hydrogen is attracted by rather strong forces to two atoms instead of only one, so that it may be considered to be acting as a bond between them. This is called a hydrogen bond.

Hydrogen bonds form between functional groups A–H and atoms B such that the proton usually lies on a straight line joining A–H---B. The distance between the nuclei of the A and B atoms is considerably less than the sum of the van der Waals (i.e., "hard core") radii of A and B and the diameter of the proton, i.e., the formation of the hydrogen bond leads to a contraction of the A–H---B system. The atoms A and B are usually only the most electronegative, i.e., fluorine, oxygen and nitrogen. Chlorine is as electronegative as nitrogen, but because of its large size only forms weak hydrogen bonds. Hydrogen bonds involving sulphur, some C–H groups and the π electrons on aromatic rings have also been invoked, but these are much weaker, they are less easily identified, and the evidence for their presence is often ambiguous.

The strength of hydrogen bonds also depends upon the atoms to which A and B are attached, i.e., the nature of the molecule or molecules involved. For example, highly charged species can give rise to very short and hence very strong hydrogen bonds. For the most part the strength of typical hydrogen bonds lies in the range 1–10 kcal/mole, which can be compared to the 50–70 kcal/mole characteristic of covalent bonds or the roughly 0.2–1.0 kcal/mole characteristic of dispersion or weakly polar forces. As a result, they cannot be

---

* J. G. Kirkwood, *J. Chem Phys*, **7**, 911 (1939); *Trans. Farad. Soc.*, **42A**, 7 (1946).
** L. Pauling, *The Nature of the Chemical Bond*. Third Edition. Cornell University Press, Ithaca, New York, 1960.

considered to be "cross-links", but in the melt there is a dynamic situation, with hydrogen bonds constantly breaking and reforming at the urgings of thermal motion.

Typical examples of hydrogen bonds found in polymers are illustrated in figure 7.2. Polymers containing amide, urethane and alkyl or aromatic hydroxyl groups associate in the form of linear chains. Cyclic hydrogen bonded structures are favored in molecules containing carboxylic acid and urazole functional groups. Carboxylic acids have been studied extensively and it is well known that the six membered cyclic rings favored by pairs of these functional groups form particularly strong, linear hydrogen bonds (some linear open chain structures may also occur at very high concentrations of acid groups). Hydrogen bonds can also form between "unlike" groups. Of particular importance in the study of polymer blends are mixtures where one component self-associates (i.e., has functional groups of the type described above), while the second does not, but has a functional group capable of forming a hydrogen bond with a A–H group of the self-associating polymer. Examples of this second type of functional group are ethers, esters and nitrogen containing heterocyclic rings (e.g., pyridine).

Finally, in our tour of the types of intermolecular interactions found in polymers we will conclude with ionomers.

These are usually copolymers where the minor component ($\approx 5\%$) has a functional group that can form strong ionic interactions. In ethylene-methacrylic acid copolymers, for example the proton of the carboxylic acid can be exchanged to form a salt: We have represented the structure as a dimer that would act to cross-link the chains, but the structure of ionomers is actually far more complicated and at present incompletely understood. The crucial point here is that the ionic domains phase separate into some form of cluster, as illustrated in figure 7.3, and these do indeed act as cross-links, as these ionic bonds are considerably stronger than hydrogen bonds (although ionic interactions can be of the order of 100 kcal/mole, the small clusters found in ionomers presumably makes the interaction strength far less than this).

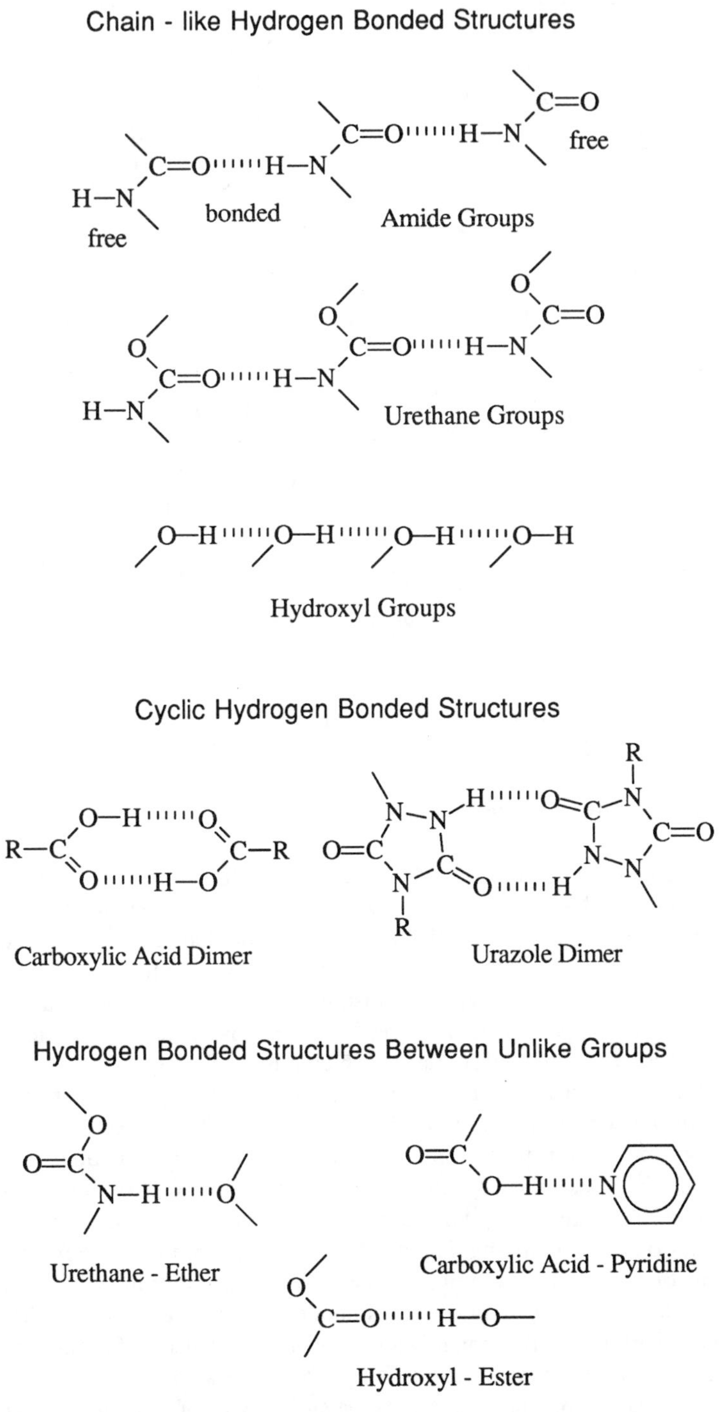
Chain - like Hydrogen Bonded Structures
free
bonded
free
Amide Groups
Urethane Groups
Hydroxyl Groups
Cyclic Hydrogen Bonded Structures
Carboxylic Acid Dimer
Urazole Dimer
Hydrogen Bonded Structures Between Unlike Groups
Urethane - Ether
Carboxylic Acid - Pyridine
Hydroxyl - Ester

Figure 7.2 *Typical examples of hydrogen bonds found in polymers.*

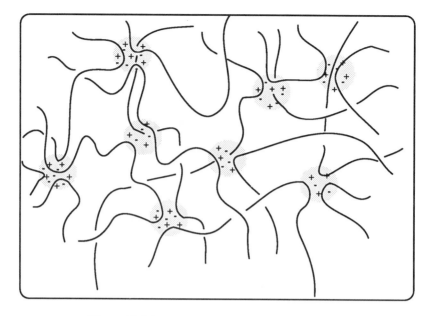

Figure 7.3 *Schematic diagram of clusters in an ionomer.*

C. THE CONFORMATIONS OR CONFIGURATIONS OF POLYMER CHAINS

Some Comments on Definitions

If we could look at an isolated polymer chain, frozen in space, and specify the position of each of its atoms, then we would have a full description of what is called the *conformation* of the chain. Except that this is also called a *configuration*, which can be very confusing because this term is one we used earlier to describe geometric and stereoisomers. This is something you just have to put up with in this subject, however, and one must take the definition of the word configuration from the context in which it is used.

Obviously, if a chain sits in a regular zig-zag or helical type of shape then it is in an *ordered* conformation or configuration. If we bend it to something that looks like an extraordinary long piece of cooked spaghetti, then it is in a *disordered* (or random) conformation or configuration. Clearly a long flexible chain can hypothetically be bent into an enormous number of such configurations, each distinguishable from the others in the sense that the positions of the atoms in the chain relative to one another are different. The origin of this flexibility is in rotations about the chemical bonds of the chain backbone. Unfortunately, certain types of *local* arrangements of groups around a bond are also called conformations, so we have the same word being used to describe a local arrangement as well as the entire shape of a chain. Nobody said this subject was going to be easy. Hopefully, we will make clear how we are using these words in what follows.

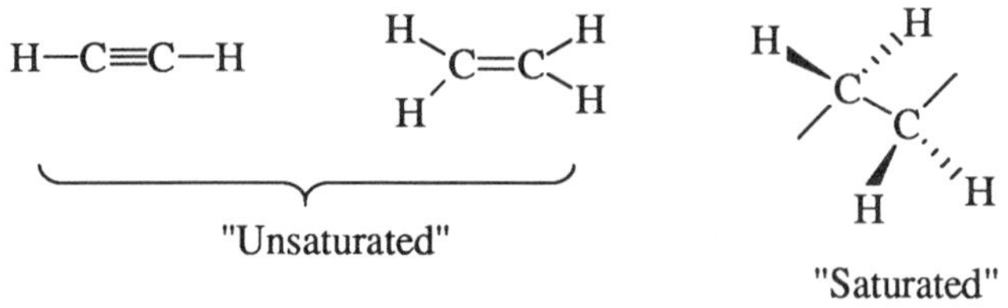

"Unsaturated"

"Saturated"

Figure 7.4 *Examples of unsaturated and saturated molecules.*

Bond Rotations

As we have stated earlier in this text, it is assumed that the student has at least some passing familiarity with physical chemistry and that this would include the nature of the covalent bond. Our concern here is the freedom of rotation about such bonds and as a reminder for those students that have trouble recalling what they have covered in previous courses (surely a very small minority!), we have shown in figure 7.4 examples of *unsaturated* hydrocarbon molecules containing triple and double carbon/carbon bonds, and an example of a *saturated* hydrocarbon, a portion of a polyethylene chain where the backbone atoms are linked by single bonds.

We will not consider these in terms of orbitals and such like, but simply note that for triple and double bonds there are very high barriers to rotation and motion is largely confined to simple torsional vibrations about these bonds. The single (sigma) bonds found in the polyethylene backbone are spherically symmetrical, however, and *in the absence of interactions between non-bonded atoms,* free rotation about such bonds is allowed. Such free rotation would permit the chain to take on an enormous range of shapes or configurations and if we are to relate molecular structure to macroscopic properties we have to figure out a way to describe this disordered state. We will come back to this problem shortly, because even though the assumption of free rotation is clearly unrealistic, it nevertheless allows us to obtain some important insights and to construct a useful model.

In reality, of course, there are various interactions that result in restrictions on rotations around bonds and to certain arrangements of bonded atoms, or groups of atoms, being favored energetically over others. This can be understood using a very simple approach, where atoms are considered to be hard spheres and barriers to rotation are then simply a result of steric repulsions between nonbonded atoms. Thus, for a molecule of ethane, shown in figure 7.5, the conformation where the hydrogen atoms on the different carbon atoms are closest together (eclipsed conformation) is much less favored than the conformation where they are furthest apart (staggered). This is because in this situation, where atom-atom distances are short, repulsive forces dominate.

We can represent these differences by means of a plot of potential energy against bond angle rotation, as shown in figure 7.6. If we arbitrarily let one of the staggered conformations define our starting point (i.e., where the bond angle

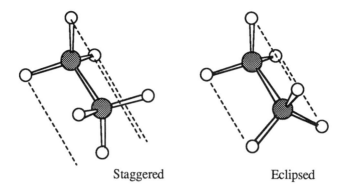

Staggered   Eclipsed

Figure 7.5 *Staggered and eclipsed conformations of ethane.*

rotation is equal to zero), then as we rotate around the bond we must go over an energy barrier to get from one staggered conformation to the next, equivalent, staggered conformation.

The positions of maximum energy correspond to the eclipsed positions (rotations of ±60°) where the forces of repulsion are a maximum. In reality, the forces between nonbonded atoms are far more complicated than this simple model would indicate and involve various interactions between the nuclei, electrons, etc. The simple model gives us an answer that is of the right form, however, which is all we require for a qualitative understanding. It's just that we could not use such a model to calculate the height of the energy barrier or the details of the shape of the potential energy function with any degree of accuracy.

The height of the energy barrier can be estimated from various spectroscopic measurements and for ethane it is about 2.9 kcal/mole. This is far greater than

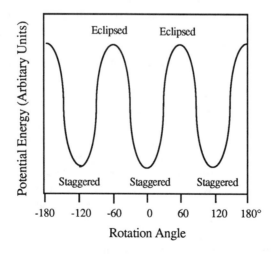

Figure 7.6 *Schematic diagram of the potential energy of ethane as a function of rotation angle.*

the value of thermal energy at room temperature ($\approx$0.6 kcal/mole), so that the staggered position is strongly favored over the eclipsed conformation. Nevertheless, a potential energy barrier of this height is not sufficient to prevent rotations between staggered positions, so that at ambient conditions we can imagine that the C–C bond in a particular molecule of ethane not only oscillates around a mean position, corresponding to one of the "wells" or potential energy minima, but every now and again will gather enough thermal energy to surmount the potential energy barrier and go to one of the other two minimum energy positions.

There are three things about this simple picture that need further comment:

1) The most crucial point is that we must not treat a conformation as a static thing like an object in a photograph, but keep in mind that above the absolute zero of temperature there is always motion, ranging from simple vibrations about a mean position to larger scale rotations between energy minima. The type and range of motion present will depend upon temperature, the chemical nature of the molecule, whether or not it is in a crystal, etc.

2) We stated above that every "now and again" a particular molecule would "gather enough energy" to surmount the rotational barrier. A better description would be in terms of a collection of molecules where the energy is allocated or partitioned between the molecules according to a (Boltzmann) distribution, such that at any instant of time a given fraction of the molecules has enough energy to rotate between minima. (We are edging towards statistical mechanics here, but don't panic!)

3) "Every now and again" is a very short time period in *macroscopic* terms. The lifetime of a *rotational isomer*, which is what we call each of the conformations we wish to identify, is of the order of 10^{-10} secs. This is a long time compared to the frequency of a molecular vibration ($\sim 10^{-12}$ to 10^{-14} secs), however, which is often the time scale that is important at the molecular level.

If we now turn to the simplest saturated hydrocarbon polymer, polyethylene, we can apply the same simple arguments and find that certain conformations are favored over others. Thus if the chain is laid out in the fully extended (planar zig-zag) conformation shown in figure 7.7, we find that atom/atom distances between protons on adjacent carbons are at a maximum and repulsive forces minimized. Each bond is in a so-called *trans conformation*. (Which should not be confused with the *trans* stereoisomers in dienes described in Chapter 1. It is

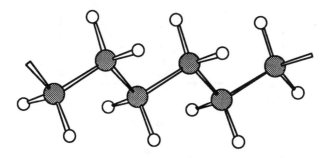

Figure 7.7 *All trans conformation of polyethylene.*

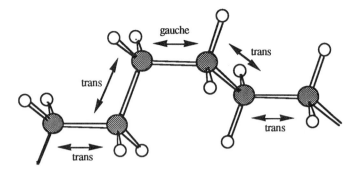

Figure 7.8 *Schematic showing a gauche conformation in polyethylene.*

unfortunate that trans is the most appropriate word for describing both types of arrangements.)

Rotation about one of the bonds would then bring the hydrogens on adjacent carbons to an eclipsed position where the potential energy is a maximum. Further rotation would bring one of the hydrogens to a position that is best described as being between the two hydrogens that are on the adjacent carbon atom, which will be another energy minimum and corresponds to a so-called *gauche* conformation, illustrated in figure 7.8. Here the hydrogens are not as close as when they are eclipsed, but not as far apart as when they are in the trans conformation, so this local energy minimum has a higher energy than that corresponding to a trans arrangement. The form of the potential energy curve for bond rotation in polyethylene is shown in figure 7.9, and it can be seen that there are two gauche conformations, one obtained by rotating 120° one way and the other corresponding to a rotation of 120° in the other direction.

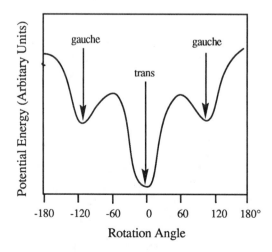

Figure 7.9 *Schematic diagram of the potential energy of a polyethylene chain as a function of rotation angle.*

Unlike ethane, where we had 3 equivalent minimum energy conformations, we now have a single conformation that is favored over all others and two other minima that have higher energies and are somewhat less favored. Why doesn't the molecule just sit in the fully extended zig-zag shape corresponding to all the bonds being trans? Well it does, but only those parts of a chain that are in a crystal. Trapped in a crystal lattice, the chain is only permitted this one shape because of the demands of regular packing, whereas in the melt or in solution it takes on a variety of shapes or conformations, corresponding to a state of higher entropy.

Regular close packing in the crystal can maximize the force of attraction between the chains so that crystallization will depend upon the balance between maximizing interactions (favoring order) and maximizing entropy (favoring disorder). This balance will depend upon temperature, chain flexibility, the strength of intermolecular forces, and so on. We are starting to get ahead of ourselves here, however, and we will discuss the thermodynamic properties of polymers and crystallization later. The point that we are again making is that even though one conformation might be favored over certain others, as the trans is over the gauche in polyethylene, there will still be a given fraction of bonds in a chain that will be in the less favored state at any moment in time and providing that the energy barrier corresponding to the eclipsed position is not too high, then a particular bond will rotate between these different local conformations. It's just that in a polyethylene chain a particular bond will tend to spend a longer period of time (on average) close to and oscillating about a trans position than about a gauche position. Alternatively, if we were to look at the whole chain at a particular instant of time we would see a larger fraction of bonds close to the trans conformation than close to the gauche.

Obviously, the conformations allowed to a chain depend upon the nature and size of the groups that are attached to the polymer backbone. In isotactic polypropylene, for example, the methyl groups are all along the same side of the chain and in the extended conformation these groups strongly repel one another. By forming a 3_1 helix (i.e., three chemical repeat units per turn of the helix), illustrated in figure 7.10, the chain manages to get the methyl groups as far away from one another as possible and this is the minimum energy conformation of this polymer (corresponding to a TGTG' sequence of conformations is successive bonds, where T is trans and G and G' are each of the two gauche arrangements). As one might guess, this is also the conformation found in the crystal (see below).[*]

This is about as far as we want to go in our discussion of bond rotations. It should be clear that there is a conformation of the whole chain that corresponds to a minimum energy, because each of its bonds is in a minimum energy (local) conformation. There is a complication here that we haven't mentioned, but this normally arises when we consider disordered conformations. The complication is that we cannot treat the local arrangement of groups in one bond as independ-

---

[*] One should not automatically assume that the minimum energy conformation of the isolated chain would necessarily be that found in the crystal (although it usually is). Forces of attraction between regularly packed chains might distort favored conformations or even favor another conformation altogether, if the total energy was thereby minimized.

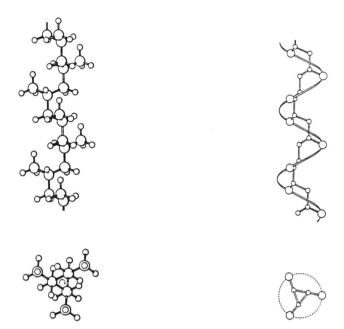

Figure 7.10 *The 3_1 helical structure of isotactic polypropylene. The chain on the left hand side shows all the atoms, while the one on the right hand side has the hydrogen atoms removed, for clarity of presentation. The figures at the bottom show the projection obtained from looking down the chain; note how the methyl groups are as far apart as they can get. Reproduced with permission from G. Natta and P. Corradini, Nuovo Cimento, Suppl. to Vol. 15, 1, 40 (1960).*

ent of the next, as certain combinations of successive conformations can bring atoms that are usually distant into close proximity. We will mention this again when we talk about the rotation isomeric states model, but not in any detail. The curious should consult Flory's second book*.

D. RANDOM WALKS, RANDOM FLIGHTS AND DISORDERED POLYMER CHAINS

In the preceding section we have seen that we can construct an ordered conformation for a polymer chain by allowing the bonds to sit in their minimum energy positions. We will defer a discussion of the extent to which such ordered arrangements are found until we consider polymer crystals, but clearly we can proceed in a logical fashion from describing ordered conformations, to describing the packing of such chains in crystals, to the overall shape and appearance of these crystals (which we will call morphology) and ultimately to attempts to relate these various facets of structure to properties.

---

* P. J. Flory, *Statistical Mechanics of Chain Molecules*, Hansen Publishers, Reprinted Edition, 1989.

To accomplish the same thing for polymers that for whatever reason have a disordered or random chain conformation might at first sight seem a much more formidable and perhaps even impossible task, because of the enormous range of conformations or configurations available to such a chain. Even if each bond were to be restricted to three possible positions (say trans and the two gauche positions), and if for simplicity we assume that each of these is of equal energy, there would be $3^{10,000}$ conformations (equal to $10^{4,771}$) available to a chain consisting of 10,000 bonds!* It is this huge number of possibilities that saves us, however, because it allows a statistical approach, which results in extraordinary insight into the nature of these materials and their properties, because such properties must ultimately depend upon an average of contributions from all the different chains, together with effects due to their interactions.

In his classic book on rubber elasticity, Treloar** presented a picture of the general form of a disordered or random coil or, if we state it even more precisely, a statistical conformation of a polymer chain, which is reproduced in figure 7.11. It was obtained by using polyethylene as a model polymer and allowing the valency angles to have their normal values. The C–C–C bond angle is thus fixed at a value close to 109°, but rotations *around* each bond are allowed,

Fixed   Allowed

but only to the extent that they could take up one of six equally spaced rotations defined by starting at the trans position and successively rotating 60°. Each of these positions was given equal probability and the actual value used for each bond was then determined by the throw of a dice. Figure 7.11 is just a single example of the enormous number of conformations available, but is typical of most of the conformations that would be found. Some "untypical" (or, more accurately, less probable) examples would be a fully stretched out chain, or one that is tightly compressed into a ball. It is not that they could not occur, but there are far fewer such conformations. In other words, if we were to look at a specific chain over the course of a long period of time, where as a result of thermal motion it went from one conformation to another through bond rotations, it would not spend an appreciable fraction of this time in any of these highly stretched out or tightly compressed conformations, but for most of the time would look something like the picture shown in figure 7.11. Equivalently, if we were to examine a collection of a large number of chains at some instant in time and freeze their motions, most of them would look like figure 7.11, and only a few would be very stretched out or highly compact in their shapes.

---

* Given by all the different combinations of rotations available to the chain. One bond has 3 different possibilities. Two together have 3^2 different combinations. A molecule with three bonds has 3^3 different combinations of arrangements, and so on.
** L. R. G. Treloar, *The Physics of Rubber Elasticity, Third Edition*, Clarendon Press, Oxford, 1975.

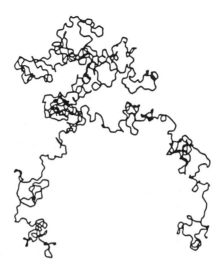

Figure 7.11 *One conformation of a 1000 link polyethylene chain. Reproduced with permission from L. R. G. Treloar,* The Physics of Rubber Elasticity, Third Edition, *Clarendon Press, Oxford, 1975.*

The knowledgeable student probably realizes that in presenting this argument we are indulging in some statistical mechanics, but we neglected to say so in order not to scare off those whose minds are filled with terror at such a prospect.

We must now delve deeper, however, and make this picture more quantitative. Fortunately we can keep the arguments at a fairly elementary level and arrive at a beautifully simple and important result. What we will need to accomplish this is a measuring stick or parameter of some kind that tells us something about the shape of the chain. For this purpose we will use the distance between the ends of the chain, the end-to-end distance, which we will call R. Obviously, the distance between the ends will be equal to the chain length if the chain is fully stretched out, or would be close to zero if it is compacted into a tight ball, but would be somewhere between these limits for the types of configurations shown in figure 7.11. We want to know both the distribution of end-to-end distances and the average value taken over all the possible configurations or shapes of the chain. Obtaining the average value is much easier than the entire distribution and we will do that first.

This type of problem is similar to a number of others in physics and physical chemistry that can be related to so-called random walks or random flights, most notably Brownian motion (more of this in a minute). The random walk problem was posed explicitly by Pearson in a paper published in *Nature* in 1905;

> A man starts from a point O and walks l yards in a straight line; he then turns through any angle whatever and walks another l yards in a second straight line. He repeats this process n times. I require the probability that after these n stretches he is at a distance between r and $r + dr$ from this starting point O.

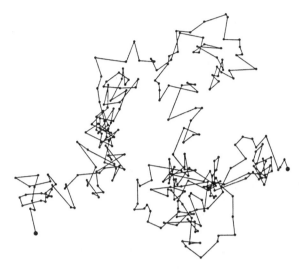

Figure 7.12 *The random flight observed by Perrin. Reproduced with permission from J. Perrin,* Atoms, *English translation by D. L. Hammick, Constable and Company, London, 1916.*

This random walk is two dimensional and implicitly assumes that the walk takes place on a flat surface. We require the equivalent problem in three dimensions, which should therefore be called a *random flight*. This is illustrated in figure 7.12, which shows the random meanderings or Brownian motion of a particle in a fluid.

This motion is a consequence of collisions with the molecules of the fluid and was observed by Perrin back in 1916*. Experimental observations were made at specific intervals of time and the positions of the particles at each chosen point in time were linked by straight lines (the particle actually took a random path to get from point to point). If we let the distance between each observation be a "step length" l, it is obvious from figure 7.12 that in this case the steps are not of constant length, but vary. Our polymer problem will be simpler in one sense, in that we will relate the step length to the bond length, which we will make constant. However, the path of Brownian diffusion can cross through positions in space that were previously traversed, while a polymer chain cannot. But we will proceed by ignoring this difficulty. In effect, we are going to make two assumptions that are both completely wrong; first, there is not only totally free rotation around the bonds of the chain, but the chain is freely jointed (i.e. the valency bond angle is no longer fixed but can take any value). The second assumption is that the chain can pass through regions of space that are already occupied by other bits of itself (at this point we are only considering a

---

* The motivation for this work was provided by some of Einstein's first papers, which were on the theory of Brownian movement. These related the diffusion of a particle to its size thus allowing microscopic dimensions to be determined from macroscopic measurements. We will use similar types of measurements to determine polymer molecular weights later in this text.

hypothetical isolated chain. Interactions with other chains, or solvent, will be introduced later). Nevertheless, the result we will obtain will be of the right form for a polymer surrounded by other chains of the same type! This is because the effect of restricted valence angles and bond rotations can be accounted for by a "prefactor" in the equations that does not affect the form of the simple main result, and because in a collection of chain of the same type various interactions cancel out, so that the chain behaves as if it could "pass through itself"!

Although the result we will require will be for a three dimensional random flight, it is easier to grasp some subtleties concerning the average end-to-end distance by first considering a one dimensional random walk. Here steps in a "forward" or "backward" direction are allowed with equal probability and we let distances from the starting point in the "forward" direction be defined as positive, while those in the other direction are negative:

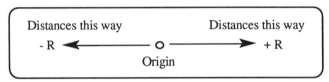

It should be immediately obvious that if steps in the forward and backward direction are equally probable, then *on average* a person making such a random one dimensional walk would end up where he or she started. *But that is on average.* If the person took one thousand steps in one such walk, they might end up say 30 steps from the origin in the plus direction. If the person started out for a second time they might end up say 30 steps in the minus direction. And so on. The final position of the walker averages out to be at the origin, because for a large number of walks final positions in the plus direction cancel out with equivalent final positions in the minus direction (we say that the distribution of final positions is symmetric about the origin). What if we want to know the number of steps, or distance (assuming all the steps are of the same length and equal to 1 unit, say a yard, just to annoy our French friends) that on average a person ends up from the origin *regardless* of whether or not it is in the plus or minus direction. Then it is obviously no use devising some method to determine the *average value* of the distance between the start and the finish of the walk, which we will designate by using special brackets, $<R>$, because we have already seen that this will be zero (our arguments were qualitative and intuitive, but they can be made rigorous). The value of R^2 for any specific walk will always be positive however, so if we calculate its average value for many, many walks, $<R^2>$, we will have a measure of the distance we require. This still isn't quite what we want, because we have a distance squared, so we will take the square root of this average, $<R^2>^{1/2}$. *That* is the thing we want to find and it is called the *root mean square end-to-end distance* (i.e., the square root of the mean or average of the squares of the end-to-end distances of a whole lot of walks). There is a particularly easy method for determining $<R^2>^{1/2}$ that is described by Feynman[*] and we will give that first before proceeding to more conventional derivations.

---

[*] R. Feynman, *Lectures on Physics, Volume 1*, Addison-Wesley, Menlo Park, CA, 1963.

We let $\langle R_N^2 \rangle$ be the average of the squares of the end-to-end distance of a one dimensional walk of N steps. We let each step be of length one unit, for simplicity. After 1 step the value of R can only be ± 1, so that R^2 is equal to 1^2. The average value of R^2 for walks of one step is, of course:

$$\langle R_1^2 \rangle = 1 \tag{7.5}$$

Now we make a jump and observe that we can obtain the allowed values of R_N if we know R_{N-1}, because we only have two possibilities:

$$R_N = R_{N-1} + 1 \tag{7.6}$$

or:

$$R_N = R_{N-1} - 1 \tag{7.7}$$

Then for the squares:

$$R_N^2 = R_{N-1}^2 + 2R_{N-1} + 1 \tag{7.8}$$

or:

$$R_N^2 = R_{N-1}^2 - 2R_{N-1} + 1 \tag{7.9}$$

Now, on average, we should obtain each of these values half the time, because there is a fifty/fifty chance of a + or a - step. Hence, the *expected value* of R_N can be obtained by adding these last two equations and dividing by 2. Identifying the average value with the expected value we get:

$$\langle R_N^2 \rangle = \langle R_{N-1}^2 \rangle + 1 \tag{7.10}$$

We already know that:

$$\langle R_1^2 \rangle = 1$$

so that:

$$\langle R_2^2 \rangle = \langle R_1^2 \rangle + 1 = 2 \tag{7.11}$$

$$\langle R_3^2 \rangle = 3 \text{ etc.} \tag{7.12}$$

Hence:

$$\langle R_N^2 \rangle = N \tag{7.13}$$

or, the answer we're looking for:

$$\langle R_N^2 \rangle^{1/2} = N^{1/2} \text{ (for steps of length l)} \tag{7.14}$$

The only bit that is troubling here might be the identification of the expected value with the average value. We have actually been through this before when discussing chain statistics, but if you need reminding think of it this way. For each toss of an unbiased coin you would *expect* to get a head 50% of the time and a tail the other 50% of the time. If you toss a coin 1000 times you won't often get exactly 500 heads and 500 tails, however, but perhaps 490 of one and 510 of the other. If you don't know probabilities and statistics, perhaps you *expected* to get exactly 500, but that will only be rarely obtained. But if you repeat this coin tossing a lot of times *on average* you will end up with 500 of each (because the next time you may have got 510 of the first and 490 of the second, and so on). We identify the expected value with the average obtained over a large number of events, or in our problem, chains.

The result we obtained, that the root mean square end-to-end distance is proportional to the square root of the number of steps in a random walk, is important and tells us a lot, as we will see shortly. First, however, we must jump from a one dimensional random walk to a three dimensional random flight. We obtain the same result. In order to show this we will consider a polymer chain to be a set of vectors, each representing a bond, as illustrated in figure 7.13. Thus the first bond or vector goes between the first two atoms in the chain and will have a certain length, l and a certain direction in space. We will represent such a vector in bold type, $\mathbf{l}_1$. The distance between the ends of a chain is then simply the sum of the vectors representing all the bonds:

$$\mathbf{R} = \sum_{i=1}^{N} \mathbf{l}_i \qquad (7.15)$$

Where there are N bonds in the chain we are considering. $\mathbf{R}$ is now a vector also, because it has both size and direction. Of course, if we know the direction of each bond with respect to a set of Cartesian or rectangular coordinates we could calculate the scalar quantity R, the distance between chain ends regardless of direction, by laborious application of trigonometry. For example, the distance between atoms 0 and 2 in figure 7.13 can be obtained by drawing a line between these atoms and using the known bond lengths ($R_{0 \to 1}$ and $R_{1 \to 2}$) and angles between the bonds to calculate $R_{0 \to 2}$. Then the distance $R_{0 \to 3}$ is calculated and so on until we eventually get $R_{0 \to N}$, which we usually just call R. We would then have to repeat this for all configurations and then find the average value. The reason we introduce vectors is so that we don't have to go through all of that. The quantity we require is the distance between ends regardless of direction, so as before we consider R^2 rather than R, which is obtained from:

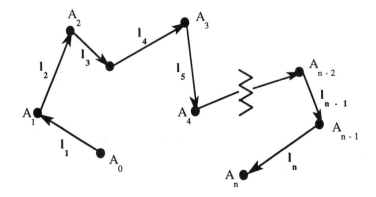

Figure 7.13 Diagram of a set of vectors depicting random flight.

$$R^2 = \mathbf{R} \cdot \mathbf{R} = \left(\sum_{i=1}^{N} l_i \right) \cdot \left(\sum_{j=1}^{N} l_j \right) \tag{7.16}$$

or:

$$\mathbf{R}^2 = \sum_{i=1}^{N} l_i^2 + 2 \sum_{i<j}^{N} l_i l_j \tag{7.17}$$

where all the terms involving $i = j$ in equation 7.16 have been taken out and put in the first summation of equation 7.17 and all the rest left in the second term. We now want the average value of R^2 taken over all possible configurations:

$$<R^2> = \sum_{i=1}^{N} <l_i^2> + 2 \sum_{i<j}^{N} <l_i l_j> \tag{7.18}$$

The second term in this equation can be written as:

$$<l_i l_j> = l_i l_j \cos \theta \tag{7.19}$$

where θ is the angle between the vectors. If all angles of θ are equally probable, which they would be for a freely jointed chain, then its average value is equal to zero (positive values cancel out with minus values in the sum), so we obtain:

$$<R^2> = \sum_{i=1}^{N} <l_i^2> \tag{7.20}$$

If all the bonds are of the same length, l, then the average value of the square of the length of ith bond is simply l^2, so we have:

$$<R^2> = \sum_{i=1}^{N} l^2 = Nl^2 \tag{7.21}$$

Hence:

$$<R^2>^{1/2} = \sum_{i=1}^{N} l^2 = l N^{1/2} \tag{7.22}$$

Again, the root mean square end-to-end distance has been shown to be proportional to $N^{1/2}$, but we now include the bond length l, where before we let the step length (equivalent to the bond length) be equal to one. This equation tells us that if we had a chain consisting of 10,000 bonds, then the average distance between the ends of the chain is just 100 bond lengths! Accordingly, if we could grab such an "average" chain by its ends and pull, it would follow the applied force by simple bond rotations and rearrangements of its conformations. Only when it had been stretched a hundred times its initial end-to-end distance would the backbone bonds experience a tension, as at this point the chain would be fully extended. Stretching the chain takes it from a more probable to a less probable state, so that upon releasing its ends there is an *entropic* driving force that would result in the chain regaining its most probable condition (in a freely rotating chain thermal motion would allow this). This is the basis for rubber elasticity. Clearly it is a fundamental property of disordered polymer chains. Once more, we are getting ahead of ourselves, this time talking about aspects of the thermodynamics of rubber elasticity and most probable states before we have considered important preliminaries. We have to defer a discussion of rubber elasticity until later and simply conclude this section by making some observations concerning what we have called most probable states.

It should already be clear that a simple determination of the average end-to-end distance has already provided considerable insight into the molecular basis for the properties of certain types (non-crystalline) of polymer chains. The model is simple and leaves out important details, such as the effect of steric restrictions on bond rotations, intra and intermolecular interactions, etc. We will get to these, if only qualitatively, in the next section. We will see that at least in the solid state the fundamental result, $<R^2>^{1/2} \propto N^{1/2}$, still holds. However, by just looking at the average of the end-to-end distance we are compressing an awful lot of information concerning the range of conformations into a single parameter. A broader description of the range of conformations available to a chain is through a *distribution function*, which for a polymer chain can be expressed in terms of a plot of the probability of finding a chain with an end-to-end distance, R, P(R) against R, as shown in figure 7.14. The general shape of this plot is easy to understand. We would expect it to be symmetric about the origin, simply by extending our arguments concerning one dimensional walks to three dimensions. If a very large number of walks are considered, then there will be as many end positions x,y,z from the origin (0,0,0) as there are -x,-y,-z positions and so on. We would also expect the probability distribution to trail off the further we got from the origin (i.e., large values of R), and to be zero at

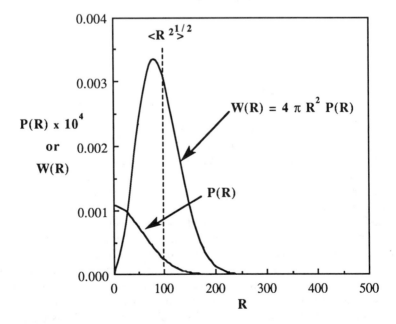

Figure 7.14 Distribution functions for the end-to-end distance R of a chain of 2,500 units each of length 2 Å.

values of R greater than the stretched out length of the chain, Nl. Accordingly, a bell shaped curve, half of which is shown in figure 7.14, is what we would intuitively expect. However, in formulating a molecular theory of rubber elasticity (amongst other things) we will need a more precise description than this. The simplest and most widely used result, which we will not derive here (see Flory's books, cited previously), is that the probability distribution takes the form of a Gaussian, with:

$$P(R) = A \exp(-BR^2) \qquad (7.23)$$

where:

$$A = \left(\frac{2\pi}{3}\right)^{-3/2} <R^2>^{-3/2} \qquad (7.24)$$

and:

$$B = \frac{3}{2} <R^2>^{-1} \qquad (7.25)$$

In order to obtain this result it is necessary to make a number of additional assumptions (i.e., over and above a freely hinged or jointed chain and neglect of self-intersection), one of which is that $N \rightarrow \infty$. Accordingly P(R) still has finite (but very small) values at values of R greater than Nl (the chain length). Nevertheless, for many purposes this is an extremely good and useful approximation. We just have to be aware that in certain circumstances the assumptions will no longer be valid (e.g., very short chains, long chains at high extensions, etc.) and we must expect deviations from theoretical predictions.

We can also obtain a distribution function that describes the probability of finding an end a distance R from the origin (i.e., where we fix the other end) regardless of direction by using equation 7.23 to obtain the probability that this end is located in an element of volume $4\pi R^2 dR$. This is the radial distribution function and is simply given by $W(R) = P(R) \cdot 4\pi R^2$ and is also plotted in figure 7.14. The maximum in this curve is the most probable value and is not equal to $<R^2>^{1/2}$, but $(2/3)^{1/2} <R^2>^{1/2}$. The second moment of this curve is actually $<R^2>^{1/2}$, shown as the vertical dotted line in figure 7.14.

The root mean square end-to-end distance (equation 7.22) can also be formally obtained from the distribution function, but this is far more statistics than we actually need for a simple understanding, so we will conclude this section by summarizing the important parts that you will need later on. These are:

1) A polymer chain can take on an enormous range of configurations as a result of bond rotations.

2) These configurations or conformations can be described statistically, with the end-to-end distance R being a useful parameter for doing this.

3) The average value of R taken over all possible conformations can be expressed in terms of its root mean square value $<R^2>^{1/2}$, which is proportional to the square root of the number of bonds, $N^{1/2}$.

4) The distribution function P(R) takes the form of a Gaussian curve, *to a first approximation*.

The Effect of Restricted Rotation, and Short and Long Range Interactions

We will now consider the effect of various restrictions, starting with the effect of restricted bond angles and ascending in difficulty until we reach the excluded volume problem, which is a very difficult problem indeed.

In the preceding derivation we have assumed a freely jointed chain and the first step towards reality is to now fix the *bond angles* at their usual values, which for saturated polymers such as polyethylene are tetrahedral. Free rotations around the bonds will still be allowed, however. The effect of this restriction is to introduce a correlation between one bond direction and the next, so that the average value of cos θ in equation 7.19 is no longer zero. It can be shown (see Flory's second book) that in the limit of large N (remember that N is the number of bonds in the chain):

$$<R^2> = Nl^2 \left(\frac{1 + \cos \theta}{1 - \cos \theta} \right) \qquad (7.26)$$

so that $<R^2>^{1/2}$ is still proportional to $N^{1/2}$.

The next step is to consider the effect of restricted rotation. If there were no correlations between rotations of one bond and the next, then we could obtain a second correlation factor that would obviously depend upon the potential function describing the barriers to bond rotation. This problem has been described by Volkenstein* and Flory (cited earlier) and it can be shown that the second correction term has the same form as that given in equation 7.26, with a factor cos φ replacing cos θ, providing that the potential function is symmetric and bond rotations are independent of one another;

$$<R^2> = Nl^2 \left(\frac{1 + \cos \theta}{1 - \cos \theta} \right) \left(\frac{1 + \eta}{1 - \eta} \right) \qquad (7.27)$$

where η = cos φ, the average value assumed by the bond rotation angle φ. (The factor η can take different forms, depending upon the potential function.)

In general, the exact shape of the potential energy function is not known, so that cos φ cannot be calculated without making a simplification of some sort. Because information concerning the heights of the barriers to rotation is usually available, one such simplification is the use of the *rotational isomeric states* model, where it is assumed that a bond sits in one of its minimum energy conformations, which for polyethylene would either be the trans or one of the two gauche conformations, for example. Fluctuations around these positions are ignored. The fraction of bonds in each conformation can then be calculated by weighting each conformational state in terms of the energy difference between them (so for polyethylene more bonds are in the trans than the gauche forms). This allows an average value of cos φ to be calculated.

These various correction factors can now be incorporated into a parameter C_∞ defined by:

---

* M. V. Volkenstein, *Configurational Statistics of Chain Molecules*, Interscience Publishers (1963).

$$\langle R^2 \rangle_0 = C_\infty N l^2 \qquad (7.28)$$

where C_∞ accounts for the difference between the actual dimensions of a given polymer chain, given by $\langle R^2 \rangle_0$ and that obtained from the freely jointed chain model, $N l^2$. It is possible to measure $\langle R^2 \rangle_0$ experimentally, $N l^2$ is known, so the experimentally derived value of C_∞ can be compared with that calculated from those quantities given in the square brackets of equation 7.27, using the rotational isomeric states model to determine η. It has been found that there are major discrepancies, so there is a big problem somewhere.

The problem is not with the rotational isomeric states model, as such, but rather the assumption that bond rotations are independent. In polyethylene, for example, it can easily be shown using models that if a given bond is in a gauche rotational state, then neighboring bonds are unlikely to be in a gauche rotational state of the opposite sign, as this would bring certain atoms much too close together. In other words, there is a dependence of the conformations of bonds upon those of their neighbors. Unfortunately, this correlation does not admit of a solution that can be expressed in a simple form. Flory (second book) describes various matrix methods that can be used, however and these allow a calculation of the factor C_∞ for real polymer chains.

The crucial point in the above discussion is that short range interactions that affect bond rotations do not affect the dependence of $\langle R^2 \rangle^{1/2}$ on $N^{1/2}$, but can be handled by a separate term C_∞, or even C_N if we modify the equation to account for the finite length (N bonds) of the chain. This allows us to approximate a real polymer chain by an equivalent freely jointed chain. We no longer consider individual bonds, however, but a sufficient number of adjacent bonds that the combinations of their various possible rotations give a result that is equivalent to a freely jointed unit. We won't pursue this further and now turn our attention from short range restrictions to interactions of longer range, by which we mean interactions between groups that are a distance apart along the chain, but because of the bends and loops in a typical coiled conformation become close to one another in space. In other words, we are considering the problems involved with assuming the chain can "pass through itself", or self intersect, when constructing a random flight.

One approach that at first sight might seem appealing is to construct a statistical model based on self-avoiding walks. By this we mean the chain is not allowed to pass through parts of space already occupied by other bits of itself. In constructing random walk models of this type we essentially say that the walker remembers where he or she has been and there is an "excluded volume" consisting of those parts of space occupied by other bits of the chain.

If this self-avoiding walk were confined to one dimension, for example, then the walker could only go one way, because a step in the reverse direction to which he or she started would not be allowed (see Figure 7.15). If each step is length l, the distance from the origin would always be Nl (ignoring direction). The end-to-end distance depends on N rather than the value of $N^{1/2}$ obtained when crossing its own path was allowed. Clearly, the effect of self-avoidance is to expand the end-to-end distance. This carries through to three dimensions.

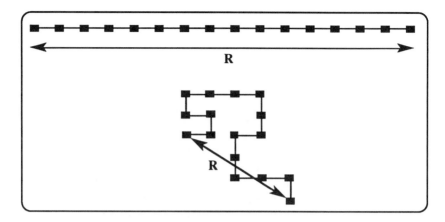

Figure 7.15 Self-avoiding lattice walks of 15 steps in one and two dimensions.

Although the problem of self-avoiding walks is a difficult one[*] it is possible to show that:

$$\langle R^2 \rangle_{SAW}^{1/2} = \text{Constant } N^\upsilon \qquad (7.29)$$

where:

$$\upsilon = \frac{3}{d+2} \qquad (7.30)$$

and d is the dimensionality of space. The subscript SAW stands for self-avoiding walk. If $d = 1$ then $\upsilon = 1$, a result we have already obtained, and if $\upsilon = 3$ then $d = 0.6$. Accordingly, you would look at this and say that because a self-avoiding walk model appears to be a much more realistic model for a polymer coil, then $\langle R^2 \rangle$ should not be proportional to $N^{0.5}$, but to $N^{0.6}$. But in the solid state it isn't, the random flight model gives the correct answer! It is only in dilute solution that the self-avoiding walk result is obtained and $\langle R^2 \rangle$ is proportional to $N^{0.6}$! (These results have been firmly established by neutron scattering experiments.) This seems really strange and it is typically of the genius of Flory that he anticipated this result quite early in the evolution of polymer physical chemistry ($\approx$1949). His derivation of the chain expansion factor rests on some drastic assumptions, but nevertheless appears correct in its essentials. He preferred to use a factor α describing the expansion of the chain in a good solvent relative to its "unperturbed" random flight form, i.e:

$$\langle R^2 \rangle^{1/2} = \alpha \langle R^2 \rangle_0^{1/2} \qquad (7.31)$$

and the dependence on dimensionality in equations 7.29 and 7.30 was subsequently made more explicit by Fisher[**], but in its essentials remained the Flory treatment.

---

[*] See the discussion in R. Zallen, *The Physics of Amorphous Solids*, Wiley, 1983.

[**] M.E. Fisher, comment on *J. Phys. Soc. Japan*, **26**, supplement, p. 44 (1969) (appended at the end of cited paper).

Why is a polymer chain in the solid state "ideal" (i.e., random flight), but in a good solvent expanded? The problem is in the simple way we have used a statistical model to account for the variety of chain conformations. We have ignored intermolecular interactions. The presence of such interactions serves to "weight" the configurations such that some would be more probable than others. If we only consider a hypothetical chain floating in a vacuum, which is in effect what we were doing, then polymer chain segments that came too close to one another would "repel". The essence of the Flory argument is that in the solid state a segment of a chain is surrounded by other segments. It does not know whether or not these segments are other parts of the chain to which it belongs, or are bits of other chains. All the interactions are the same and the polymer coil gains nothing by spreading out. It therefore maintains "ideal" (random flight) dimensions. In a good solvent, however, the chain gains free energy by expanding, because any contacts between different parts of the same chain are then replaced by more favorable interactions with the solvent (we use the term "good" solvent to mean that the interactions between it and the polymer segments are preferred to polymer segment-segment interactions). What if the solvent is not so good? Will the chain "collapse" (i.e., coil up on itself)? More on this later (in solution thermodynamics)!

All this might seem difficult to grasp at first and is indeed difficult to describe simply without involving more complicated arguments and equations. To really appreciate this problem you must travel deeper into this subject than the scope of this text allows. Hopefully, from these discussions you will have obtained a feel for the difficulties that you will encounter if you decide to take this journey.

E. POLYMER MORPHOLOGY

Morphology is a word we have stolen from biologists and botanists, where it means the study of form and structure. It has a corresponding, although perhaps not precisely the same, meaning when applied to polymers; the study of order. As such, it most obviously refers to the study of crystallinity in polymers and many discussions of polymer morphology are confined to this subject. Some polymers (actually block copolymers) that are incapable of crystallizing can also form ordered structures, but we will not discuss these in this introductory text.

Crystallinity in Polymers

Polymers are semi-crystalline. In certain ways polymers that crystallize are distinctly different in their characteristics to equivalent low molecular weight materials. Two of the most revealing experimental aspects of these differences concern the types of patterns observed in X-ray (and electron) diffraction experiments and the range of temperatures over which polymer materials melt.

We will assume that at some point in your studies you have encountered the basic principles of X-ray diffraction. If atoms are arranged in a regular three dimensional array in space, then they will scatter X-rays to give a regular pattern of spots that can be observed on a photographic film, as illustrated in figure 7.16. The structure of the crystal can be reconstructed from the pattern and

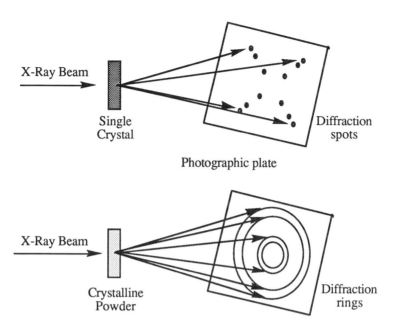

Figure 7.16 Schematic representation of a simple X-ray diffraction experiment.

intensities of these spots, but in order to obtain the full wealth of information that this technique allows, you have to grow a crystal of macroscopic size that can be placed in the diffractometer.

Now consider the situation where we cannot obtain one large crystal, but only a collection of randomly arranged smaller crystals, say in the form of a powder. If this powder is placed in a diffractometer, the crystals would all lie at different angles to the incident X-ray beam, so that the individual discrete spots from each crystal trace out a set of concentric rings. Employing an instrument that can scan this pattern along a line stretching radially from the center of the film and measure the intensity of the scattered radiation (a densitometer), we would see something like the pattern of peaks or bands shown in figure 7.17, with the height of the peaks being a measure of intensity. If we had a collection of relatively "large" crystals (say 1 µm in diameter) the lines would be fairly narrow, as illustrated by the "diffractogram" shown at the top [figure 7.17(a)]. For a collection of much smaller crystals, we would find that the width of these circles is broader than that obtained from the larger crystals, by an amount that depends inversely upon their relative size. The smaller the crystals, the broader the diffraction peaks, as illustrated in figure 7.17(b). Finally, if we now take the crystals and melt them, the X-ray diffraction pattern does not completely disappear, but instead a very broad line is obtained in our densitometer scan, superimposed upon which weak secondary lines are sometimes observed, as shown in figure 7.17(c). This is because in amorphous materials there is still local order, in the sense that there is a reasonably well-defined average number of nearest neighbors and scattering from these gives the first peak.   The average

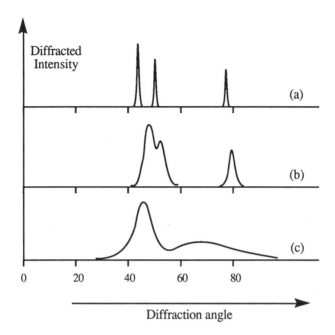

Figure 7.17 *X-ray diffraction patterns of (a) a sample consisting of crystals that are about 1 μm in diameter, (b) the same sample prepared in the form of smaller crystals (≈ 5 nm across) and (c) the same sample in the amorphous state.*

number of second nearest neighbors is much less well-defined in the amorphous or liquid state, and third neighbors even less so. Consequently, you don't usually see much in the way of additional peaks.

For most non-polymeric materials you see crystalline patterns below the melting point, broadened by size effects and crystalline defects, while above the melting point you see the amorphous pattern. *In crystallizable polymers you see both at the same time.* This is illustrated in figure 7.18, which shows the pattern obtained from a low molecular weight paraffin* in the crystalline state (a), in the liquid or amorphous state above its melting point (b), and finally, the pattern obtained from polyethylene (below its melting point), which shows crystalline peaks (quite broad ones, compared to the lines obtained from the paraffin) superimposed upon the broad scattering from amorphous material (c). So polyethylene must consist of small crystals (broad crystal peaks) that coexist and are somehow embedded in a matrix of amorphous material.

Turning our attention to crystallization and melting temperatures, we again see that polymers behave very differently to their low molecular weight analogues. Let us compare the melting of polyethylene to that of some paraffins. Figure 7.19 shows measurements of volume taken as a function of temperature, using a dilatometer. The close-packed crystalline state naturally has a smaller

---

* The n-paraffins have the same chemical structure as polyethylene, their chains are just very short.

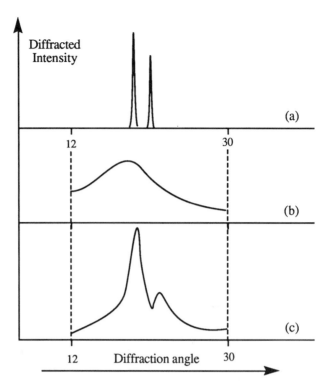

Figure 7.18 *Schematic illustration of the X-ray diffraction pattern of (a) a low molecular weight paraffin in the crystalline state, (b) the same paraffin heated above its melting point and (c) the x-ray defraction pattern of polyethylene.*

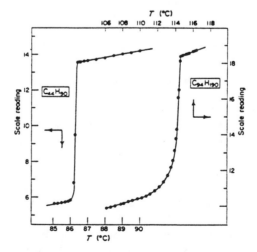

Figure 7.19 *Volume as a function of temperature for the n-hydrocarbons $C_{44}H_{90}$ and $C_{94}H_{190}$. Reproduced with permission from L. Mandelkern,* Comprehensive Polymer Science, *Vol. 2, Pergamon Press, Oxford, Chapter 11, 1989.*

volume than the amorphous state and for the paraffin, $C_{44}H_{90}$, melting occurs over an extremely narrow range. This is exactly the type of behavior you would expect from the melting of large, more-or-less perfect crystals.

If we now consider a longer chain paraffin, $C_{94}H_{190}$, the melting range is now broadened somewhat, from a fraction of a degree in $C_{44}H_{90}$ to about 1.5 to 2°C. Upon going to polyethylene this melting range is broadened even further, as shown in figure 7.20. The filled circles in this figure show an ordinary polyethylene sample that has a broad range of molecular weights. The sample was crystallized under carefully controlled conditions, because fast or slow cooling, etc., can also dramatically affect the melting point range. Nevertheless, the melting range is still very broad, indicating that polymer crystals come in a range of sizes and with various degrees of perfection. As we will see, one reason for this is that the presence of long chains results in a complex crystallization process, but if there is a range of chain lengths there will be additional problems.

In a paraffin crystal all the chain ends can get in "register" in the crystal (i.e., line up like sardines in a can), because all the chains are the same length, but this cannot happen in a polymer with a distribution of molecular weights. This is not to imply that if the polymer chains were all of the same length, then we would obtain crystals consisting of fully extended, lined up chains. For reasons that we will discuss in Chapter 8, polymers cannot crystallize in this fashion. What actually happens is that chains of different lengths crystallize at different *rates* and the morphology of polymer crystals is determined by crystallization kinetics. Accordingly, if we examine the melting behavior of a narrow molecular weight fraction of polyethylene, shown as the open circles in figure 7.20, then the melting range narrows considerably, but is still broader than the paraffins.

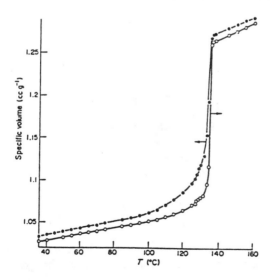

Figure 7.20 *Specific volume-temperature relation for linear polyethylene samples.* ●, *unfractionated polymer;* ○, *fractionated. Reproduced with permission from R. Chiang and P. J. Flory, JACS,* **83**, *2857 (1961).*

From all of this, we can conclude at this point that polymers form mixtures of small crystals and amorphous material and melt over a broad range rather than at a fixed melting point. Thus, as Hoffman et al.* put it, "They had laid upon them the curse of not obeying thermodynamics", because according to the phase rule, a single component mixture should be *either* crystalline (neglecting defects) *or* amorphous at a particular temperature (not both at the same time) and the transition between these states should be sharp and first-order (rather than over a range of temperatures). However, this assumes the material has achieved equilibrium, which for various reasons polymer crystals do not, as we will discuss more fully in the next chapter when we discuss crystallization kinetics. Before getting to this we need to first consider polymer morphology, i.e., What do polymer crystals look like?, How are their chains arranged relative to one another?, and Where are the amorphous bits? We will start at the beginning and consider why some polymers crystallize and others do not.

Packing in Polymer Crystals—Why Some Polymers Crystallize and Others Do Not

In principle, X-ray diffraction allows the complete structure of a crystalline material to be determined. This information is expressed in terms of the arrangements of the atoms or molecules of a material in a so-called unit cell. A schematic picture of individual atoms arranged in a very simple unit cell is shown in figure 7.21. The basic idea is that once we know the structure of the unit cell, then we can obtain a picture of the entire crystal by simply packing many of these unit cells adjacent to one another along the crystallographic axes. Of course, many real unit cells appear far more complicated than that shown in figure 7.21, but the basic idea remains the same.

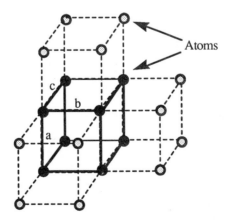

Figure 7.21 *A schematic picture of atoms in a simple cubic unit cell (heavy lines). The rest of this crystal can be built by stacking identical unit cells next to one another along the x, y and z axes.*

* J. D. Hoffman, G. T. Davis, and J. I. Lauritzen, Jr., in *Treatise on Solid State Chemistry, Vol. 3* (Editor: Hannay, N. B.), Plenum Press, New York, p.497, 1976.

There are three things about the picture obtained from X-ray diffraction experiments that concern us here. First, for simple, low molecular weight materials, it is often possible to obtain good single crystals that provide a wealth of data (diffraction spots) from which an accurate picture of the arrangements of atoms and molecules can be determined. Synthetic polymers cannot be obtained in the form of large single crystals and it is necessary to obtain a diffraction pattern from a drawn or stretched fiber, where the chains are aligned to some degree (depending on the amount of stretching) in the fiber direction. Only a few spots are observed (usually broadened into arcs) and the process of determining structure is much more difficult. Second, even low molecular weight materials form crystals that contain defects and these can profoundly affect properties. We shouldn't think that crystalline materials have a completely perfect structure. (Anybody who buys a diamond usually learns this very quickly.) Finally, in low molecular weight materials a molecule is smaller than the size of a unit cell. In polymers this is not so and individual chains pass through many unit cells.

The details of the determination of polymer crystal structure are beyond the scope of what we want to cover here. We will simply observe that X-ray crystallography played a key role in early studies of polymers and the establishment of the macromolecular hypothesis and proceed to describe two or three typical examples of polymer unit cell structures. This will make it immediately clear why some polymers crystallize and others do not.

Figure 7.22 shows a representation of the unit cell of polyethylene. There are three things you should notice. First, as we have just mentioned, only a small part of each chain lies in a unit cell. Accordingly, a knowledge of the arrangement of chains in the unit cell is a sort of local knowledge, in the sense that we do not know what sections of the rest of the chain are doing. Are all the segments also in the crystal or are some in those amorphous regions that we know are also present in polymers? Second, the chains are in the preferred, minimum energy, all trans or zig-zag conformation. This is generally, but not always, the rule for polymer crystals, particularly if there are several conformations of almost equal energy. Finally, the crystal structure is close packed, as one might expect if intermolecular attractions are to be maximized. This means that defects, such as short chain branches, generally cannot be accommodated in a crystal lattice (some small defects are occasionally incorporated into certain polymer crystals, but these naturally distort the lattice). Accordingly, just small amounts of branching in polyethylene serve to reduce the *degree of crystallinity*. In general, only the straight parts of the chains crystallize and the branches are banished to the amorphous domains. At high degrees of branching there would be no ability to crystallize at all.

If we now turn to isotactic polypropylene we observe a similar general pattern, as can be seen from figure 7.24. Here, the preferred conformation of the chain is the 3_1 helix, as we discussed previously. The chains are arranged in a regular close packed manner in the crystal, as illustrated in the projection shown in the figure, again to maximize intermolecular interactions. Clearly, an atactic polypropylene would not be able to do this, in so much as it could not form a regular conformation, and even if it could, the irregular protrusion of the methyl groups from the main chain would preclude ordered, symmetrical packing.

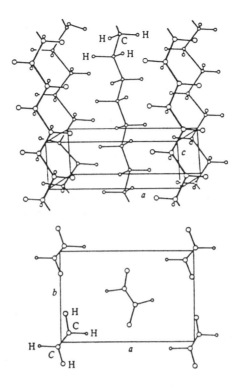

Figure 7.22 *The orthorhombic crystal structure of polyethylene. Reproduced with permission from C. W. Bunn,* Fibers from Synthetic Polymers, *R. Hill, Ed., Elsevier Publishing Co., Amsterdam, 1953.*

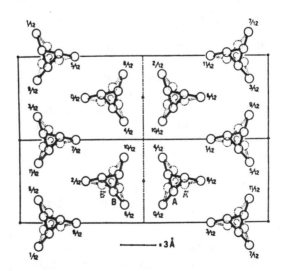

Figure 7.23 *A view (projection onto {001} plane) of the crystal structure of the monoclinic unit cell of isotactic propylene. Reproduced with permission from G. Natta and P. Corradini,* Nuovo Cimento, *Suppl. to Vol. 15, 1, 40 (1960).*

Our general principle is thus a very simple and obvious one. In order to be able to crystallize, a homopolymer must have an ordered chain microstructure; i.e., it must be linear, stereoregular, or whatever else is required by its chemical nature*. If it is a copolymer than there must be a regular arrangement of units, in an alternating or block form (and the block size must be large enough to allow crystals to form). We would emphasize here, however, that regularity of chain structure makes a particular polymer *capable of crystallizing*. This does not mean that it will. Natural rubber (*cis*-1,4-polyisoprene), for example, has a very regular, linear chain structure, but at room temperature in its unstretched state, it is completely amorphous. This is because in the disordered state there are numerous configurations available to the chain and it has a large entropy. Upon crystallization a chain will sit in a single preferred conformation and there is a significant loss of entropy relative to the disordered state. Crystallization will therefore only occur if there is a sufficient gain in energy as a result of maximizing interactions through ordered close packing. In natural rubber the possible modes of packing at room temperature do not provide sufficient energy to overcome the loss of entropy associated with ordering the chains, so that crystallization only occurs if the entropy is first reduced by application of an external force to stretch the chains.

In addition to the thermodynamic factors that can prevent crystallization, there are also kinetic factors. The process of crystallization can be prevented by fast cooling (quenching) from the melt to a temperature that is below the glass transition, where there is insufficient mobility for crystallization.

Finally, we will consider one more example of a polymer crystal structure, that of nylon 66, as this allows us to mention *polymorphism*, the ability to exist in more than one crystalline form (although we did mention this briefly in Chapter 6). One of the crystal forms of nylon 66 is shown in figure 7.24. Here, the chains are aligned so as to maximize the number of hydrogen bonds (which you will recall are relatively strong interactions) that can form between amide groups on adjacent chain segments, shown as dotted lines in this figure. This minimizes the overall energy. The hydrogen bond prefers a linear arrangement of the atoms involved, in this case, the N–H---O atoms of adjacent amide groups, but here other factors come into play. The proper packing of the CH_2 groups and the arrangement of the chains in their preferred extended conformation will also minimize the total energy. Sometimes all of the packing and conformational factors cannot be accommodated simultaneously, and as a result, there are different crystal structures of nearly equal energy. Which form is obtained will depend upon the conditions of crystallization.

---

* Perversely, there are always exceptions to the rule, and atactic poly(vinyl alcohol) crystallizes extensively.

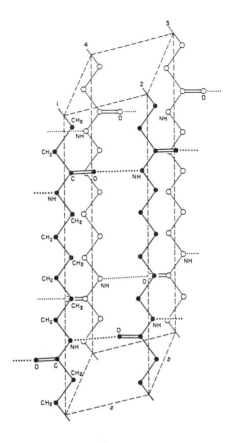

Figure 7.24 *The triclinic crystal structures of nylon 6,6, α-form. Reproduced with permission from C. W. Bunn and E. V. Garner,* Proc. Roy. Soc. (London), *189A, 39 (1947).*

From Fringed Micelles to Single Crystal Lamellae

We have now established that in the depths of its crystalline domains a polymer is arranged in a regular ordered form in the same manner as small molecules, but taken as a whole, the crystals co-exist with amorphous material. We have noted that this is quite unusual and we must now consider the form or arrangement of the crystalline and amorphous domains (and later, how this comes about).

In studies of polymer morphology, the year 1957 stands as a divider. Before this time the most widely accepted picture of the structure of semi-crystalline polymers was provided by the *fringed micelle model*, illustrated in figure 7.25. The crucial feature of this model is that individual chains traverse regions of both order and disorder, going from a small crystallite, to an amorphous region, into another crystallite, and so on. We will see that this pattern is an important feature of the structure of melt crystallized polymers, but it is not the only path that can be taken by the chains.

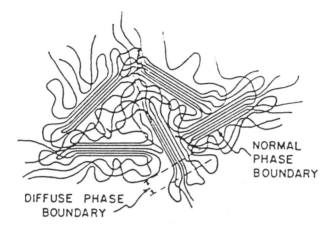

Figure 7.25 *Schematic representation of the fringed micelle model. Reproduced with permission from J. D. Hoffman, T. Davis and J. I. Lauritzen,* Treatise on Solid State Chemistry, Vol. 3, *Chapter 7, Plenum Press, New York, 1976.*

In 1957 Keller, Till and Fischer, independently and almost simultaneously, succeeded in growing so-called single crystal lamellae by cooling hot, dilute solutions of polyethylene in xylene*. At a certain temperature crystallization starts to occur, but because the chains are not tangled up with one another, as they are in the melt, then a particularly simple crystalline form is obtained. Elec-

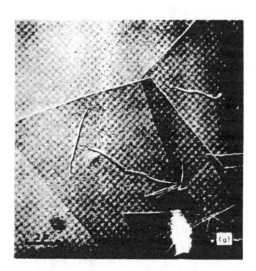

Figure 7.26 *Electron micrograph of a polyethylene single crystal grown from tetrachloroethylene solution. Reproduced with permission from P. H. Geil,* Polymer Single Crystals, *Robert E. Krieger Publishing Company, Huntington, New York, 1973.*

* Single crystal lamellae had been obtained prior to this date, but their significance had not been recognized.

tron micrographs showed that these crystals appeared to have a flat diamond type shape as shown in figure 7.26. Later, it was demonstrated that in solution these crystals are actually more like hollow pyramids and that the observed flat shape is due to a collapse upon drying. Also, polymer single crystals of different shape or habits can be obtained. We will not describe these here, as we only need to consider the one crucial feature of the lamellae that forms the primary focus of our discussion.

Electron diffraction experiments demonstrate that the polymer chain axes are essentially perpendicular to the large flat faces and parallel to the thin faces (think of the crystal as a sheet of paper that is many times longer in two dimensions than it is in the third, "thickness," direction). Because the thickness of the crystal is only of the order of a 100 Å or so, while the length of the chains used in the original experiments were of the order of 2000 Å, it was concluded by Keller and O'Connor, in a paper presented at the 1957 Discussions of the Faraday Society meeting, that "as the parallel alignment of the chains is almost perfect, we are forced to conclude that the molecules must sharply fold on themselves." This was controversial and Morawetz, in his book on the history of polymer science (cited previously), reports that E. W. Fischer encountered such ridicule of his paper, which also suggested chain folding, that he feared losing his job. The experimental facts were incontrovertible, however, and it was soon realized that chain folding also occurs in melt crystallized samples, but we will come to that soon enough.

The concept of chain folding suddenly allowed an understanding of various previously puzzling observations and insight into the mechanical properties of polymer materials, as we will discuss later in this book. The subject remained controversial for many years, however, and can still incite the occasional heated

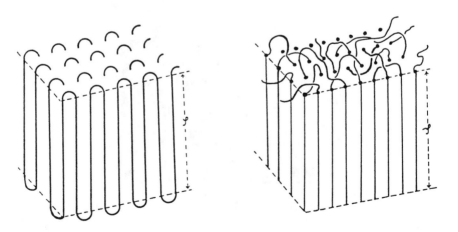

Figure 7.27 *The adjacent reentry and switchboard models for polymer single crystals. Reproduced with permission from P. J. Flory, JACS, 34, 2857 (1962).*

exchange. It was not the concept of chain folding as such that excited subsequent controversy, but questions concerning the nature of the fold surface. Keller proposed that folding occurred in a tight, regular manner, such that a chain emerges from the crystal then immediately reenters at an adjacent position, as shown schematically in figure 7.27. Flory and his co-workers maintained that the fold surface should be essentially disordered, with chains entering and leaving at random and this was called the switchboard model, as also illustrated in figure 7.28. (This is a really good analogy, but if you were born sometime in the last twenty years you may have no idea what an old telephone switchboard looks like. We can only sympathize and advise you to watch more old movies.)

The arguments concerning the relative merits of the two models reached polemical proportions, culminating in a now famous meeting of the Faraday Society in 1979. Many stories have been told about this meeting, some no doubt apocryphal, but it is our understanding that Flory felt he was "ambushed" and the proponents of the "ordered" reentry model considered themselves victorious (the arguments predominantly revolved around the requirements of steric packing at the interface and the interpretation of neutron scattering data). As a result, Flory reportedly came back from this meeting seriously displeased and proceeded to work on a statistical mechanical model of crystal surfaces and interphases (see figure 7.28). This lattice model allowed the calculations of the number of adjacent reentry sites in single crystals, which Flory determined to be less than 40%. The precise value of this number depends (amongst other things) on assumptions concerning chain flexibility, however, and this figure has been revised upwards to a value as high as 80% for very flexible chains[*] , so one could argue that everybody was right to some degree (or equally wrong, if you take a more jaundiced view). Certainly, the controversy appeared to have provoked Flory into developing a new and powerful method of analysis and it

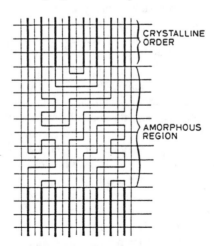

CRYSTALLINE
ORDER

AMORPHOUS
REGION

Figure 7.28 A schematic representation of the interphase in lamellar semi-crystalline polymers. For single crystals all the chains must return to the same crystal. Reproduced with permission from K. A. Dill and P. J. Flory, Proc. Nat. Acad. Sci., 77, 3115 (1980).

---

[*] S. K. Kumar and D. Y. Yoon, *Macromolecules*, **22**, 3458 (1989).

seems safe to conclude that for single crystals all chains reenter within (about) three lattice sites of the surface. For polyethylene this corresponds to a surface region of about 15 Å, in good agreement with experimental measurements of the extent of the surface regime.

Crystallization from the Melt

The crucial principle that you should bring with you from the previous discussion is the concept of chain folding. At first it was uncertain if this should apply to melt crystallized samples and Bunn, a leading figure in X-ray studies of polymer morphology, commented that "if it occurs in crystals grown from a melt, this is still more surprising because we suppose a melt to be a tangle of long-chain molecules." It soon became apparent that chain folding also occurs in melt grown crystals but the elucidation of their fine structure is a more difficult problem.

The most striking feature of the crystallization of many polymers from the melt is that it results in the formation of so called spherulites that are large enough to be seen in the optical microscope (i.e., they are a lot larger than single crystal lamellae). As the name suggests, these are spherical objects. They are not wholly crystalline, but contain some sort of arrangement of both crystalline and amorphous parts. If viewed in an optical microscope under crossed polarizers*, they appear as in figure 7.29, which shows the growth of spherulites from the melt. There are two things you should notice about the spherulites shown in this figure. First, they are only really spherical at the initial stages of growth. Later, they impinge upon one another to form straight or hyperbolic boundaries. Second, the use of cross-polarizers reveals a pattern which is related to the birefringent properties of these crystalline polymers.

It will be assumed that you know roughly what this means, but if you don't it is sufficient to understand that the polymer crystals are anisotropic in their optical properties (there is a difference in the refractive index for light polarized parallel to the chain axis relative to light polarized perpendicular to the chain axis). Accordingly, the observed pattern says something about the underlying arrangement of the crystalline domains. Figure 7.29 shows the characteristic and commonly observed Maltese cross, which indicates some sort of radial order. This type of observation, together with X-ray diffraction experiments, indicated that the chains are oriented in a direction that is perpendicular to the radius of the spherulite. This was at first puzzling and hard to understand until the idea that chain folding also occurs in melt crystallized samples took hold. Furthermore, the arrangement of the crystalline and amorphous parts becomes more apparent when we consider the results of some very clever experiments by Keith and Padden, which involved crystallization of mixtures where one component could not crystallize. A micrograph is shown in figure 7.30. Here, 10% isotactic poly-propylene was crystallized from a mixture containing 90% of the atactic polymer. This latter material was then removed by solvent (the crystalline parts

---

* The incident light is polarized in one direction and after passing through a thin film of the sample, is viewed through a second polarizer placed at 90° to the first. In this position, no light would be transmitted unless there is some interaction with the sample.

Figure 7.29 *Spherulites of (a) nylon 6,10 during crystallization and (b) polyethylene oxide after completion of growth, observed through crossed polaroids. Note that the featureless background in (a) which is amorphous material and the radiating fine structure and Maltese Cross pattern in (b). Reproduced with permission from F. Khoury and E. Passaglia, in* Treatise on Solid State Chemistry, *N. B. Hannay, Ed., Vol. 3, Chapter 6, Plenum Press, New York, 1976.*

Figure 7.30 *Growth of isotactic polypropylene spherulites in the presence of 90% atactic polypropylene. Note the branched fibrils. Reproduced with permission from H. D. Keith and F. J. Padden, J. Appl. Phys., 35, 1270 (1964).*

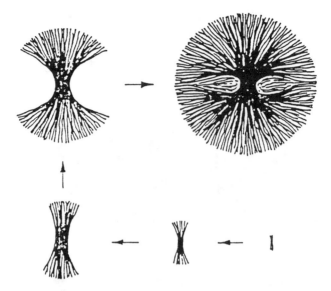

Figure 7.31 *A schematic representation of the development of a spherulite. Reproduced with permission from D. C. Bassett,* Principles of Polymer Morphology, *Cambridge University Press, 1981.*

were insoluble at ordinary temperatures) revealing a pattern of branched lamellar like structures. Figure 7.31 shows a schematic representation of how spherulites are thought to develop, from an initially formed central lamellae (sometimes confusingly called a fibril), which in the course of its growth, subsequently branches to form the spherulite. The space between the arms is filled with amorphous material.

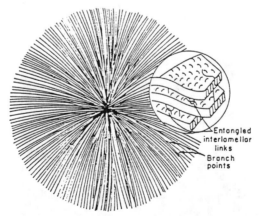

Figure 7.32 *A schematic representation of the structure of a spherulite. Note the interlamellar links between the spherulite arms. Amorphous material also sits between the spherulite arms, but is not shown here. Reproduced with permission from J. D. Hoffman, T. Davis and J. I. Lauritzen, in* Treatise on Solid State Chemistry, *N. B. Hannay, Ed., Vol. 3, Chapter 7, Plenum Press, New York, 1976.*

The lamellar arms of the spherulite consist of chain folded material, as illustrated in figure 7.32, but the nature of folding here is considered to be much less regular than in solution grown crystals. In addition, there are a certain number of chains that emerge from one lamellar arm, coil around in the amorphous interlamellar region for awhile, and then enter another crystalline lamellar arm, as in the old fringed micelle model (see the model of the interphase region shown in figure 7.28). These are known as *tie molecules* and play a key role in mechanical properties. The generally tough nature of certain polymers and their ability to undergo plastic deformation depends upon the presence of chains connecting the lamellae.

Fibers

One consequence of chain folding is that melt crystallized polymers are not as strong as they could be or as strong as we would like for certain applications. It is not our purpose to discuss mechanical properties here, but it should be intuitively clear that if we could arrange the chains such that they were perfectly stretched out and aligned and then pulled in the direction of alignment, then the strong covalent bonds of the chains would take the load. (There is a complication because of the finite length of the chains. The stress has to pass from one chain to another, but for the sake of our simple argument, we will ignore such difficulties here.) Conversely, in a melt crystallized sample containing randomly oriented spherulites, it would be the weaker forces of attraction between the chains (and also physical entanglements, etc.) that resist a load and one can envisage pulling the chains out of a crystal, in other words, unfolding them, to make a material where the chains have a more elongated form. That is exactly what we do when we make a fiber. If a sample is "drawn" (stretched) at a temperature above T_g but below T_m, a morphology is produced where there is a preferred orientation of the chains parallel to the draw direction. In fact, a material such as polyethylene can be drawn to many times its original dimensions at room temperature, as illustrated in figure 7.33, and this process of drawing, if done properly, does confer increased strength and modulus in the draw direction.

The question for us here, is "What is the morphology of drawn samples?" The answer is we don't really know, at least in comparison to the type of detailed knowledge that has been built up for single crystal lamellae, spherulites, and some other morphologies which we have neglected here because they are less common or do not illustrate additional general principles.

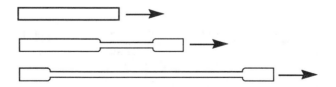

Figure 7.33 A schematic diagram depicting the drawing of a polymer.

In producing fibers we would like to obtain completely extended chains. Nature accomplishes this to a great extent in utilizing cellulose, which is synthesized in plants in sets of about 30 chains which form long or extended fibrils. In general, for many synthetic polymers, this level of orientational order is not achieved using ordinary processing methods. Perhaps the most convincing model of the morphology of drawn fibers is the fibrillar structure proposed by Peterlin, and the formation of these structures is illustrated schematically in figure 7.35. Here, the chains are partly elongated, but not completely, and they still remain somewhat folded.

There are two fairly recent developments that have allowed the development of fibers with much more extended structures and hence, enhance mechanical properties. First, polymers with very stiff "backbones" have been synthesized. These rod-like molecules have liquid crystalline properties, one aspect of which is that they become aligned in solution at concentrations above a certain critical level. Clearly, it is much easier (in principle!) to obtain highly oriented fibers with extended chain structures from such polymers (in practice, there are a lot of problems). Second, new processing methods, based on drawing polymers such as polyethylene from a gel (so as to minimize the effect of chain entanglements during the orientation process) have resulted in extremely high strength, high modulus fibers. But this takes us out of the area of morphology and into processing and mechanical properties. We will save such delights until later in this book. We will conclude this chapter with a brief aside on size and microscopy.

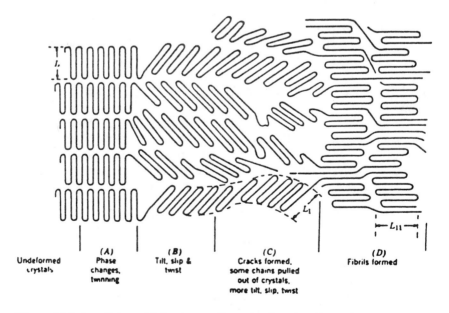

| (A)
Undeformed
crystals | (B)
Phase
changes,
twinning | (B)
Tilt, slip &
twist | (C)
Cracks formed,
some chains pulled
out of crystals,
more tilt, slip, twist | (D)
Fibrils formed |

Figure 7.34 Peterlin's model for the transformation from lamellar to fibrillar structure. Reproduced with permission from A. Peterlin, J. Polym. Sci., **C9**, 61 (1965).

F. FINAL WORDS—A BRIEF COMMENT ON SIZE

Although we have presented various micrographs in the preceding section, we have said nothing about microscopy. There is tremendous skill and craft, particularly in sample preparation, that is required in order to obtain good pictures (and not be fooled by artifacts!), so this is a technique that requires a separate and more advanced treatment. To round out this chapter, however, we think it is useful to briefly comment on the relative sizes of the structures we have discussed. These are illustrated schematically in figure 7.35, which on the left hand side shows two chains arranged in a crystal. The interchain distances are of the order of 5 Å (5×10^{-8} cm) (and obviously vary from polymer to polymer). If we now go to the single crystal lamellae, these measure approximately 10 μm (1 $\times 10^{-3}$ cm) along the sides of their diamond-like plates (the figure shows a measurement across the diagonal) and so are approximately 30,000 times larger than interchain distances along this dimension (i.e. a lot of chain stems can be stacked in single crystals). The thickness of the lamellae is only about 150 Å, however, so compared to their long dimensions these crystals are very thin indeed ($\approx 1/1000$) and can be likened to a sheet of paper. Finally, spherulites, in the middle of their growth phase before they impinge on one another, are approximately 100 μm (1 x 10^{-2} cm) in diameter (they start out smaller, obviously, and end up larger).

This difference in size determines which technique can be used to study their morphology. The limiting factor is (roughly) the wavelength of light, in that features that are smaller than this cannot be resolved (however, we have recently

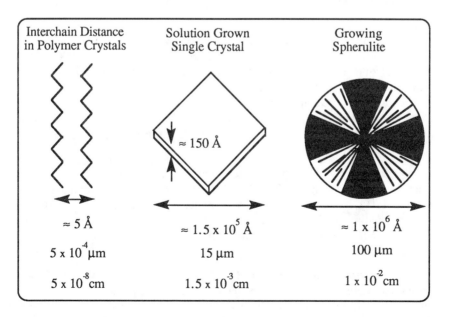

Figure 7.35 *A schematic representation of the size associated with various features of polymer crystals.*

read that optical microscopists have learnt some new tricks that might allow an improved resolution). If we assume that we are using visible light in the 5000 Å wavelength range, then spherulites can be observed, but no details of the structure of single crystals can be resolved. It is necessary to use electron microscopy in order to observe these latter structures.

The field of microscopy is presently experiencing something of a revolution, with new techniques (scanning tunneling microscopy, surface forces microscopy) and improvements to older methods becoming available. We don't suppose that the old controversy about the nature of chain folding will ever be resolved to everybody's satisfaction until some enterprising microscopist figures out a way to "see" a single crystal surface, so we await developments in this field with interest.

G. STUDY QUESTIONS

1. In this chapter we discussed minimum energy conformations for just two polymers, polyethylene and isotactic polypropylene. Go to the literature and find out the minimum energy conformations of the following polymers:

> (a) Syndiotactic polypropylene
> (b) Nylon 6
> (c) Poly(ethylene terephthalate)

Discuss these conformations in terms of the simple steric repulsion approach given in this chapter.

2. Discuss the limitations of the freely jointed and rotating bond model for a disordered polymer chain

3. (a) Given that C_∞ for polyethylene is 6.8, what would be the root mean square end-to-end distance of a chain of degree of polymerization 100,000, given that the chain is in a melt of other high molecular weight polyethylene chains ? (Assume the projected length l of an ethylene unit is ≈ 1.78Å, calculated from tetrahedral bond angles and usual C-C bond lengths).

(b) Assuming that all constants of proportionality cancel out, what is the ratio of the root mean square end-to-end distances of the same chain in a good solvent compared to that in the melt ?

4. You are given two polyethylene samples, A and B. Sample A has an average molecular weight of 200,000, a density of 0.92 and a crystalline melting point of 110°C. Sample B has the same molecular weight (and practically the same molecular weight distribution), but its density is 0.96 and its melting point is 133°C. What is the difference in the chain structure of these two materials and how does it cause these variations ?

5. Poly(ethylene terephthalate) (PET), which has glass transition (T_g) and crystalline melting temperatures of 69 and 267°C, respectively, can exist in a number of different states depending upon temperature and thermal history. Thus it is possible to prepare materials that are semi-crystalline with amorphous regions that are either glassy or rubbery and amorphous materials that are glassy,

rubbery or melts. Consider a sample of PET cooled rapidly from 300°C (state I) to room temperature. The resulting material is rigid and perfectly transparent (state II). The sample is then heated to 100°C and maintained at this temperature, during which time it gradually becomes translucent (state III). It is then cooled to room temperature, where it is again observed to be translucent (state IV).

(a) Identify each of the states.

(b) In state IV PET has a higher modulus and is less flexible than in state III. Why is this ?

(c) Why is the PET transparent in state II, yet translucent in state IV ?

6. (a) A sample of linear polyethylene is dissolved in a large excess of xylene at 130°C and the dilute solution is then slowly cooled. After a couple of days a fine white suspension is obtained. What does this suspension consist of ?

(b) If the suspension is filtered and dried, then heated to 170°C and slowly cooled to room temperature, a different type of structure is obtained. Compare and contrast the two different types of structures.

7. Discuss the relationship between chain microstructure and crystallinity.

H. SUGGESTIONS FOR FURTHER READING

(1)   D. C. Bassett, *Principles of Polymer Morphology,*
       Cambridge University Press, Cambridge, 1981.

(2)   P. J. Flory, *Principles of Polymer Chemistry*,
       Cornell University Press, Ithaca, New York, 1953.

(3)   P. J. Flory, *Statistical Mechanics of Chain Molecules*,
       John Wiley & Sons, New York, 1969.

(4)   P. H. Geil, *Polymer Single Crystals*,
       Robert E. Kreiger Publishing Company, Huntington, New York, 1973.

(5)   H. Tadokoro, *Structure of Crystalline Polymers*,
       John Wiley & Sons, New York, 1979.

(6)   L. R. G. Treloar, *Physics of Rubber Elasticity,*
       Third Edition, Claredon Press, Oxford, 1975.

CHAPTER 8

Crystallization, Melting and the Glass Transition

Thermodynamics " . . . a pretty gloomy topic,
the part of science that tells us that the world is running down,
getting more disordered and generally going to hell in a handbasket."
—James Trefil, 1991[*]

A. OVERVIEW AND GENERAL APPROACH

In the next few chapters we will be considering properties (thermal, solution, certain mechanical properties like rubber elasticity) that are best described in terms of the language of thermodynamics and statistical mechanics. Pedagogically, this presents some problems of knowing where to start and how much fundamental material to assume as a common core of knowledge. Furthermore, most students regard thermodynamics as a pretty painful subject, involving the apparently fearsome, intangible and mysterious (to many) concept of entropy. In part, some of these difficulties are a residue of the historical development of this subject, which started before the general acceptance of the atomic nature of matter and was concerned with things like the efficiency of heat engines. Classical thermodynamics, of course, only describes relationships between macroscopic properties (pressure, volume, temperature, etc.), but if we view concepts such as entropy from the viewpoint of statistical mechanics, which provides the link between atomic or molecular structure and macroscopic properties through a consideration of the statistical properties of very large numbers of atoms or molecules, they are much easier to understand. Unfortunately, these subjects are often taught separately and it has been our experience that many students don't always make the necessary connections and bring the required fundamentals to their studies of polymers. As a result, as soon as we start mentioning free energy or entropy, it is sometimes possible to see visible manifestations of unease pass like a wave through the class. We will therefore start our discussion of what can be thought of as thermodynamic properties with a brief review of certain essential features of thermodynamics and statistical thermodynamics.

We will only be dealing with that part of the subject known as equilibrium thermodynamics. For processes which are controlled by the rate at which they occur, such as crystallization, kinetic theory has provided the most insight. There

---

[*] Description given by James Trefil in a review of the book, *The Arrow of Time*, by P. Coveney and R. Highfield, *The New York Times* Book Review Section, Sunday, June 23, 1991.

are also certain properties that we do not exactly know how to classify, particularly the glass transition temperature. Is this a true second order phase transition (defined in more detail later in this chapter), or are we looking at a purely kinetic phenomenon? (Even if there is a true underlying thermodynamic transition, kinetics certainly determines where the T_g will be observed in a particular material, depending upon experimental factors like the rate of cooling.) Accordingly, the first few sections of this chapter may seem in the beginning a little out-of-place. Nevertheless, even in our discussion of kinetically controlled phenomena like crystallization, we will be using thermodynamic quantities, such as free energy and entropy. Equilibrium considerations are still crucial for kinetically controlled phenomena, as they tell us where the system would like to go, even if it cannot quite get there, and this can provide an important level of understanding.

B. SOME FUNDAMENTALS

Thermodynamics—An Elementary Review

We will start with the laws of thermodynamics, which most students vaguely recall as having something to do with temperature, energy and entropy and whose confusing chronology is beautifully described by Atkins[*]:

> There are *four* Laws. The third of them, the *Second Law*, was
> recognized first; the first, the *Zeroth Law*, was formulated last;
> the *First Law* was second; the *Third Law* might not even be a law
> in the same sense as the others.

The zeroth and third law deal with temperature, the zeroth law essentially acting as a definition by stating that if there is no net heat flow between two objects, then they are at the same temperature. The third law is the "you can't get there from here" law, which says you cannot reach the absolute zero of temperature in a finite number of steps or operations. We will not say anything more about these laws except to make the point that because temperature is a thing of everyday experience, most students think they understand what it is, but they usually don't. They can usually reach the point of stating that it measures the "degree of hotness" of a body, but what does this mean? How are the molecules behaving in a hot body that makes it different to a cold body? We will briefly come back to the molecular basis of temperature below.

The first law, concerning the conservation of energy, presents few problems, as it makes intuitive sense to most people, at least those who believe that you don't get something for nothing. The big difficulty for most students is the second law, which can be stated in various ways, two of which are:

1)   You cannot convert heat completely into work.
2)   Heat doesn't flow from a "colder to a hotter ".

---

[*] P. W. Atkins, *The Second Law*. Scientific American Library. W. H. Freeman and Co., New York (1984).

These are not precise and rigorous statements and don't have the necessary caveats, but our aim here is only to remind you of some basic laws that you should have learned elsewhere. The crucial point is that both of these statements can be shown to be equivalent once we introduce the concept of entropy. The problem with understanding entropy, in our view, is that it is usually first described in terms of the dissipation of energy in an irreversible process, or in terms of the "energy tax" one must pay in order to operate a cyclic heat engine. As revision we will quickly examine both of these ideas and then proceed to a molecular interpretation, which is much more easily grasped because it allows you to see a mechanism. We can then introduce some concepts and equations which we will use extensively in describing various thermodynamic properties of polymers.

We will first consider an irreversible process, using the example given in the classic textbook by Lewis and Randall.* This starts out in a very simple way. A weight is attached to a pulley and this weight is also in contact with a reservoir which is a source or a sink of heat (to reiterate, in all of what follows we leave out the usual caveats of isolating the system, doing things very slowly and so on, as our only aim here is to remind you of some key points). The weight is allowed to fall and gives up energy in the form of heat to the reservoir through some frictional process, and also through impact as it thuds to a halt on the floor. We know from the first law that this energy is still around, but we also know from everyday experience that the weight cannot be lifted again without the intervention of an *external directed force*. We can take all the heat we want from the reservoir and put it into the weight, but it won't jump up in the air, it just gets hotter! So, the first thing we can say is that there is some sort of quality or organization to energy and something has been lost that is a measure of the irreversibility of this process of falling under gravity.

We can now make this experiment a bit more complicated by having two heat reservoirs, one hot and one cold, instead of just one. We let the falling weight first give up some energy to the hot reservoir, which is our first irreversible process. This energy is then transferred from the hot reservior to a cold reservoir. One of the statements of the second law given above tells us that this is also an irreversible process, in that heat will not *spontaneously* flow from the cold reservoir to the hot. Don't get confused here. We're still talking about the transfer of the same amount of total energy in each step, the first law is still being enforced, but there is something different about its "form" or "quality". It should be obvious that the total "degradation" of the quality of the energy in the process,

$$\text{Frictional Energy} \xrightarrow{\quad Q \quad} \text{Hot Sink} \xrightarrow{\quad Q \quad} \text{Cold Sink}$$

must be more than each of the two steps separately, because the total process is, after all, irreversible. This implies that *there must be an inverse dependence of the "irreversibility" on temperature*. If we let the quality Q/T be a measure of this

---

* G. N. Lewis and M. Randall, *Thermodynamics*. Revised by K. S. Pitzer and L. Brewer, Second Edition, McGraw Hill, (1961).

irreversibility, where Q is the heat transferred in each step, then we get a parameter than would be of the right form, because the transfer of "irreversibility" from the weight to the hot reservoir is then less than the transfer directly to the cold:

$$\frac{Q}{T_{hot}} < \frac{Q}{T_{cold}}$$ (8.1)

because, of course, $T_{hot} > T_{cold}$. This quantity, Q/T, the heat transferred to the reservoir divided by the temperature of the reservoir, is related to the entropy, but for an irreversible process is not equal to it. In fact, we can write for such a process:

$$\Delta S > \frac{Q_{irrev}}{T}$$ (8.2)

and recall that yet another statement of the second law (there are an awful lot of them) is that in an irreversible or spontaneous process the entropy always increases.

The problem that most students have with this thermodynamic approach is that it describes something that is called entropy without providing a physical picture or mechanism by which it works. It doesn't tell you in what manner a system with a high entropy differs from one with a low entropy. If you don't know, or have forgotten, or never understood it, we will maintain the suspense a little longer, because first we wish to examine a reversible process. This is not because we wish to inflict more thermodynamics on you than we have to (although this is doubtless good for your soul), but because it will allow us to define entropy in terms of a mathematical equality and also introduce the concept of free energy.

The definition of a reversible process is in some ways subtle, and you should check your old physical chemistry notes to see if you grasp it. In the reversible transfer of heat, for example, you can imagine two bodies at exactly the same temperature, then make the temperature of one infinitesimally smaller or greater than the other so that heat flows between them. The direction of heat flow can be reversed by changing which of the two bodies is infinitesimally hotter. In this process the transfer of heat must be "smooth" so as not to generate hot spots or introduce other complications into the analysis. For an isothermal (constant temperature) process the entropy change can then be *defined* as:

$$\Delta S = \frac{Q_{rev}}{T}$$ (8.3)

In order to consider the effect of a temperature change in a system the process can then be divided into a sequence of infinitesimal steps, each occurring at a different but fixed temperature and these steps are summed (actually integrated) to give the change in entropy.

This brings us to the analysis of a piston moving in a cylinder, as illustrated in figure 8.1. (Pistons moving in cylinders seem to be a ubiquitous part of classical thermodynamics, but you have to remember that this subject had its

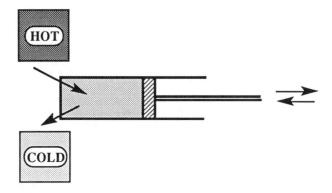

Figure 8.1 *Schematic diagram of a piston that can move in a cylinder attached to a hot reservoir and a cold reservoir.*

birth in the age of steam engines.) Imagine that this piston is initially cold, but is then attached to a hot reservoir so that a transfer of heat occurs. This causes the gas to expand an amount dV. If the pressure on the gas provided by the piston is a constant, P, then the amount of work done in moving the piston head is PdV. Now, we cannot obtain any more energy out of this system unless the piston returns to its initial position, but we must expend exactly PdV energy units in order to do this (if there are no frictional losses, etc.). Accordingly, there is no net production of work in this cycle (i.e., we cannot make something happen, like turn a wheel). If we cool the cylinder after it has expanded, however, by now attaching it to a cold reservoir, then it becomes easier to push the piston back to its original position (i.e., requires less energy) and we can get some overall work out. We are, of course, crudely describing a Carnot cycle, leaving out a proper description of all the individual infernally confusing stages of isothermal and adiabatic expansion, compression, etc. The crucial point is that in order to get work out of the hot reservoir we must throw some heat away into the cold reservoir:

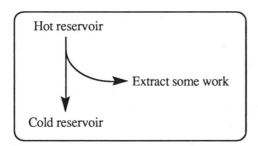

To use an analogy described by Adkins[*], think of the energy flow from hot to cold as a waterfall from which energy can be extracted to turn a turbine. You cannot obtain all the energy, because the water must have enough kinetic energy

[*] C. J. Adkins, *An Introduction to Thermal Physics*, Cambridge University Press, (1976).

to exit the plant after you have done with it. But the higher the waterfall (hence the greater the potential energy of the water) the more energy is available for extraction. Similarly, the greater the temperature difference between hot and cold "sinks" (reservoirs) the more energy can be obtained, or the greater the efficiency of the engine. At this point, thermodynamic texts usually look at this problem in more detail and then move on to the third law. That is not our concern. We just want you to have an idea that there is only a certain amount of energy in a system that is available to do work and later on we will call this the free energy. We will consider equations that will describe this quantitatively, but first we must go back to entropy. Clearly this "energy tax" that has to be paid in order to obtain useful work from heat must also be related to the "quality of energy", or, more precisely, its distribution, a thing we have called entropy, so now its time to look at the molecules and obtain a clearer idea of the nature of this intangible beast. To finish this section, however, we cannot resist using a quote from Eddington[*] concerning the inexorable requirements of the second law (hence why you should make every effort to understand its implications);

> The law that entropy always increases – the second law of thermodynamics – holds, I think, the supreme position among the laws of Nature. If someone points out to you that your pet theory of the universe is in disagreement with Maxwell's equations – then so much the worse for Maxwell's equations. If it is found to be contradicted by observation – well, these experimentalists do bungle things sometimes. But if your theory is found to be against the second law of thermodynamics I can give you no hope; there is nothing for it but to collapse in deepest humiliation.

Some Basic Statistical Mechanics

We will look for a molecular interpretation of entropy by first examining why heat flows from a hotter to a colder body, followed by our falling weight example. First we need to take a brief look at what we mean by temperature. We know from the zeroth law that if there is no net heat flow between two bodies, then they are at the same temperature, but how can this be interpreted in terms of what the molecules of each body are doing? In our opinion the easiest way to understand this is to go back and look at your old notes on the kinetic theory of gases. There it is demonstrated that temperature is a measure of the average kinetic energy of the particles (atoms or molecules). If we take two gases, one of which is hot so that its particles are on average moving fast (i.e., have a lot of K.E. $= mv^2/2$) and mix it with one that is cold (smaller values of $mv^2/2$), then through a process of random collisions some of the kinetic energy of the fast particles is transferred to the slow ones so that a new average somewhere between the initial two states is eventually obtained. Unlike a gas, a solid has kinetic energy by virtue not of translational motions but vibrations about mean positions, but the principle is the same. The vibrations get faster as

---

[*] A. S. Eddington, *The Nature of the Physical World*, Cambridge University Press, (1928).

we heat it up (and eventually get large enough to break up, or melt, the lattice). This doesn't explain why heat doesn't flow from a colder body to a hotter body, however. Why don't the cold molecules just spontaneously slow down a bit and the fast ones speed up? It turns out that in principle they can, its just that the probability of them doing so is so incredibly small that it never happens.

Consider a representation of a hot reservoir just as it is placed in contact with a larger cold reservoir, represented by the shaded and unshaded squares, respectively, in figure 8.2(a). (Here we are using a representation given by Atkins' book on the second law, cited above.) Now in reality there would be a distribution of energies in the hot reservoir and a distribution in the cold, but to give a simple picture we will just consider the average of each, so the squares can either be shaded or unshaded. Now as energy is transferred from the hotter to the colder, we should introduce some intermediate shadings to represent our new average in-between temperatures, but it's easier to make our point by assuming that heat is only transferred in chunks of one square at a time, so that the cold reservoir gets warmed up when a shaded square swaps placed with an unshaded one, as shown progressively in figures 8.2(b) and 8.2(c). The advantage of this unrealistic and somewhat convoluted model for heat transfer from our viewpoint is that in addition to describing temperature differences and the exchange of energy, we can use the same representation or model to describe the spontaneous mixing of two gases or liquids, initially separated by a barrier. We will use so-called lattice models to calculate the entropy change in this process in chapter 9.

To return to figure 8.2, as a result of random swaps, all the shaded and unshaded squares eventually become randomly distributed, as illustrated in figures 8.2(d) and 8.2(e). By this we mean that if we subdivide our container into regions of equal volume (or area, in the two dimensional representation shown in figure 8.2), then on average there will be the same number of hot and cold particles in each and the system will be at equilibrium. The parts will still be moving around, however. Accordingly, for heat to flow from a colder to a hotter body is equivalent to asking what is the probability that *as a result of random swaps* all of the shaded squares in one of the representations of the random (or most probable) state, figure 8.2(d) or 8.2(e), we can obtain the region representing the initially hot region of space, figure 8.2(a). For macroscopic systems this probability is about the same as a snowball's chance in hell! Once the energy has been dispersed, it stays that way. We can immediately see that entropy is a measure of the way energy (and also matter) is distributed in a system, and equilibrium is reached when the system reaches its "most probable" state[*].

Taking this idea that the entropy is related to the number of ways energy and matter can be distributed in a system, we now make an immediate jump to the fundamental equation of statistical mechanics (which is carved on Boltzmann's tomb[**]):

---

[*] Note that there are fluctuations about this most probable state (e.g., the distribution of black and white squares in figure 8.2), but in macroscopic systems these fluctuations are extremely small.

[**] There is a wealth of anecdotes concerning the strange things people have carved on their tombstones. Our favorite is, "I told you I was sick".

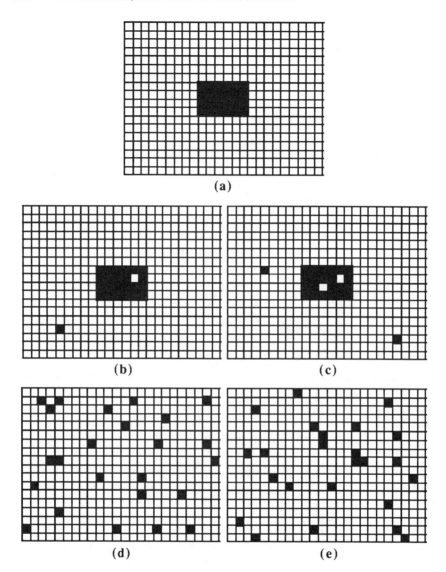

Figure 8.2 *Schematic diagrams representing an irreversible process of heat transfer (or the mixing of gases).*

$$S = k \ln W \qquad (8.4)$$

where S is the entropy, k is a constant and W is a measure of the number of arrangements available to the system. (It is now common to use the symbol Ω instead of W, but it is the latter that is carved on Boltzmann's tomb.)

This is one of those equations that just seem to come out of thin air. They cannot be derived from any other fundamental equation, but as far as we have ever been able to tell describe the way the world works. If you accept this and then go back to figure 8.2(a), it can be seen that only one arrangement corresponds to all the particles in the central box being shaded. (Shaded boxes can be switched with one another without changing the overall "hotness" or temperature of the initial area. The same applies to switching unshaded boxes with one another. We say that boxes of the same type are indistinguishable and swapping the positions of identical types of boxes does not create a new arrangement.) There are a very large number of random arrangements of the type shown in figures 8.2(d) and 8.2(e), however. If we made the system larger say about 10^{24} boxes, roughly equivalent to the number of molecules in a mole, the probability of obtaining this singular arrangement through random swaps would obviously be negligible.

We can make the same arguments concerning the falling weight. Here there is an added variable because all the molecules of the weight have some initial coherent movement downwards, so energy is not only dissipated from the weight to the table, but the direction of motion of its molecules becomes randomized after impact. To spontaneously jump in the air the weight would not only have to gather energy from the table, but there would have to be a decrease in entropy corresponding to all the molecules suddenly moving in the same direction. It doesn't happen because to obtain this type of concerted action *as a result of the random redistribution of energy*, is so improbable. That's why heating the weight doesn't raise it into the air. It has more energy, *but its distribution is not in the right form (*it stays randomly distributed).

As we have noted in passing the same arguments apply not just to energy but to particles. If figure 8.2 now represents two different gases, initially separated, then there is clearly a large entropic driving force for them to mix. Furthermore, if we go back and look at the number of conformations available to a flexible chain, discussed in the preceding chapter, we see that obtaining the fully stretched out conformation as a result of random bond rotations is inherently unlikely, because it is only one of numerous other possibilities. Even better, if we could devise a way to count the number of configurations, then we could use equation 8.4 to calculate the entropy. This is the heart of statistical mechanics, finding ways of counting the arrangements available to a system, subject to the constraints imposed by the total energy of the system and number of particles present. For many problems something called a partition function is used instead of equation 8.2, but this is simply an alternative description of the way energy and matter is distributed in a system and is just an extension of the same approach and arguments.

Some Pertinent Equations and a Discussion of Free Energy

Having provided some background to remind you of key fundamentals we will now simply extract from the subject of thermodynamics some equations we will use in this and the following chapters to describe the properties of polymers. We start with the first law of thermodynamics, which states that energy is conserved. The law can be expressed in the following form:

$$dE = dQ + dW \qquad (8.5)$$

This equation states that the difference in the thermal energy transferred to a system (Q) and the work done on that system (W) must go somewhere, and it changes what is called the internal energy (E) (or sometimes simply called the energy) of that system. The internal energy is simply the sum of the kinetic and potential energies, so that energy changes reflect the motions of its particles (KE) and/or their interactions with one another (PE).

The type of work (W) we will be considering is related to the change in dimensions of the system, so recalling the definition of work as force times distance we can write:

$$dW = f\,dl \qquad (8.6)$$

where f is the force and l is the change in dimensions of the system. This is the form of the equation we will use in describing rubber elasticity, but most thermodynamic texts deal with problems involving the work done by a piston against a constant pressure P. The two descriptions are equivalent, because if A is the area of the piston, then we can write:

$$f\,dl = \frac{f}{A}A\,dl = P\,dV \qquad (8.7)$$

where force per unit area is, of course, pressure (P) and A dl represents the change in the volume (V) of the system dV.

We can now rewrite the first law:

$$dE = dQ - P\,dV \qquad (8.8)$$

where the negative sign is introduced because in this case we are discussing work done *by* the system as the piston is moved. We combine this with the equation that comes out of the second law:

$$dS = \frac{dQ}{T} \qquad (8.9)$$

we obtain:

$$dE = T\,dS - P\,dV \qquad (8.10)$$

We will use this equation in a couple of ways. First in a direct form when we discuss rubber elasticity (Chapter 11). If we substitute fdl for −PdV, because we are now discussing work done *on* the system (chain) in order to stretch it:

$$dE = T\,dS + f\,dl \qquad (8.11)$$

or:

$$f = \left(\frac{dE}{dl}\right) - T\left(\frac{dS}{dl}\right) \qquad (8.12)$$

which says that when a force is applied to a system it can change the internal energy and the entropy. We will see that for materials such as metals the first term dominates, because the bonds are stretched directly, while in a lightly cross-linked rubber the second term dominates, because the initial response to the load is a change in the distribution of conformations and hence entropy of the system.

Less transparently, we will also use this equation as a stepping stone to the concept of free energy. Equation 8.10 can be rewritten:

$$dS = \frac{dE + P\,dV}{T} \tag{8.13}$$

It will be recalled that at equilibrium entropy is a maximum and if we can count the number of arrangements available to a system and hence obtain S, then equation 8.13 can be used to link statistical mechanics to thermodynamic properties. That is not our concern here, however, and we simply note that experimentally the variables E and V are difficult to work with as a pair. It is much easier to work with temperature (T) and volume (V), as a pair, or T and pressure (P) as a pair. This has led to the definition of two additional thermodynamic properties, the Helmholtz free energy (F) and the Gibbs free energy (G), defined by:

$$F = E - TS \tag{8.14}$$

and

$$G = (E + PV) - TS \tag{8.15}$$

where V and T are the natural variables for F, while P and T are the natural variables for G. Defining the enthalpy of the system H as:

$$H = E + PV \tag{8.16}$$

we have:

$$G = H - TS \tag{8.17}$$

It is easier to work with some variables (i.e., P,T or V,T) for one problem and others for another. In discussing mixing and equilibrium melting points we will use G, but in discussing rubber elasticity we will use F.

The formal definitions of free energy given as equations 8.14 and 8.17 do not give a feel for the meaning of free energy. One way of thinking about this quantity is to recall from our initial discussions that only a certain amount of thermal energy can be converted into work in a reversible process. The free energy is simply a measure of the maximum work that can be obtained, as might be intuitively grasped by noting that it is equal to an energy minus an entropy term. The Gibbs free energy is a measure of the capacity to do non-expansion work (at constant T and P) while the Helmholtz free energy is related to the capacity to do isothermal work. However, there is (for our purposes) an even better way of defining the free energy[*]. If we consider as our example the question of whether or not two liquids will mix, we can use as our starting point a model system similar to that shown in figure 8.2, where the two liquids are initially separated by a barrier that is then removed. The second law tells us that the two liquids will mix if the entropy change is positive, $\Delta S > 0$. This would suggest that all liquids will mix, because it should be self-evident from the sequence of figures also shown in figure 8.2 that there are more arrangements available to a randomly mixed system than one that remains separate. Its entropy will therefore be larger than the sum of the entropies of the initially separate

---

[*] Atkins, P.W. *The Elements of Physical Chemistry*, W. H. Freeman and Co., New York, (1993).

liquids (recall equation 8.4). However, we also know that oil and water do not mix. (Actually they do just a little bit, but what we are really saying is that they are phase separated; we'll explain this in chapter 9!) So, where have we gone wrong? The problem is that we have only considered a part of the entropy change that would occur if the liquids were to mix. Mixing involves a heat exchange with the surroundings, the direction of energy transfer depending upon whether the process is exothermic or endothermic. As a result of this energy change there is also a change in the entropy of the surroundings (ΔS_{surr}). For a spontaneous process to occur, (i.e., for the liquids to mix) the second law tells us that $\Delta S > 0$, but this inequality refers to the *total* entropy change ($\Delta S_{tot} > 0$), not just the entropy change in the system (ΔS_{sys}). We must therefore write:

$$\Delta S_{tot} = \Delta S_{sys} + \Delta S_{surr} \qquad (8.18)$$

and impose the condition that $\Delta S_{tot} > 0$ for mixing to occur; but how do we calculate ΔS_{surr}? (We will assume that we can construct a model to calculate ΔS_{sys} and our aim is to calculate whether or not $\Delta S_{tot} > 0$ to predict if two liquids will or will not mix.) What we would really like is to be able to confine our attention to the properties of the system alone, because we can usually construct models that at least allow us to estimate the quantities that we are trying to determine. Fortunately, if the process we are considering is reversible we can write:

$$\Delta S_{surr} = -\frac{\Delta Q}{T} = -\frac{\Delta H}{T} \qquad (8.19)$$

at constant pressure (where $\Delta Q = \Delta H$): we have put in a minus sign to indicate that we are discussing heat transferred from the system to the surroundings, the heat being the heat of mixing, a property of the system (resulting from interactions between the molecules). We can now substitute equation 8.19 into equation 8.18 and multiply through by -T to obtain:

$$- T\Delta S_{tot} = \Delta H - T\Delta S_{sys} \qquad (8.20)$$

If we now *define* the Gibbs free energy ΔG to be equal to $-T\Delta S_{tot}$:

$$\Delta G = - T\Delta S_{tot} \qquad (8.21)$$

we have:

$$\Delta G = \Delta H - T\Delta S \qquad (8.22)$$

where this equation refers only to changes within the system and has the conditions that P and T are constant. (The equivalent equation for F has V and T being constant.)

In our discussion of polymer properties we will use these equations in a couple of different ways, as a criterion for the direction change and as a criterion for equilibrium. We can say that if a process is to spontaneously occur (i.e., $\Delta S > 0$) then the change in free energy must be negative (i.e., $\Delta G < 0$, see equation 8.21). We will use this criterion to decide whether or not a polymer will dissolve in a solvent or mix with another polymer. Second, at equilibrium the entropy reaches a maximum and hence the free energy is a minimum (remember, they have the opposite sign) and no further spontaneous change is possible. In other words, once equilibrium has been reached $\Delta G = 0$. We will use this in discussing crystalline melting points.

Derivatives of the Free Energy and the Order of Phase Transitions

In the following sections we will deal with transitions between the liquid (melt) and solid state in polymer systems. As we have already mentioned in our preliminary discussion of states of matter in polymer materials, crystallization is accompanied by a discontinuous change in volume at the crystallization temperature. There are also discontinuities in the entropy and enthalpy at this point (i.e., there is a latent heat involved in the process) and transitions such as these are known as first order transitions (the condensation of a gas to form a liquid has the same characteristics). The reasoning behind this classification can now be appreciated if we consider a schematic representation of free energy as a function of temperature, as illustrated in figure 8.3. For a crystalline material (top left hand side) we can draw two lines, one representing the free energy of the melt and the other representing the free energy of the crystal.

In the crystalline state the configurational entropy of the chains is small, because they are arranged in an ordered conformation and are locked in position. Conversely, this regular close packed arrangement maximizes the attractive forces between the chains. At low temperatures (below the melting point), this latter term dominates, the overall free energy of the crystal is lower than that of the melt and is therefore the stable state. As the temperature increases this balance between the enthalpic and entropic contributions to the free energy changes. At the melting point the free energy of the crystal and melt are equal, but at higher temperatures the free energy of the liquid melt is now lower, so that this becomes the preferred or stable state. Clearly the free energy remains continuous as a function of temperature, but there is a shift from one curve to another at the melting point. It is simply a matter of recalling elementary calculus to realize that this shift will result in a discontinuity in the first derivative of the free energy with respect to temperature, because the slope of each of the curves that meet at the melting point is different. Recalling the fundamental thermodynamic relationship:

$$\left[\frac{\partial G}{\partial T}\right]_P = -S \qquad (8.23)$$

We can see that there must be an abrupt and discontinuous change in S at the melting point. The same is true of the enthalpy because:

$$\left[\frac{\partial (G/T)}{\partial (1/T)}\right]_P = H \qquad (8.24)$$

and the same holds true with respect to other variables:

$$\left[\frac{\partial G}{\partial P}\right]_T = V \qquad (8.25)$$

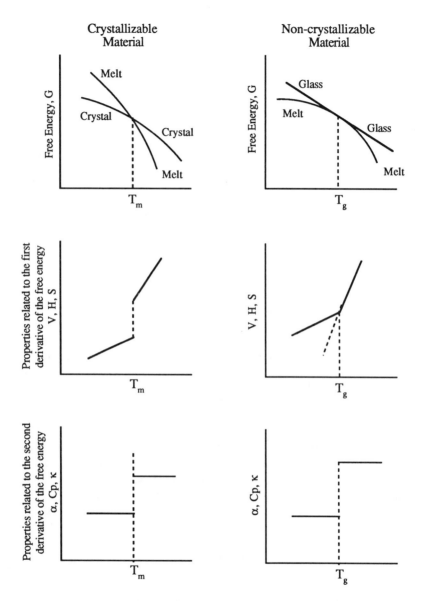

Figure 8.3   *Schematic representation of the free energy as a function of temperature. The left hand side is for a crystallizable solid and there is an abrupt change in the slope of the free energy/temperature line at the the melting temperature, T_m. This gives discontinuities in V, H and S. For a non-crystallizable solid, there is no abrupt change in first derivative quantities.*

Because the quantities S, H and V are related to the first derivative of the free energy, this type of transition is called first order. In the same way, transitions where S, H and V are continuous, but their derivatives change abruptly, as also shown on the right hand side of figure 8.3, are called second order. This is because these quantities must, of course, be related to the second derivatives of the free energy:

$$-\left[\frac{\partial^2 G}{\partial T^2}\right]_P = \left[\frac{\partial S}{\partial T}\right]_P = \frac{C_p}{T} \tag{8.26}$$

$$\left[\frac{\partial^2 G}{\partial P^2}\right]_T = \left[\frac{\partial V}{\partial P}\right]_T = -\kappa V \tag{8.27}$$

$$\frac{\partial}{\partial T}\left[\left[\frac{\partial(G/T)}{\partial(1/T)}\right]_P\right]_P = \left[\frac{\partial H}{\partial T}\right]_P = C_p \tag{8.28}$$

and:

$$\frac{\partial}{\partial T}\left[\left[\frac{\partial G}{\partial P}\right]_T\right]_P = \left[\frac{\partial V}{\partial T}\right]_P = \alpha V \tag{8.29}$$

where C_p is the heat capacity, κ is the compressibility and α is the thermal expansion coefficient. Polymer crystallization and melting can be considered true first order phase transitions; the glass transition temperature *may* be related to an underlying second order transition, but this remains an open question, as we will see. Nevertheless, it can be detected by measuring the specific heat (using differential scanning calorimetry (DSC), where there is a (more or less) abrupt change at T_g).

C. CRYSTALLIZATION AND MELTING: SOME EQUILIBRIUM CONSIDERATIONS

Having spent some time reviewing the basic fundamentals of equilibrium thermodynamics, we are going to consider in the rest of this chapter phenomena, such as crystallization and the glass transition, that are largely governed by kinetic rather than equilibrium considerations. We are not abandoning thermodynamics completely, however, as by trying to determine what would constitute the state of lowest free energy we obtain considerable insight. It's just that the chains can't get there. In effect, on cooling from the melt the system changes so that the free energy decreases as quickly as possible. This leads it into a non-equilibrium state where it is trapped by energy barriers that cannot be scaled.

Nevertheless, equilibrium considerations tell us the direction in which the system would like to move. We will first describe some of the crucial experimental observations that can be accounted for in this fashion, focusing most of our attention on aspects of crystallization. Melting will be considered in more detail later in this chapter, but we will introduce the concept of the equilibrium melting temperature here, as it is a fundamental parameter in theories of crystallization kinetics, which will be discussed in the next section.

As we have already mentioned in chapter 7, semi-crystalline polymers melt over a range of temperatures and we defined the melting point as the temperature at which the last measurable vestige of crystallinity disappears. This melting point will vary somewhat with thermal history for a given sample, but let us start by assuming that we can define some arbitrary conditions whereby we obtain a sample with a fairly well defined melting temperature, T_m. If the sample is taken above this temperature into the melt and then cooled, crystallization does not occur when T_m is reached, but at much lower temperatures (anywhere from 15°C to 50°C, depending upon the polymer).

For most polymers crystallization is also a relatively slow process and the growth of spherulites can be observed and their rate of growth often measured using an optical microscope. This is in stark contrast to the crystallization of small molecules, which occurs rapidly at temperatures just slightly below the melting point or, as it is usually put, at low degrees of undercooling. Crystallization depends upon concentration fluctuations within the melt (or solution) such that at some critical temperature these result in the formation of what is called a primary nucleus that is stable and can then grow. There is an energy barrier in all materials that prevents this happening at T_m, but clearly the formation of a stable nucleus is for some reason much more difficult in polymers.

If we are considering lamellae crystallized from solution instead of the melt, then again crystallization occurs at appreciable undercoolings from the melting or dissolution temperature obtained when held in the same solvent, but both these temperatures are lower or depressed from those observed in the absence of solvent, for reasons we will discuss later in this chapter.

The second experimental factor that we need to account for is the chain folded nature of polymer crystallization. Furthermore, the crystal thickness or fold period is a fairly well-defined quantity and varies inversely with $T_m - T_c$. This is regardless of whether we are examining melt crystallized samples or those grown from solution, providing the latter are kept in contact with the same solvent in both the crystallization and melting experiments. Although in solution both T_m and T_c are depressed, it is the difference or degree of undercooling that determines fold period. Plots of this latter quantity against $(T_m - T_c)^{-1}$ fall on the same curve for both solution and melt grown samples, getting larger as the undercooling gets smaller. This indicates that the fold period should approach that of the fully extended chains as T_c gets closer to T_m, assuming we started with a monodisperse sample (more on molecular weight effects below). Experimentally, however, a polymer will not crystallize at all if the undercoolings are too small. Nevertheless, this suggests that a polymer might prefer to crystallize in an extended chain form, but there is some barrier preventing this from occurring. In fact, if a polymer is "annealed" by heating to a temperature above its crystallization temperature, T_c, but below its melting temperature, T_m, then the crystal irreversibly refolds to give a longer fold period. This indicates that the crystalline form grown at the original T_c is not the equilibrium structure, because if it were, the fold period would be expected to change reversible as the temperature is raised and lowered between T_c and the annealing temperature.

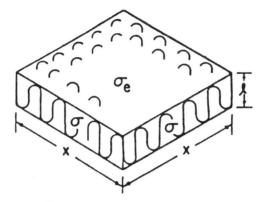

Figure 8.4 *Schematic representation of a lamellar single crystal showing dimensions. Reproduced with permission from J. D. Hoffman, G. T. Davis and J. I Lauritzen,* Treatise on Solid State Chemistry, *N. B. Hannay, Ed., Vol. 3, Chapter 7, 1976.*

This brings us to a discussion of what constitutes the equilibrium structure of a polymer crystal and the question of why the chains cannot get there. Consider as a starting point a single crystal lamella, which you no doubt recall as having the form of a thin plate with folds on the top and bottom surfaces, as illustrated in figure 8.4. Shown in this figure are the symbols σ and σ_e, representing the free energies per unit area of the fold surface and the edges of the crystal, respectively. The free energy of the crystal as a whole can now be written as the sum of the free energy of the bulk or the "inside" of the crystal and the surfaces:

$$\Delta G_{cryst} = (4 \times \ell)\, \sigma + (2\, x^2)\, \sigma_e - (x^2\, \ell)\, \Delta g \qquad (8.30)$$

$\underbrace{}$ $\underbrace{}$ $\underbrace{}$
area of   area of   volume
sides   of folds   of
of crystal   crystal

where Δg is the bulk free energy of fusion and for simplicity it has been assumed that the lateral dimensions of the crystal are the same along each face (x).

The chain segments that lie within the bulk of the crystal are usually in their minimum energy conformation and stack so as to maximize their interactions (they actually try to arrange themselves so as to minimize their overall free energy, the sum of the conformational and interactional contributions). In contrast, the segments on the fold surface are usually in some higher energy conformational arrangement (for example, in polyethylene this will be some sequence of *trans* and *gauche* conformations) and there is also some additional free energy associated with the fact that all surface segments are at an interface, so on their "inside" they contact the crystal while on their "outside" they contact the surrounding medium. The surface free energies in equation 8.30 can therefore be considered "excess" terms (i.e., defined with respect to the free energy of a segment in the bulk of the crystal) and the overall free energy written as a difference between surface and bulk terms. Accordingly, the crystal can minimize its free energy by reducing its surface area and maximizing the number of segments within the bulk. This is more explicitly demonstrated if we rewrite equation 8.30 as:

$$\Delta G_{cryst} = 2 x^2 \left[\frac{2\ell}{x} \sigma + \sigma_e \right] - x^2 \ell \Delta g \qquad (8.31)$$

Because $x \gg \ell$ (i.e., the crystal is very thin compared to its lateral dimensions) to a first approximation:

$$\Delta G_{cryst} = 2 x^2 \sigma_e - x^2 \ell \Delta g \qquad (8.32)$$

This tells us that the crystal can get to a lower free energy by becoming thicker. It follows that the minimum free energy is obtained for a crystal where the chains are fully extended. Indeed, if the crystal is annealed at a temperature higher than its crystallization temperature it *irreversibly* refolds to form a crystal which is thicker (a higher value of ℓ), as we have already mentioned.

At this point you should ask why the polymer did not initially fold with this longer fold length; well, that's where kinetics comes in, but we'll get to that shortly, after we obtain two more pieces of information from equation 8.32. The first has to do with melting and we will use the criterion for equilibrium described in the preceding section:

$$\Delta g = 0 \qquad (8.33)$$

(Note that we are using lower case letters to indicate that this is a free energy per unit volume of the crystal, not an overall free energy.)

This state of equilibrium should occur at the melting point of the (theoretically) very large perfect crystals formed from fully extended chains, where we can neglect surfaces. This we will call the equilibrium melting temperature, T_m^o, to distinguish it from the melting point of (real) crystals, T_m. Assuming Δh_f is independent of temperature then at T_m^o:

$$\Delta g = \Delta h_f - T_m^o \Delta s_f = 0 \qquad (8.34)$$

It turns out that we can measure the enthalpy of fusion, Δh_f and there are also extrapolation methods to obtain T_m^o, so we can use the equilibrium condition to express Δs_f in terms of things we know:

$$\Delta s_f = \frac{\Delta h_f}{T_m^o} \qquad (8.35)$$

If we now consider a temperature T where $\Delta g \neq 0$, then we can substitute for Δs_f to obtain:

$$\Delta g = \Delta h_f - T \Delta s_f = \Delta h_f \left(1 - \frac{T}{T_m^o} \right) \qquad (8.36)$$

Substituting into equation (8.32):

$$\Delta G_{cryst} = 2 x^2 \sigma_e - x^2 \ell \Delta h_f \left(1 - \frac{T}{T_m^o} \right) \qquad (8.37)$$

Assuming that $\Delta G_{cryst} = 0$ also applies to the state of metastable equilibrium found at the melting point $T = T_m$ we obtain:

$$2\,\sigma_e - \ell\,\Delta h_f \left(1 - \frac{T}{T_m^o}\right) = 0 \qquad (8.38)$$

or:

$$T_m = T_m^o \left(1 - \frac{2\,\sigma_e}{\ell\,\Delta h_f}\right) \qquad (8.39)$$

This equation is called "Thompson's rule", or the Thompson-Gibbs equation and tells us that the actual melting temperature of a polymer, T_m, is always less than the equilibrium melting temperature by an amount that depends inversely on the polymer crystal thickness[*].

It also explains why polymer samples melt over a range of temperatures, one of the peculiarities of polymer crystals that we mentioned in our discussion of morphology. Samples crystallized over a range of temperatures (and also those with a range of molecular weights, see below) will have a distribution of fold periods and hence melting points, because of the dependence of T_m on ℓ.

The final insight we can obtain from equation 8.32, is an expression for the minimum fold period, ℓ_{min}:

$$\ell_{min}^{*} = \frac{2\,\sigma_e}{\Delta h_f}\left[\frac{T_m^o}{T_m^o - T_c}\right] \qquad (8.40)$$

This is the shortest stable fold, corresponding to $\Delta G_{cryst} = 0$, that can be obtained at an undercooling of $\Delta T = T_m^o - T_c$, where T_c is the temperature of crystallization. Note the inverse dependence of the fold period on ΔT. For crystallization temperatures equal to T_m^o, only crystals with infinite fold periods could grow, while at temperatures just below T_m^o the shortest stable fold would correspond to the fully extended chain. The probability that as a result of random fluctuations a collection of fully extended chains could form an initial or primary nucleus is, of course, vanishingly small. (As we mentioned in Chapter 7, the probability of finding even one flexible chain in the fully extended state is minuscule.) Undercoolings have to be sufficient that as a result of random fluctuations a fold period that is stable to further growth is obtained. This fold period must be longer than the minimum given by the equilibrium condition expressed in equation 8.40 (which being in equilibrium would remelt as fast as it would crystallize). We will come back to this point shortly in our discussion of crystallization kinetics.

---

[*] Because there are experimental methods for measuring ℓ, this equation can also be used to determine T_m^o and $2\sigma_e$ by plotting T_m as a function of ℓ using data obtained from samples crystallized at different temperatures.

D. THE KINETICS OF POLYMER CRYSTALLIZATION

If the degree of crystallinity of a polymer sample is measured as a function of time during crystallization from the melt at a constant T_c, then a sigmoidal shaped curve is obtained, as illustrated in figure 8.5. This shape remains the same as T_c is varied, but is shifted to shorter times as the undercooling is decreased (providing that we are considering a fairly narrow temperature range so T_c does not approach the glass transition). It can be seen that there is an initial induction period (where primary nuclei are formed) followed by a period of fast spherulite growth. At a point where the sperulites start to impinge, the rate of crystallization slows down. Even after the volume of the sample has become completely filled with spherulites, crystallization continues, albeit very slowly. This latter process is called *secondary crystallization*, while the initial stage of radial growth of the spherulites is called *primary crystallization*. These two phenomena should not be confused with primary and secondary nucleation, which we will discuss below.

The initial stages of crystallization, where the spherulites are growing independently, has often been interpreted or modeled in terms of the Avrami equation

$$\phi_c(t) = 1 - \exp{-(kt)^n} \tag{8.41}$$

originally developed on the basis of geometrical arguments concerning the statistics of how a sample volume becomes filled by growing objects of different shape. The quantity of $\phi_c(t)$ is a measure of the overall degree of crystallinity and

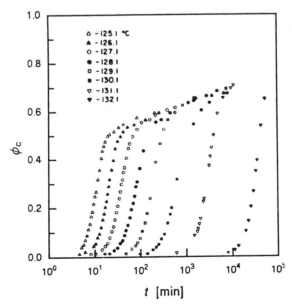

Figure 8.5 *A plot of the degree of crystallinity of polyethylene, $\phi_c(t)$, as a function of time (log scale). Reproduced with permission from the data of E. Ergoz, J. G. Fatou and L. Mandelkern,* Macromolecules, 5, 147 (1972).

the Avrami exponent, n, is related to the nature of crystal growth and should be a whole number. When applied to polymer crystallization it seldom is, however, and the Avrami equation is now usually regarded as no more than a useful empirical way of representing the data.

Over the last thirty years the dominant theory of polymer crystallization kinetics is that due originally to Lauritzen and Hoffmann, modified in various ways over the years and discussed in a detailed review by Hoffman et al.* This approach describes the rate of radial or lateral crystal growth (i.e., before the spherulites impinge), as a function of the degree of supercooling, $\Delta T = (T_m^o - T_c)$.** A full discussion of this theory is beyond the scope of this introductory overview, but essentially considers secondary nucleation. We have mentioned primary nucleation already, essentially noting that below the melting point there is a temperature dependent critical nucleus size and crystals that are larger than this are stable to further growth. In a pure polymer melt this process is referred to as homogeneous nucleation. In practice polymer melts are seldom pure and primary nucleation is heterogeneous, starting on the surface of foreign particles. Once a stable primary nucleus is formed, however, it becomes much easier to add new units, as there is now there is a surface to "sit on", and this process is called secondary nucleation. Because adding chains to an existing nucleus is easier, it occurs more rapidly than primary crystallization and is the rate limiting step that determines the crystal growth rate, so this is where attention has focused. Before discussing secondary nucleation in polymers in a little more detail, however, it is important to note that the formation of nuclei depends upon two terms, an energy of activation for nucleus formation and a transportation term accounting for the movement of the molecules from the melt or solution to the nucleus site. This latter term, in turn, depends upon viscosity (amongst other things). As the temperature is lowered nuclei form more easily, but in polymer melts, in particular, viscosity can also increase significantly with decreasing temperature. Hence, at low temperatures nucleation can decrease dramatically. Obviously there must be a temperature at which nuclei formation, both primary and secondary, is a maximum and this is illustrated schematically in figure 8.6. This figure demonstrates that at lower crystallization temperatures a large number of small crystals are formed (primary nucleation > secondary nucleation). Crystallization at higher temperature will lead to a smaller number of larger crystals, because the rate of primary nucleation is slow, but the formation of secondary nuclei follows at a much faster rate.

The kinetic theory of crystallization developed by Hoffman et al. results in an expression for the initial thickness of the crystal that is given by:

---

* J. D. Hoffman, T. Davis and J. I. Lauritzen, Jr., *Treatise on Solid State Chemistry*, Vol. 3, p. 497. N. B. Hannay, Editor, Plenum Press, New York, (1976).
** Experimentally, overall crystallinity, the type of data used in the Avrami equation, is measured by techniques such as dilatometry, which are based on the difference in density of the crystalline and amorphous domains. Radial growth rates can be measured directly by optical microscopy.

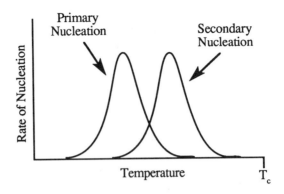

Figure 8.6 *Primary and secondary nucleation rates as a function of crystallization temperature T_c.*

$$\ell_g^* = \frac{2\,\sigma_e}{\Delta h_f}\left[\frac{T_m^o}{T_m^o - T_m}\right] + \delta\ell \qquad (8.42)$$

which is the same form as equation 8.40 but with an added term, $\delta\ell$, to account for the fact that the initial fold period must be longer than the minimum (equation 8.40) in order to be stable to further growth. The $\delta\ell$ term is usually of the order of 10–40 Å. As mentioned above, this equation is obtained by assuming a secondary nucleation mechanism and this can take various forms, described in terms of crystallization regimes. In regime I a stem is laid down and the rest of the row quickly fills up, as illustrated in figure 8.7, while in regime II new secondary nuclei are formed before rows are completed (also shown in figure 8.7). Regime I occurs at low supercoolings and regime II at higher supercoolings (see below). There's also a regime III, where prolific multiple nucleation occurs, but we'll ignore that here. The growth rate in these regimes is described by a general equation of the form:

$$G = G_o \exp(-\Delta E/kT_c)\,\exp(-\Delta F^*/kT_c) \qquad (8.43)$$

where T_c is the crystallization temperature and G_0 is a constant. As in the formation of primary nuclei, there are two terms, one of which has the quantity ΔE, representing the activation energy for transporting crystallizing units from the liquid to the crystal face, while ΔF^* is the free energy of formation of the nucleus. The growth rate obviously depends on the balance between the transport or diffusion term and the nucleation term. The crucial points to keep in mind are:

1) ℓ_g^* is kinetic in origin and represents the thickness that allows the growing crystal to be stable.
2) Crystals with $\ell > \ell_g^*$ also form, *but reach the stable region more slowly.*
3) Crystals with $\ell_g^* = \ell_{min}$ cannot reach the stable region at all.

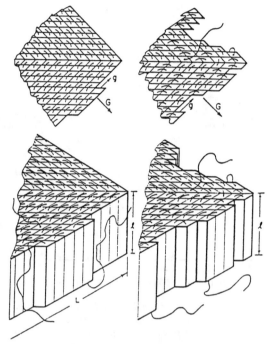

Figure 8.7 *Schematic diagrams representing crystallization regime I (left) and II (right). In regime I a single nucleus (chain stem) forms on the surface and rapidly completes a layer. In regime II new layers are formed before previous layers are complete. Reproduced with permission from J. D. Hoffman et al., cited previously.*

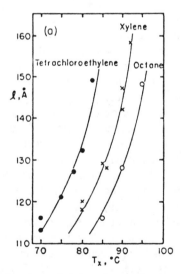

Figure 8.8 *Fold length plotted as a function of crystallization temperature for polyethylene in various solvents. The dissolution or melting temperature of polyethylene in each of these solutions is different (see later) and the data shows the expected dependence of ℓ upon ΔT. Reproduced with permission from J. D. Hoffman et al., cited previously.*

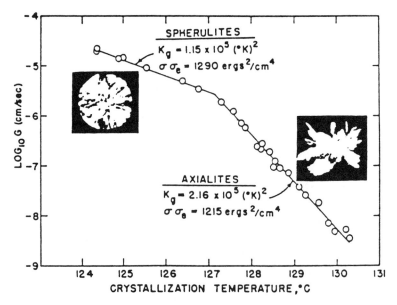

Figure 8.9 *The crystal growth rate in melt crystallized polyethylene as a function of crystallization temperature. The change in slope from regime I to regime II is accompanied by a change in morphology (inserts). Reproduced with permission from J. D. Hoffman et al., cited previously.*

This theory accounts for the thin character of polymer crystals. A stable crystal with a fold period ℓ_g^* is formed and this grows laterally. There is no crystallization on the high energy fold surface. It also predicts that the fold period should be proportional to $K_g/\Delta T$ (where $\Delta T = T_m^o - T_c$), a relationship we mentioned in the preceding section that is good up to high supercoolings, as shown in figure 8.8. The K_g term takes different forms depending upon the crystallization regime being considered. In many ways, this theory appears to work very well. A plot of the log of the growth rate vs. crystallization temperature should be linear with a slope that depends upon K_g and hence the crystallization regime. Figure 8.9 shows that this is indeed so.

In spite of this success, the Lauritzen/Hoffmann theory and its modifications have been under increasing critical scrutiny in recent years. Furthermore, it essentially does not treat phenomena such as crystal thickening and molecular fractionation. Although we have not mentioned this latter phenomenon, it is important and has become the basis for a new separation technique (Temperature Rising Elution Fractionation, or TREF). When a polymer is crystallized by slow cooling from the melt, the low molecular weight component crystallizes at lower temperatures, in separate subsidiary lamellae or at the boundaries of the spherulites formed by the higher molecular weight material. Similarly, those chains containing defects, such as branches, also crystallize at lower temperatures. We won't go into the details of these segregation effects any further, as it takes us outside the boundaries of the broad-brush overview we are trying to give. Nevertheless, it is an important effect and you should keep it in mind as one of the fundamental characteristics of polymer crystallization.

E. THE CRYSTALLINE MELTING TEMPERATURE

Characteristics of the Crystalline Melting Point

Melting is a familiar phenomenon to most people, who usually associate it simply with the conversion of a solid to a liquid by heating. We must be far more precise in our definition, however. For example, a polymer such as atactic polystyrene is a fairly rigid solid at room temperature, but as the temperature is raised above 100°C it first becomes "leathery", "tacky" and then liquid-like. It would therefore, seem to "melt" in the everyday general sense of the word. But, there is no sharp transition from solid-like to liquid-like properties and instead changes occur over a fairly broad range of temperature. When we use the word "melting" we will mean something far more specific and well-defined than this; the transition from an ordered crystalline phase to a disordered liquid phase, usually at a well-defined temperature. One characteristic of this type of transition is that the sharp rings observed in an x-ray diffraction powder pattern disappear at temperatures above the melting point, where all that remains is the diffuse halo characteristic of amorphous materials. Atactic polystyrene cannot "melt" in this sense, because it never crystallizes. (It becomes more "liquid-like" in its properties at higher temperatures for reasons we will discuss in the following section.)

There are other characteristics of the change from crystalline order to liquid-like disorder that are a consequence of this being a first order transition. As we discussed earlier, such transitions are accompanied by a discontinuity in quantities such as volume and enthalpy. This allows for various experimental measurements of melting. An instrument called a dilatometer, for example, can be used to measure the volume as a function of temperature. Or, a differential scanning calorimeter (DSC) can be used to measure and plot the heat supplied to a sample, again, as a function of temperature. Such thermograms, as they are called, show a peak associated with the latent heat of fusion at the melting point, as illustrated in figure 8.10. We will not go into this type of experiment any further, but simply refer back to specific volume (i.e., volume per unit mass) measurements to make our points. This data was presented in Chapter 7 (figure 7.19) in our discussion of the semi-crystalline character of polymer materials. The key point is that for low molecular weight materials the transition is usually sharp and well-defined. For polymers there is a not quite so well-defined transition that is accompanied by an obvious melting of some portion of the sample at lower temperatures; i.e., the melting range is broad. This breadth must obviously be associated with the distribution of the size and degree of perfection of the polymer crystals, which, in turn, must depend upon the regularity of the polymer microstructure (e.g., whether or not there is a small degree of branching, etc.), the molecular weight and molecular weight distribution, and so on. As discussed in the preceding section on crystallization, the degree of crystallinity and the size and perfection of the crystalline domains are also significantly affected by the rate at which a sample is cooled from the melt.

Because the range over which a polymer melts is often fairly broad, it is usual to take the temperature at which the last crystals disappear as the melting

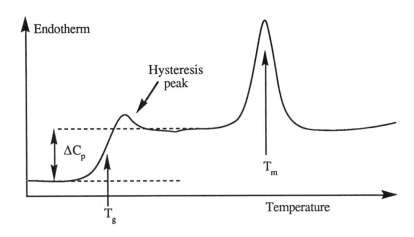

Figure 8.10 *Schematic representation of a DSC plot showing a change in the specific heat (ΔC_p) at the glass transition temperature (T_g) and an endothermic peak at the melting temperature (T_m).*

point. In dilatometer experiments this would be defined by the point where the almost vertical part of the specific volume vs. temperature curve, illustrated earlier in figure 7.19, sharply changes direction and becomes a gently sloping straight line. If we were to now repeat the experiment with samples of a polymer crystallized at different temperatures, however, we would find that the experimentally determined melting point would vary! The higher the temperature of crystallization, the higher the melting point of the polymer crystals that are obtained. The reason for this is simple and we have already explored it in the preceding two sections. Polymers crystallize in the form of thin "platelets" or lamellae (these are usually associated with solution-grown crystals, but we can think of the "arms" of the spherulites formed from the melt as consisting of a form of such lamellae). The forces of attraction between the chains in the well-ordered depths of the crystal are greater than at the surface, so that thicker crystals (higher "bulk" to surface ratio) have higher melting points. Crystalline domains grown at higher temperatures* are thicker (have a longer "fold period") and therefore also melt at a higher temperature. If crystallization occurs by simply letting a polymer cool from the melt, rather than holding it at a specific temperature, crystallization will obviously occur over a range of temperatures, resulting in a material with crystalline domains that has a distribution of fold periods and hence melting temperatures.

For practical purposes the fact that the melting temperature of a particular sample of a polymer varies with thermal history is usually not that critical. We can talk about polyethylene having a melting point of about 135°C, for example, and this is usually good enough. If we are concerned with fundamental studies, however, we need something more precise, the equilibrium melting temperature.

---

* Or smaller degrees of undercooling defined as T_m - T_c, where T_m is the melting temperature and T_c is the crystallization temperature.

This can be defined as the temperature at which an essentially extended chain crystal is in equilibrium with its melt. Hoffman and Weeks[*] demonstrated how this could be measured by determining the melting temperature (T_m) of samples crystallized at various temperatures (T_c). A plot of T_m vs. T_c is made and extrapolated to the $T_m = T_c$ line (there can be experimental problems here). The intersection point gives the equilibrium melting temperature. We are getting beyond the scope of what is appropriate for this broad overview of the subject, however, so those curious readers who wish to understand and explore this part of the subject more completely should go to the original literature.

Factors That Affect the Crystalline Melting Temperature

Having considered some fundamental aspects of polymer melting, we can now turn our attention to some very interesting questions. The first of these is basic. Why does one polymer, say Kevlar®, have a very high (for a polymer) melting temperature, about 370°C, while the T_m of polyethylene is only of the order of 135°C or so? This is the type of question that not many students think about, tending to accept that certain materials have a high T_m as a matter of course or divine imposition. Strength is another property that is often considered in this same manner. Because of the experiences of everyday life most people usually accept without question that steel is stronger than garbage-bag polyethylene. However, "it ain't necessarily so," in the words of the classic old song, as we will see later when we consider mechanical properties. There are things we can do to "pump-up" the tensile strength (and modulus) of polyethylene.

Although we cannot "pump-up" the melting point of a polymer with a given chemical structure to any appreciable extent, can we design a new polymer to have a high T_m? Also, how is the T_m affected by copolymerization, molecular weight, the presence of diluents, etc.? Most of these questions can be answered by using some simple thermodynamics and statistical mechanics, which we will consider here, but in a largely qualitative manner, pointing to more advanced texts or original papers for a fuller treatment.

The Effect of Chemical Structure

The simplest way to consider the effect of structure on the melting temperature is to start by writing an equation for the change in free energy upon melting, which we will call ΔG_f:

$$\Delta G_f = \Delta H_f - T \Delta S_f \qquad (8.44)$$

This is immediately discouraging to many students who even after our brilliant lectures can remain uncomfortable for some time with basic thermodynamics. Relax and keep in mind that we use the concept of free energy to explore just a couple of basic things. First, we can use it to decide whether or not a particular process is allowed by thermodynamics (e.g., Will *this* polymer dissolve in *that* solvent?). One necessary condition for this is that the free energy

---

[*] J. D. Hoffman and J. J. Weeks, *J. Res. Natl. Bur. Stand.*, **66A**, 13 (1962).

change should be negative. The second way we will use the free energy applies to the problem we now have before us. Once equilibrium has been reached, any further change in free energy must be zero, by definition. In other words, if kinetic factors would allow us to obtain polymer crystals in equilibrium with a melt, then there would be as many chains crystallizing as there are melting in a given time period, so there would be no overall change in free energy. We therefore have:

$$\Delta G_f = 0 \qquad (8.45)$$

and it immediately follows that:

$$T_m = \frac{\Delta H_f}{\Delta S_f} \qquad (8.46)$$

Those readers with good memories will recall that we have already been through this argument (we used lower case symbols before, Δg, Δh_f etc., as we defined these quantities in terms of a unit volume of crystalline material and we were separating factors involving the bulk of the crystal from those involving the surface). Our purpose there was to obtain equations relating the fold period to the melting point. Here we will take a different path. Now that we have simply established that the melting temperature is related to the ratio of the enthalpy change to the entropy change upon melting, can we obtain some insight into why one polymer would have a higher melting temperature than another? The answer is yes, but this insight remains largely qualitative. Nevertheless, it is very useful, as in any approach that provides fundamental understanding.

Equation 8.46, as it stands, is no use at all, however. We must relate the enthalpy and entropy changes to aspects of molecular structure in order to obtain insight. We'll start with the enthalpy. We will assume that this is simply related to the force of attraction between the chains. Accordingly, ΔH_m must be related to *the difference in* the forces of attraction between polymers packed in a regular array in the crystalline domains and the forces between those same chains when randomly intertwined in the melt. Obviously, regular packing in the crystalline domains allows attractive forces to be maximized. One might therefore conclude that those polymers with strong intermolecular forces would have a higher melting point than those where the attractions are weaker. For example, comparing the melting temperatures of polyethylene and nylon 6:

(where we use the symbol ≈ to mean "about"). The forces of attraction between the simple hydrocarbon segments found in polyethylene are weak dispersion forces (≈ 0.2 kcal/mole). In contrast, nylon 6 contains the amide group, which forms hydrogen bonds that are an order of magnitude stronger (≈ 5 kcal/mole). This is certainly a factor in the large melting point difference between these two polymers, but we have to be careful here. ΔH_f is the *change* in enthalpy on going from the crystal to the melt. All of the amide groups are hydrogen bonded in the crystalline phase, but *these hydrogen bonds are not all broken in the melt.* Although the melt is dynamic (i.e., the segments are always moving about), hydrogen bonds are still present. If we could focus on an individual amide group, we would see it form a hydrogen bond with another such group, hang around for a while, "break" and become free for a while, form a new hydrogen bond with some other group, and so on. At any instant of time there is an equilibrium distribution of hydrogen bonds. The quantity ΔH_m is related to the difference in the number of hydrogen bonds in the crystal and the *average number* in the melt. This is not as big as you might at first think, so although intermolecular attractions play a role, they are by no means the only factor or even necessarily the dominant factor. Nevertheless, *other factors being equal,* one would conclude that in general polymers containing polar functional groups would have higher melting points than non-polar polymers, and polymers that hydrogen bond or that attract each other because of the presence of ionic species would have the highest melting points of all (refer back to Chapter 7 if you need to remind yourself about intermolecular forces).

Another factor that can affect the enthalpic term is the presence of bulky groups, as these may prevent close packing of the chains. Obviously, this will depend upon the details of the structure and we can make no general observations. Bulky groups can also act in the opposite sense and *raise* the melting temperature, however, through their effect on the conformations allowed to the chain. This brings us to the entropy term.

In the crystalline state a polymer chain is in a single ordered conformation (e.g., the planar zig-zag of polyethylene). Upon melting the chain escapes the cage of the crystalline lattice and now has the freedom to sample all the conformations available to it. In other words, through bond rotations it will constantly change its shape from one random coil type conformation to another. Recalling that the entropy is given by:

$$S = k \ln \Omega \qquad (8.47)$$

it should be immediately obvious that for a flexible polymer chain there is a large change in entropy on going from the crystalline phase to the melt. If we confine our attention to the entropy associated with chain conformations, then in polymer crystals, Ω, the number of configurations available to the chains is small, as large portions of the chains (those within the bulk of the crystal) are restricted to the same conformation. In contrast, a flexible polymer has a large number of configurations available to it in the melt, so that the entropy *change*:

$$\Delta S_f = k (\ln \Omega_{melt} - \ln \Omega_{cryst}) \qquad (8.48)$$

is large. If the polymer chain is stiff, because of the presence of bulky groups in the backbone, for example, then the change in entropy associated with melting will be less than for a more flexible polymer and this will tend to make the melting temperature higher ($T_m = \Delta H_f / \Delta S_f$; if ΔS_f is smaller, T_m is larger).

In general, oxygen groups in the backbone of the polymer (ether and ester functional groups) tend to make it more flexible, while benzene rings make it "stiffer". This can be illustrated by examining the melting temperatures of the following three polymers, where the intermolecular forces are roughly the same (dispersion, weak polar):

	T_m
Poly(ethylene oxide)	$\approx 65°C$
Polyethylene	$\approx 135°C$
Poly(p-xylene)	$\approx 400°C$

The presence of bulky groups attached to the side chain can also raise the melting temperature, as bond rotations are inhibited by steric hindrance and the number of configurations available to the chain becomes limited:

	T_m
Polyethylene	$\approx 135°C$
Isotactic Polypropylene	$\approx 170°C$
Isotactic Polystyrene	$\approx 225°C$

To summarize, those polymers that have stiff chains and the type of functional group that results in strong interchain attractions have the highest melting temperatures, while simple hydrocarbon polymers, such as polyethylene, where the chain is relatively flexible and interchain forces are relatively weak, have low melting temperatures.

The Effect of Diluents, Copolymerization and Molecular Weight

We will not discuss the effect of diluents, etc. in any depth here, for two reasons. First, the details of the analysis start to take us beyond the scope of what we can cover in this introductory text. Second, a proper analysis requires that we first introduce some solution thermodynamics. Accordingly, we will confine our discussion to some simple qualitative observations and arguments.

It is quite easy to obtain a qualitative understanding of the effect of diluents on the basis of very simple arguments of the type introduced in the preceding section. The melting temperature is inversely related to the entropy change on going from the crystalline state to the melt for a pure polymer. If there is a diluent present, say a good solvent*, then the liquid state is a solution, rather than a melt. A solution of polymer and solvent has a larger entropy than a pure polymer, as we argued qualitatively in the introduction to this chapter and will demonstrate more explicitly later. Accordingly, in the presence of a diluent the melting point is lowered.

As might be expected, the *random* inclusion of comonomer units into a polymer chain also acts to reduce and broaden the crystalline melting point. Indeed, if the concentration of comonomer units is high enough, then the sample will not crystallize at all. For example, in random ethylene-co-propylene copolymers there are no detectable crystalline regions of ethylene units once the propylene content reaches (roughly) 20-25%. Even at small propylene contents the degree of crystallinity and the crystalline melting point are significantly reduced. Again the reason is simple. Propylene units cannot fit into the polyethylene crystalline lattice and are thus excluded from the crystalline domains. Only (fairly) long sequences of ethylene units can crystallize and as the propylene content increases the number of such sequences decreases and at some comonomer content these are insufficient for crystallization to occur. Also, even at low comonomer concentrations there is a distribution of sequence lengths, resulting in a distribution of crystalline fold periods and hence broad melting temperatures. This is for random copolymers, of course. Block copolymers of crystallizable units will behave differently, with the degree of crystallinity depending upon the molecular weight of the blocks, amongst other things.

The effect of molecular weight can also be considered in the same way as random copolymerization. This is because the end units on a chain are chemically different. They are usually bulkier than the repeating segments and therefore excluded from the lattice. Obviously, as the chain length (molecular weight) of the polymer is increased the number of end-groups is proportionally

---

* We'll discuss what is meant by a "good" solvent later; for now just think of it as one which readily dissolves the polymer.

decreased and the melting temperature increases. This is not a linear increase, however, and the melting point approaches an asymptotic limit at high molecular weights.

Flory[*] has derived equations describing the depression of melting point due to these factors. On the basis of the qualitative arguments presented above it should not be too surprising that the equations have roughly the same form, and to a first approximation the melting point depression is proportional to the amount of "impurity" present, whether that "impurity" be an end-group, a comonomer unit or a diluent. If we again let T_m be the melting temperature of the sample under consideration and T_m^o be the melting temperature of a the pure, infinite chain length polymer, then for a high molecular weight polymer in the presence of diluents:

$$\left[\frac{1}{T_m} - \frac{1}{T_m^o} \right] \propto \left(\Phi_s - \Phi_s^2 \chi \right) \tag{8.49}$$

where Φ_s is the volume fraction of diluent or solvent that is present and χ is a parameter that is a measure of the interactions between the polymer and this solvent (see next chapter).

For a random copolymer where the mole fraction of the crystallizable component is X_2;

$$\left[\frac{1}{T_m} - \frac{1}{T_m^o} \right] \propto - \ln X_2 \tag{8.50}$$

If the end-groups are treated as an impurity, then it can be shown that for polymers with the most probable molecular weight distribution;

$$\left[\frac{1}{T_m} - \frac{1}{T_m^o} \right] \propto \frac{1}{\overline{M}_n} \tag{8.51}$$

where $\overline{M}_n$ is the number average degree of polymerization.

F. THE GLASS TRANSITION TEMPERATURE

Characteristics of the Glass Transition Temperature

If you were to stop a person on the street and ask them to explain the nature of glass, they would most likely mumble something along the lines of it being an optically transparent material that you stick in windows (if they bothered to answer you politely). Just as in our use of the word melting, we are going to be considerably more specific when we say a material is glassy. We will mean that the material possesses two fundamental characteristics of the glassy state; first, that there is no long range order, only the disorder characteristic of the liquid state; second, unlike the liquid state, where the molecules are moving around at

---

[*] P. J. Flory, *Principles of Polymer Chemistry*, Cornell University Press (1953).

random and can access all parts of its contained volume, in the glassy state the positions of the molecules are essentially frozen, although they still vibrate about their mean positions and there can be other types of local motions. These characteristics confer various physical properties. Because glassy materials are homogeneous they can be optically clear, as long as they don't contain functional groups that absorb visible light (note that coal, for example, is a glassy solid). Glassy solids are also relatively rigid and brittle, and we will discuss this when we get to mechanical properties.

Just as a crystalline or semi-crystalline material can be heated and become liquid-like in its behavior, so can a glass. There are distinct differences in the nature of the transition to the liquid state, however. For a crystalline material the transition is usually fairly sharp, occurring at a well-defined temperature and is accompanied by an abrupt change in volume and a latent heat. In other words, there is a discontinuity in volume and enthalpy (and also entropy) at T_m, or as we should now be able to say without too much discomfort, the transition is first order.

Upon heating a glassy solid, there is no such sharp transition in these properties. It just appears that the material "softens" in a certain temperature range until it eventually appears to be a liquid. It is a "thick", "tacky" or, more precisely, highly viscous liquid at first, but flows more easily (i.e., becomes less viscous) as the temperature is raised. If we were to keep the pressure constant and measure the volume of a sample as a function of temperature, however, we would obtain a plot similar to that shown in figure 8.11. Although there is no abrupt change in volume at any particular temperature, there is a change in the

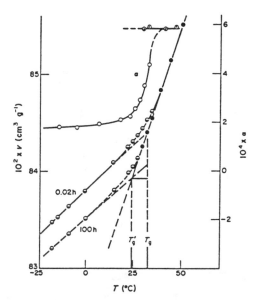

Figure 8.11 *Plots of specific volume and the coefficient of thermal expansion vs. temperature for poly(vinyl acetate). The specific volume plots show the effect of cooling rate on the measured T_g. Reproduced with permission from A. J. Kovacs, J. Polym. Sci., 30, 131 (1958).*

slope of the volume-temperature plot. The point where this occurs is what we define to be the glass transition temperature, T_g. Of course, if the slope of the volume/temperature line changes, there must be a change in the coefficient of thermal expansion (α) because:

$$\left[\frac{\partial V}{\partial T}\right]_P = \alpha V \qquad (8.52)$$

And indeed there is, as also illustrated in figure 8.11. Similarly, if we could plot enthalpy as a function of temperature we would again see a change in slope, corresponding to a change in specific heat:

$$\left[\frac{\partial H}{\partial T}\right]_P = C_p \qquad (8.53)$$

and the apparent discontinuity in C_p is illustrated in figure 8.12. (The DSC thermogram shown earlier as figure 8.9 also shows a similar discontinuity in C_p corresponding to a T_g.) If you look back in the text to our discussion of transitions you might immediately say "Ah-Ah! It's a second order transition!" But you could well be wrong. The problem is that the transition is not sharp and it can move 5 or 6°C, depending on the rate at which you do the experiment (check figure 8.11). We'll come back to this shortly, after briefly discussing what types of materials form glassy solids, and how and where polymers fit into all this.

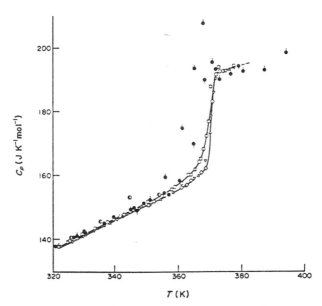

Figure 8.12 Plots of C_p vs T for polystyrene showing the discontinuity at T_g. The different types of data points represent different samples. Reproduced with permission from S. S. Chang, J. Polym. Sci., Polym. Symp., **71**, 59 (1984).

It is now thought that most materials can be obtained as a glass, providing that we can figure out a way to cool them quickly enough from the melt or condense them from a gaseous state without allowing them to crystallize. As we have seen, crystallization is governed by kinetic factors. For some materials, such as window glass, crystallization is so slow that even slow cooling from the melt results in a glassy solid. Other materials, such as metals, require extraordinary efforts in order to obtain them in the glassy state.

Because of the tangled up nature of polymer chains it is usually relatively easy to "quench" a crystallizable polymer from the melt to temperatures below the T_g^* before crystallization can start. (Remember, below the T_g the chains cannot move relative to one another and so cannot crystallize.) Even very regular polymers never crystallize completely, so that there are always amorphous regions that are either flexible (rubber-like) or rigid (glass-like), depending on whether they are above or below their T_g. This is a crucial factor in the mechanical properties of polymers, as we will see later. In addition, and in contrast to low molecular weight materials, there are numerous polymers that can *never* crystallize, because of their irregular structure (e.g., atactic polystyrene). Here, as we mentioned in our discussion of "states of matter" (Chapter 7), the glassy state is not one of metastable equilibrium in the sense that it is the result of the molecules of the material losing transitional motion before they can crystallize, but the only state allowed to the polymer at this temperature. This brings us to theories of the glass transition.

Theories of the Glass Transition Temperature

We will start this section by mentioning the famous Kauzmann paradox[**], because this sets the stage for a brief discussion of the two major approaches that have been used to explain the T_g. Kauzmann plotted values of the equilibrium thermodynamic quantities of liquids (volume, enthalpy, entropy) as a function of temperature. As we have seen, these quantities change slope at the T_g, as illustrated in figures 8.11 and 8.12 above. Kauzmann essentially investigated what would happen if the T_g did not exist by extrapolating the lines describing the properties of the liquid state to low temperatures. He found "a very startling result", the values of these properties became *less than those of the crystalline state* at temperatures above 0°K. Now, once you think about it you quickly realize that this is unreasonable and more than likely impossible. If we simply consider the entropy associated with the random arrangements and conformations of polymer chains in the liquid state, this cannot be less than that associated with the regular or ordered arrangement of chains, each having the same conformation, in the crystalline state. Similarly, how can the volume of randomly intertwined chains be less than that obtained by the usual regular close-packing in the crystal? (There are some unusual "loose" packed crystals, but these are exceptions.) The T_g "prevents" these entropy and other "catastrophes", but the question that so far remains unanswered is whether this transition is

---

[*] Polyethylene presents a particular problem because it crystallizes so quickly, but through heroic efforts even this polymer has been obtained in the glassy state.
[**] W. Kauzmann, *Chem. Rev.*, **43**, 219 (1948).

thermodynamic in origin, a manifestation of an underlying true second order transition, or whether it is purely kinetic.

There is no doubt that the quantity observed experimentally is essentially governed by kinetic factors, which can be appreciated by recalling how the rate of cooling affects the measured value of the T_g. Nevertheless, the plots of thermal expansion, α, and specific heat C_p, suggest there could well be an "underlying" (i.e., at temperatures below the observed T_g) thermodynamic transition, which can never by experimentally observed because of the intervention of kinetics. On the other hand, the observed transitions are broad, where they should be sharp. Then again, the experimentally observed crystalline melting transitions of polymers are also broad, although there is no doubt that this is a true first order transition! About all we can do in this simple introductory treatment is to note that this question remains unresolved, there are various theories and a good starting place for the interested student is the excellent review by McKenna*. We will simply observe that there is one factor that comes into both the equilibrium thermodynamic approach and various kinetic theories in one form or another, and that is *free volume*. We will discuss aspects of this in some detail because it leads us to concepts that are central to our later discussion of mechanical and rheological properties.

Free volume is not an easy quantity to define quantitatively, but it is a concept that is readily grasped. First, we must emphasize that free volume is not the same as unoccupied or empty volume. If we consider the simplest model of a solid amorphous material, hard spheres randomly packed, as illustrated in figure 8.13, then the total volume consists of the volume occupied by the spheres themselves, together with the volume associated with the inevitable gaps between the balls. However, unlike balls sitting in a pot at ambient temperature, molecules have considerable thermal motion. For a material in the glassy state

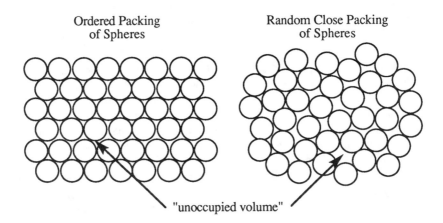

Figure 8.13 *Two dimensional representation of the packing of spheres.*

* G. B. McKenna, "Glass Formation and Glassy Behavior" in *Comprehensive Polymer Sci. Vol. 2. Polymer Props*, Edited by C. Booth and C. Price, Pergamon Press, Oxford, 1989.

we can think of the motion as the balls oscillating in a cage of their neighbors. These oscillations create some "free volume" over and above the empty space characteristic of random close packing. Clearly, this free volume increases with temperature (as the amplitude of the oscillations increases) and must be related in some manner to the coefficient of thermal expansion of the material (i.e., to the bulk expansion of the material with temperature). The free volume in the system is not equally shared between all the molecules, but fluctuates, such that at some instant of time one molecule might be trapped in a local close-packed cage of its neighbors, while another has sufficient "free volume" available to it that as a result of its random oscillations and collisions with other molecules it bounces into a new position. These possibilities are illustrated in figure 8.14. Thus, if we take a material that is a glass at some temperature and heat it, the glass transition is the point at which there is sufficient free volume to allow molecules to move (i.e., change their positions) relative to one another. Similarly, a material that is initially a liquid forms a glass upon cooling to a temperature at which the free volume has dropped to a level where the molecules can no longer move relative to one another (but still oscillate or vibrate around a mean position). This idea, that the T_g corresponds to the point where the free volume falls below a critical value was suggested by Fox and Flory[*].

There are various free volume theories of the glass transition (see the review article by McKenna), but we will briefly touch only on that due to Cohen and Turnbull, later extended by Cohen and Grest, as this allows a connection to the important Doolittle and WLF equations, the latter of which will come into our discussion of viscoelastic properties in Chapter 11. A general discussion of this model is given in Zallen's excellent introductory treatment[**] and it essentially assumes that the motion of a molecule relative to its neighbors only occurs when it acquires sufficient free volume. Using this assumption, the viscosity was related to the probability of a molecule having a free volume greater than a certain critical value and statistical mechanical arguments were used to describe the distribution of free volume in the system. We will not give the equation for the viscosity that was obtained by this treatment, but simply note that it is similar in

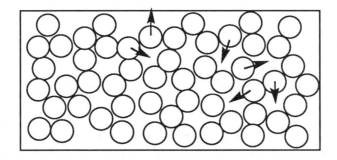

Figure 8.14 Schematic visualization of free volume allowing local translational motion.

---

[*] T. G. Fox and P. J. Flory, *J. Appl. Phys.*, 21, 581 (1950) *J. Phys Chem.*, **55**, 221 (1951) and *J. Polym Sci.*, **14**, 315 (1954).
[**] R. Zallen, *The Physics of Amorphous Solids*, John Wiley & Sons, New York (1983).

form to the empirical Doolittle equation, which was found to fit viscosity data for a wide range of systems and can be written:

$$\eta = A \exp B\left(\frac{V - V_f}{V_f}\right) \tag{8.54}$$

where V is the specific volume and V_f is the total free volume.

If we assume that V_f is proportional to $\Delta\alpha(T - T_0)$ where T_0 is the temperature which free volume vanishes and $\Delta\alpha$ is the difference between the coefficients of thermal expansion of the liquid and the glass, then we can obtain:

$$\eta \propto \exp\left(\frac{\beta}{T - T_0}\right) \tag{8.55}$$

where various terms have been bundled into the parameter β. This immediately provides a kinetic explanation for T_g. As a liquid is cooled, the temperature T approaches T_0, the viscosity increases dramatically, and the system essentially becomes frozen.

In providing a simple free volume explanation of the T_g we have made a couple of simplifications. First, we have treated our hypothetical material as if it were a collection of hard spheres. Even small molecules, let alone polymers, are nothing like this, of course. Second, in modifying the Doolittle equation to demonstrate the increase in viscosity with decreasing temperature we made a simplified substitution in order to illustrate principles as clearly as possible. We will come back to this later in Chapter 11. Here we will continue our discussion of the types of motion that become frozen as a polymer is cooled below its T_g.

The dynamics of polymer chains is an important subject, but one which is too complicated to discuss properly in an introductory treatment. We will qualitatively discuss some aspects of this subject later, when we discuss polymer melt rheology. Here we will simply note that a polymer molecule can change its shape and its center of gravity can move as a result of rotations around bonds. A rotation around a single bond, as illustrated below,

would lead to huge displacements, however, which would be forbidden by the relatively close packing of the chains in the melt (i.e., a huge amount of energy would be required to push the other chains "out of the way"). Accordingly, various types of coupled rotations of adjacent or nearly adjacent bonds have been proposed that allow fairly small displacements of segments without massive displacement of the chain as a whole. We will defer a discussion of how chains move until later, but clearly this cannot occur unless the chain segments have a certain degree of mobility. This must, of course, be related to free volume in some fashion.

This gets us to another important empirically derived relationship, the WLF equation. The initials WLF stand for Williams, Landel and Ferry, who found that mechanical relaxation times at one temperature T, could be related to those at another, say T_s, through the definition of a shift factor a_T. The relaxation time can be thought of crudely as the time taken for the polymer to adjust to a new configurational position. We will consider a more precise definition in Chapter 11, but clearly this must be related to the viscosity, the relaxation time being longer in a more viscous medium. The WLF equation was originally written:

$$\log a_T = -\frac{C_1(T - T_s)}{C_2 + T - T_s} \tag{8.56}$$

where the shift factor a_T is related to the ratio of the relaxation times at temperatures T and T_s, and C_1 and C_2 are constants for a particular polymer.

If the arbitrary reference temperature T_s is replaced by T_g (and we note that the equation only applies when $T > T_g$), then values of the constants C_1 and C_2 are often taken to be universal constants, with values of 17.44 and 51.6, respectively. Over the years variations from one polymer to another have been found, however, but that is not our concern here. Although the WLF equation was originally formulated in order to give an empirical fit to mechanical relaxation data, it can be obtained from the Doolittle equation by assuming that free volume is a linear function of temperature. We will present this derivation in Chapter 11, where we discuss relaxation behavior in more detail, but the following equation is obtained

$$\log a_T = \frac{\left(-B/2.303\, f_0\right)\left(T - T_g\right)}{f_0/\Delta\alpha + T - T_g} \tag{8.57}$$

where f_0 is the fractional free volume at T_g, V_f/V.

The WLF constants C_1 and C_2 can then be identified with the corresponding free volume terms in equation 8.57 and if we note that to a first approximation:

$$a_T \approx \frac{\eta_T}{\eta_{T_g}} \tag{8.58}$$

where η_T and η_{T_g} are the viscosities at temperatures T and T_g, respectively, then again a rapid increase in viscosity is predicted as T approaches T_g.

It has been our experience that this preliminary discussion of relaxation times and shift factors is probably not entirely clear to most students at this stage. What you should grasp here is that there is a connection between free volume theories, empirical relationships that describe the viscosity of small-molecule liquids (Doolittle equation), and relaxation behavior in polymers as described by a shift factor, a_T. You should then return and reread this section after you have worked through chapter 11, then the connections should become much clearer.

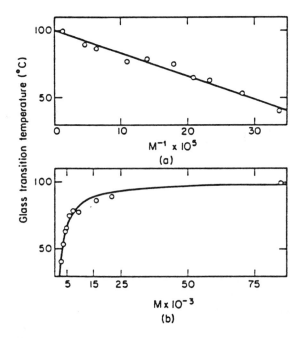

Figure 8.15 *Glass transitions of polystyrene fractions plotted as a function of molecular weight, M (bottom) and M⁻¹ (top). Reproduced with permission from T. G. Fox and P. J. Flory,* J. Appl. Phys., **21**, *581 (1950).*

Factors That Affect the Glass Transition Temperature

Just as in our discussion of the crystalline melting point, having considered the fundamental nature of the transition we can now move on to the interesting question of the factors that result in one polymer having a different T_g to another. We will see that simple free volume arguments allow a good qualitative understanding of most of these factors.

The Effect of Molecular Weight

Figure 8.15 shows the measured T_g's of polystyrene fractions plotted as a function of their molecular weight. It can be seen that for relatively short chains the T_g increases sharply with increasing molecular weight, but then levels off and appears to asymptotically approach a limiting value. The free volume approach accounts for this in an easily understood manner. We have mentioned in the preceding section that although the concept of free volume is not easily defined, it can be associated with cooperative motions. The ends of a chain have more freedom of motion than the segments in the center of a chain (think of cracking a whip—the end moves much faster than the parts in the middle and can break the sound barrier) and, crudely, can be thought of as having "more free volume". Low molecular weight chains have more ends per unit volume than long chains, hence a higher free volume, hence a lower T_g. Fox and Flory* used such simple free volume arguments to obtain the following equation:

$$T_g = T_g^\infty - \frac{K}{M_n} \qquad (8.59)$$

where T_g^∞ is the glass transition temperature that would be obtained from a polymer of infinite molecular weight, while K is a constant related to parameters describing the free volume. This relationship works well for most polymers, as figure 8.15 demonstrates, but there are exceptions and it does not explain the molecular weight dependence of the T_g of ring-like polymers (i.e., those with no ends!). The reader interested in pursuing this in more depth should consult McKenna's review article and the citations therein.

The Effect of Chemical Structure

Given that we are dealing with polymers of high molecular weight, what is the effect of chemical structure, as manifested in stiffness, intermolecular interactions, etc., on T_g? One would intuitively expect that the T_g would be affected in the same way as T_m. Stiffer chains and those with stronger intermolecular interactions would have a higher T_g. That is indeed the case, but to explain the effect on T_g we will use arguments based on free volume and molecular mobility rather than the equilibrium thermodynamic approach we applied to the T_m.

Because chain stiffness affects chain mobility, in other words the ease of rotation around the bonds of the polymer backbone, its effect on T_g is easily understood in a simple qualitative sense. If there are bulky groups, such as benzene rings, in the backbone of the polymer chain, there is a high energy barrier to rotations which then only occur at higher temperatures. (This can be related to free volume models through arguments concerning cooperative motions.) Similarly, the presence of bulky pendant groups attached to the polymer backbone also raises the T_g, through steric hindrance to bond rotations. This effect is clearly illustrated with the examples shown in table 8.1. As the pendant group gets larger, the T_g increases. There is a limit to this, however, as at some point the attached groups no longer get in the way of bond rotations as they get further and further away from the chain. Compare the T_g's of atactic polystyrene with poly(1-vinyl naphthalene) and poly(vinyl biphenyl). The effect of attaching a methyl group to the main chain of polystyrene, to give poly α-methyl styrene, shown above, is greater than increasing the size of the aromatic unit, because the close proximity of this group to the polymer backbone introduces a higher degree of steric hindrance.

The pendant groups we considered in the above examples are single species or, in the case of poly(vinyl biphenyl) very stiff units. What happens if we attach a flexible side chain to the main chain, as for example, in the methacrylates?

---

* T. G. Fox and P. J. Flory, *J. Appl. Phys.*, **21**, 581 (1950) and *J. Polym Sci.*, **14**, 315 (1954).

Table 8.1 *Glass Transition Temperatures.*

Polymer	Chemical Structure	T_g
Polyethylene	$-CH_2-CH_2-$ (n)	$\approx -80°C$ *
Atactic Polypropylene	$-CH_2-CH-$ with CH_3 (n)	$\approx -10°C$
Atactic Polystyrene	$-CH_2-CH-$ with phenyl (n)	$\approx 100°C$
Atactic Poly(α-methyl styrene)	$-CH_2-C-$ with CH_3 and phenyl (n)	$\approx 175°C$
Atactic Poly(1-vinyl naphthalene)	$-CH_2-CH-$ with naphthalene (n)	$\approx 135°C$
Atactic Poly(vinyl biphenyl)	$-CH_2-CH-$ with biphenyl (n)	$\approx 145°C$

$$-CH_2-\underset{\underset{R}{\overset{\overset{CH_3}{|}}{\underset{|}{\overset{|}{C}}}}{\overset{|}{\underset{O}{\overset{|}{C=O}}}}-\Bigg]_n$$

where:

R = Methyl —CH₃

= Ethyl —CH₂—CH₃

= Propyl —CH₂—CH₂—CH₃

= Butyl —CH₂—CH₂—CH₂—CH₃

etc.

Here the T_g *decreases* with increasing chain length, as illustrated in figure 8.16. This is again easily understood. The substituents closest to the chain, the –CH₃ (methyl) and COO (ester) group, provide the bulk of the steric hindrance. The rest of the attached side chain can "get out of the way" of motions of the main chain through rotations around side-chain bonds. Because these side chains increase the free volume through their effect on the packing of the chains (and also affect it through their mobility), the T_g is lowered.

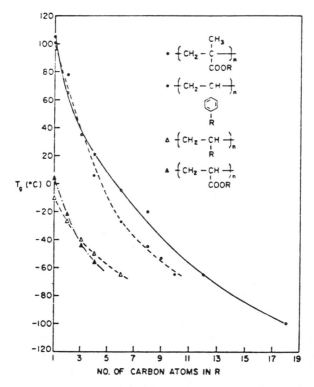

Figure 8.16 *Effect of side chain length on the glass transition temperatures of various polymers. Reproduced with permission from A. Eisenburg, in* Physical Properties of Polymers, *J. E. Mark, A. Eisenburg, W. W. Graessley, L. Mandelkern and J. L. Koenig, Eds., American Chemical Society, Washington, DC, 1984.*

Strong intermolecular attractions also act to raise the T_g. For example, if we compare the T_g's of polypropylene and poly(vinyl chloride)

		T_g
Atactic Polypropylene	$\left[-CH_2-\underset{\underset{CH_3}{\vert}}{CH}-\right]_n$	$\approx -10°C$
Atactic Poly(vinyl chloride)	$\left[-CH_2-\underset{\underset{Cl}{\vert}}{CH}-\right]_n$	$\approx 87°C$

we can argue that the chlorine atom and methyl group have approximately the same effect on bond rotations. The polar character of the Cl atom leads to stronger forces of attraction between chains, however, which in turn means that on average those groups engaging in such interactions are closer together. The free volume is less and the T_g is higher (reality is probably a bit more complicated than this, but this simple explanation gives a good feel for the major effect).

The Effect of Diluents and Copolymerization

Intuitively, one would expect that if a polymer with a T_g of about 100°C is mixed with a second component, either another polymer or a low molecular weight material with a T_g of say -50°C, then the T_g of the mixture would be somewhere in-between and would depend upon the relative proportions of the components present. This is generally what is observed, although there are one or two exceptions (e.g., blends that have very strong intermolecular interactions, where the T_g of the mixture can be higher than either of the pure components). The addition of low molecular weight materials to a polymer to make it more flexible is known as plasticization and the diluent is called a plasticizer. A key here is whether or not the system (blend of two polymers or polymer/plasticizer mixture) is miscible. By this we mean form a single phase with intimate mixing of the components at the molecular level. Clearly, if the system phase separates one would expect to observe two T_g's, the observed value of the T_g's depending upon the composition of the phase separated domains.

The T_g of a mixture can be simply rationalized on the basis of free volume arguments. For example, the addition of a low molecular weight species increases the free volume of the system and hence lowers the T_g. It is possible to go a step further than this and assume a relationship between the free volume of the mixture and the free volume of the components (e.g., that they are simply additive). This allows the derivation of equations relating the T_g of the mixture to that of its components. We won't go into this further but simply present one widely used result, the Fox equation 8.55.*

$$\frac{1}{T_g} = \frac{W_1}{T_{g_1}} + \frac{W_2}{T_{g_2}} \qquad (8.60)$$

where T_{g_1} and T_{g_2} are the T_g's of the pure components (°K) and W_1 and W_2 are the respective weight fractions present in the mixture (this equation was actually first derived to describe the T_g of random copolymers, where the same simple free volume arguments we have used for mixtures would also apply).

The Effect of Cross-linking and Crystallization

Again, it is beyond the scope of this introductory treatment to go into detail, so we will simply observe that both cross-linking and crystallinity raise the T_g. The effect of cross-linking can be crudely explained on the basis of simple free volume arguments. Cross-linking decreases free volume, because parts of the chain are tied more closely together, hence the T_g increases.

The effect of crystallization can also be qualitatively understood in terms of simple arguments. (Be careful, because some students get confused here; we are only talking about the amorphous regions of a sample when we consider the T_g). If we take a crystallizable polymer and quench it from the melt to a temperature below the T_g, so that the sample remains amorphous, we will observe a certain value of the T_g. If we now anneal this sample at a temperature above the T_g, so that crystallization occurs, the measured T_g is higher. This is presumably because the crystalline regions "tie down" and limit the mobility of the chains in the amorphous regions. The extent of the effect depends upon the degree of crystallinity and the morphology of the sample (in certain systems the T_g can even be completely masked by crystallization).

G. STUDY QUESTIONS

1. Compare and contrast the changes in structure and properties (e.g., the specific volume) that occur at the glass transition and crystalline melting temperatures.

2. Explain why polymers usually melt over a range of temperatures, unlike low molecular weight materials, which usually have a sharp melting point.

3. Consider a family of polyesters with the general structure:

---

* T. G. Fox, *Bull. Am. Phys. Soc.*, **1**, 123 (1956).

How would you expect the glass transition and crystalline melting temperatures to vary with the values of m and n? Give reasons for your answer.

4. Assuming that the T_gs of atactic polystyrene and polybutadiene are 100°C and - 60°C, respectively, calculate and plot the variation of T_g with composition for a styrene/butadiene random copolymer.

5. Consider the following:
(a) A high molecular weight 50:50 (by weight) styrene/butadiene random copolymer.
(b) A 50:50 (by weight) block copolymer of styrene and butadiene containing high molecular weight blocks.
(c) A 50:50 (by weight) immiscible blend of high molecular weight polystyrene and polybutadiene.
Using the results obtained from question 4 and with suitable explanations, at what temperature or temperatures would you expect to see glass transition(s)?

6. The authors recently published the results of a differential scanning calorimetry (DSC) study of random copolymers containing the comonomers, n-butyl methacrylate (BMA) and 4-vinyl phenol (VPh) [*Macromolecules*, **25**, 7077 (1992)].

Glass transition temperatures were determined as follows:,

Wt % VPh	T_g (°C)	Wt % VPh	T_g (°C)
0	20	65	138
10	44	80	151
28	81	100	170
50	118		

Prepare a plot of T_g versus composition and on the same plot show the curve calculated from the Fox equation (equation 8.60). What reasons can you come up with to explain why the experimental data for the copolymers does not fit the Fox equation?

7. Discuss why polymers crystallize in a chain folded fashion rather than the thermodynamically preferred extended chain form.

8. Define or briefly explain the following:
(a) Primary crystallization.
(b) Secondary crystallization.

(c) Primary nucleation.
(d) Secondary nucleation.
(e) Fold period.
(f) Equilibrium melting temperature.

9. The following data for the T_g of polystyrene as a function of molecular weight was obtained by Fox and Flory [*J. Polym. Sci.*, **14**, 315 (1954)].

M (g/mole)	T_g (°C)	M (g/mole)	T_g (°C)
85,000	100	3,590	75
19,300	89	3,041	65
13,300	86	2,600	62
6,650	77	2,085	53
4,980	78	1,675	40

What would be the T_g of a theoretically infinite molecular weight sample?

10. What is meant by *free volume*? Show how this concept may be used to rationalize the glass transition temperatures of the following high molecular weight polymers relative to polyethylene, which is believed to have a T_g of approximately - 80°C.

(a) Polydimethylsiloxane; $T_g \approx - 123°C$
(b) *cis*-1,4-Polyisoprene; $T_g \approx - 75°C$
(c) Polychloroprene; $T_g \approx - 50°C$
(d) Poly(ethyl acrylate); $T_g \approx - 23°C$
(e) Poly(methyl acrylate); $T_g \approx 0°C$
(f) Polystyrene; $T_g \approx + 100°C$
(g) Poly(phenylene oxide); $T_g \approx +200°C$

11. A paint is to be prepared from a poly(vinyl acetate) (PVAc) emulsion. PVAc has T_g of $\approx 28°C$ and at ambient temperature the dried paint is very brittle and cracks easily. Suggest methods to overcome this problem.

H. SUGGESTIONS FOR FURTHER READING

(1)   R. N. Haward, *The Physics of Glassy Polymers*, John Wiley & Sons, New York, 1973.

(2)   R. Zallen, *The Physics of Amorphous Solids*, John Wiley & Sons, New York, 1983.

(3)   J. D. Hoffman, G. T. Davis and J. I. Lauritzen, Jr., in *Treatise on Solid State Chemistry*, N. B. Hannay, Ed., Vol. 3, Chapter 7, Plenum Press, New York, 1976.

(4)   L. Mandelkern and G. McKenna, Articles in *Comprehensive Polymer Science*, C. Booth and C. Price, Eds., Vol. 2, Pergamon Press, Oxford, 1989.

CHAPTER 9

Thermodynamics of Polymer Solutions and Blends

*"The development of solution theory has resembled
the flow of the Gulf Stream across the Atlantic
—a meandering course with eddies that disappear."*
—Joel Hildebrand[*]

A. INTRODUCTION

At the start of the previous chapter we reviewed some basic principles of thermodynamics and statistical mechanics and if you have been jumping around in your use of this book you may want to go back and review this material. Here we are going to be considering the problem of mixing, which is to say that we will address the question of whether or not a given polymer will dissolve in a particular solvent or mix with a different type of polymer (i.e., form a miscible blend). Thermodynamics allows us to construct an answer to this question, in that at equilibrium the free energy must be a minimum, which in turn means that the *free energy change* upon mixing the items we are interested in, say a polymer and a solvent, must be negative. If we let the change in Gibbs free energy upon mixing be ΔG_m, then we can write this as:

$$\Delta G_m < 0 \qquad (9.1)$$

The free energy change at a given temperature is of course related to the enthalpy and entropy change by:

$$\Delta G_m = \Delta H_m - T\Delta S_m \qquad (9.2)$$

so that we need to determine if the right hand side of this equation is negative. As we will show later, this is one necessary condition for mixing to occur, but it is not the only one. The other principle criterion involves the second derivative of the free energy with respect to composition, however, so this will also require that we start from equation 9.2.

To remind you, thermodynamics is a subject that describes the essential relationships between macroscopic properties (P, V, the quantities in equation 9.2, etc.). This is all very well when describing things like the efficiency of heat engines, but for the problems we are addressing here thermodynamics only provides a starting point. To obtain a useful relationship we need to relate the

---

[*] J. H. Hildebrand, *Ann. Rev. Phys. Chem.*, **1**, 32 (1981).

quantities in equation 9.2, ΔH_m and ΔS_m, to molecular properties; the interactions between molecules and the arrangements and packing of the molecules in the mixture. This means we will have to do some statistical mechanics, but fortunately we can obtain some simple and useful results without going into too much depth.

B. THE FREE ENERGY OF MIXING

As we mentioned at the beginning of Chapter 7, entropy can be related to the number of arrangements or configurations available to the system. In order to count the configurations we need to construct a model. The simplest model of this type, and one which has proved to be extremely useful in spite of its simplicity, is *regular solution theory*, which assumes we are mixing spherical molecules of equal size. This is a reasonable place to start our discussions because, as we will see, a workable theory of polymer solutions can be obtained using most of the assumptions of this approach.

Regular Solution Theory

Although many great scientists have made significant contributions to this subject, regular solution theory is inextricably linked to the name of Hildebrand, and the 1950 version of the book he coauthored with Scott[*] is still a classic and should be a starting point for students interested in studying this subject in greater detail. This theory makes the fundamental assumption that the enthalpic and entropic parts of the free energy can be treated separately and are additive. In order to comprehend the nature of this assumption and to introduce a particularly simple method for counting configurations we will use something called a lattice model. In this approach it is assumed that the molecules can be placed on a lattice, as illustrated in figure 9.1 for a mixture of black and white "balls" (mole-

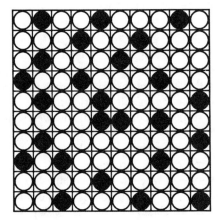

Figure 9.1 *Schematic representation of a lattice model for mixing molecules of equal size.*

[*] J. H. Hildebrand and R. L. Scott, *The Solubility of Non-Electrolytes, Third Edition*, American Chemical Society Monograph Series, 1950.

cules) of equal size. This is, of course, entirely artificial, in that real solutions do not have the degree of order implicit in this regular arrangement, but to emphasize, this model provides a useful device for counting configurations.

The representation shown as figure 9.1 should be considered as a snapshot of a situation prevailing at any instant of time. If the system is in the liquid state (i.e., not crystallized or frozen into a glass), the molecules are constantly switching positions, so that there are a large number of possible arrangements. We want to know the total number of arrangements, Ω, so that we can use:

$$S = k \ln \Omega \qquad (9.3)$$

to calculate the entropy. (In Chapter 7 we used the symbol W rather than Ω to represent the number of configurations, because that is the equation carved on Boltzmann's tomb. Many modern texts use the symbol Ω, however, as we now do here.)

The problem is that the number of configurations available to the system is constrained by the energy of the interactions (and of course by the numbers of molecules of each type present). If, for example, molecules of the same type attracted one another more strongly than molecules that are unlike, then we might expect that arrangements where black molecules are next to one another and white molecules are also adjacent to their own kind would be preferred. This greatly complicates the evaluation of the number of configurations (in fact, this is a problem that has never really been solved). The simplification used in regular solution theory, and one we will subsequently also make when treating polymer mixtures, is to assume that the energy of interaction experienced by any one molecule is simply an average taken over all possible configurations. This so-called mean field approximation corresponds to the situation where each molecule moves in a potential field (i.e., experiences forces due to intermolecular interactions) that is unaffected by local variations in composition, but is an overall average. Essentially, we assume random mixing. This assumption is good as long as the interactions between molecules are weak compared to kT, so that motion due to thermal energy essentially keeps the system random.

With this assumption we can now treat the entropic and enthalpic parts of the free energy separately. In addition, the lattice model provides a particularly simple way of calculating the configurations. If we consider a binary (two component) system such as that illustrated schematically in figure 9.1, then we can start by imagining that we take all the molecules off the lattice and then calculate the number of ways of successively placing these molecules back so as to fill up all the sites. If we have n_A molecules of type A, and n_B molecules of type B, then we have a total number of lattice sites, n_0, given by:

$$n_0 = n_A + n_B \qquad (9.4)$$

because, to reiterate, we assume that initially the lattice is filled (i.e., there are no holes; more on this later). In putting the first molecule back on the lattice, and at this point we don't care if this molecule is of type A or of type B, we can choose from any of the sites, because they are all empty, so there are n_0 different ways of putting the first molecule back on the lattice. The second molecule only has $n_0 - 1$ empty sites available to it, however, because the first molecule has occupied a site somewhere on the lattice. In the same way the third molecule has $n_0 - 2$ choices and so on. The total number of arrangements available to the molecules is thus obtained by multiplying all these choices, i.e.:

$$(n_0)(n_0 - 1)(n_0 - 2)(n_0 - 3) ----- 1 = n_0! \qquad (9.5)$$

At first you might think that this is the answer, but force yourself to re-examine figure 9.1. It should be obvious that if we swap the positions of any two of the white molecules with one another, or similarly allow any two of the black molecules to change places, the configuration is unchanged, because molecules of the same type are indistinguishable. Only if a white molecule changes place with a black molecule do we obtain a new configuration. However, in counting the ways of putting the molecules back on the lattice we counted *all* possible arrangements, as if each molecule was labelled (this one's Fred, that one's Joe, etc.) and could be distinguished from all other molecules. Accordingly, we must remove from our answer those configurations where molecules of the same type just swap places.

Using the same type of argument given above for counting the total number of configurations, it can be shown that there are $n_A!$ ways of swapping the positions of the A molecules (say the black ones) with one another and $n_B!$ ways of similarly swapping the positions of the white ones, so the total number of configurations is actually given by:

$$\Omega = \frac{n_0!}{n_A! \, n_B!} \qquad (9.6)$$

We can now immediately obtain an expression for the entropy of mixing using equation 9.3 and Stirling's approximation, which says that as long as N is large, then the log of a factorial, N!, is given by:

$$\ln N! = N \ln N - N \qquad (9.7)$$

The substitution and the elementary algebra is left as an exercise and the answer is[*]:

$$\Delta S_m = -k (n_A \ln x_A + n_B \ln x_B) \qquad (9.8)$$

where x_A and x_B are the mole fractions, given by:

$$x_A = \frac{n_A}{n_A + n_B} \qquad x_B = \frac{n_B}{n_A + n_B} \qquad (9.9)$$

If we wish to let n_A and n_B be the number of moles of A and B, rather than the number of molecules, then we can simply divide these quantities by Avogadro's number (and multiply k by the same quantity) to obtain:

$$\Delta S_m = -R (n_A \ln x_A + n_B \ln x_B) \qquad (9.10)$$

where R is the gas constant. In future we will not always specify if we are

---

[*] To obtain the entropy *change* upon mixing you actually need to construct the entropy of the A and B components in the pure state and subtract these from the entropy of the mixing calculated from equation 9.6. This is easy because the entropy of mixing A's with A's is zero.

dealing with moles or molecules, but it's easy to tell from the constant we employ (if it's the gas constant, R, our concentration variables are in moles; if it's Boltzmann's constant, k, our concentration variables are in terms of numbers of molecules).

Before proceeding you should sit back and think about equation 9.10. It is a beautiful result. We have taken a concept, entropy, that many students find difficult to grasp, and found that by using a simple model we can relate it to something we know and understand, the composition of the mixture we are considering. Of course, our lattice model of a liquid is not very realistic and as it turns out there are other entropy changes that occur upon mixing. Nevertheless, this result is a reasonable first approximation. The entropy change calculated from an estimate of the number of configurations is usually called the combinatorial entropy and in the simplest treatments any other entropy changes are neglected. We will briefly discuss some of these other factors later, but at this point we will turn our attention to the enthalpy of mixing.

In our discussion of intermolecular interactions in Chapter 7 we introduced the concept of cohesive energy density, which is equal to $\Delta E^v / V$, where ΔE^v is the heat or energy of vaporization of a volume V of a liquid. If we consider our simple mixture of A and B molecules in the pure state, then clearly the cohesive energy density must be related to the energy of interactions between like molecules. This is because it is these forces of interaction that must be "overcome" in order to extract a molecule from the liquid state, where it is surrounded by and in close proximity to other molecules of the same kind, and place it in the vapor or gaseous state, where molecules are on average much further apart (and to a first approximation we can neglect interactions between them). Accordingly, if we let C_{AA} be the cohesive energy density of the A molecules and C_{BB} be the cohesive energy density of the B molecules, then the interactions between A molecules in pure A is proportional to C_{AA} and the interactions between B molecules in pure B is proportional to C_{BB} (we have hidden some assumptions about pairwise additivity and nearest neighbor interactions here that you will have to consider if you pursue this subject in greater depth). Clearly, we could define a cohesive energy density parameter describing interactions between A and B molecules as C_{AB}.

What we want to know is the change in enthalpy upon mixing, which is related to the change in interaction energy. If we start with the interactions between just a pair of A molecules and just a pair of B molecules, then consider the change that occurs when these contacts are replaced by AB contacts, as illustrated schematically in figure 9.2, then clearly the energy change is proportional to:

$$C_{AA} + C_{BB} - 2C_{AB}$$

where the factor 2 is a consequence that two AB contact pairs are formed from a pair of AA and BB contacts (see figure 9.2)*. We know or can determine C_{AA}

---

* If you are wondering why we did not write the energy change the other way around, as $2C_{AB} - C_{AA} - C_{BB}$, which might seem more logical (i.e., mixture minus pure components), it is because the cohesive energy density is related to -E, where E is the potential energy of the interaction; see Hildebrand and Scott.

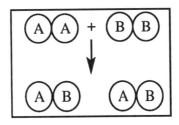

Figure 9.2 *Schematic representation of the formation of AB contacts from AA and BB contacts.*

and C_{BB} from measurements of the energy of vaporization. What we don't know is C_{AB} and the number of AB contacts in a mixture. For random mixing, however, it can be shown that the number of AB contacts is proportional to the product of the volume fractions of the two components, $\Phi_A \Phi_B$. The quantity C_{AB} is much more difficult to determine, but if we are considering mixtures where the molecules interact by dispersion forces only, then we can use the geometric mean assumption:

$$C_{AB} = (C_{AA})^{1/2} (C_{BB})^{1/2} \qquad (9.11)$$

This approximation, due originally to Berthelot, is more than just an assumption and follows from the dependence of dispersion forces on the volume of the (spherical) molecules that are interacting. It appears to work reasonably well for many non-spherical and even weakly polar molecules and allows the energy change per pair of molecules to be written as a perfect square by substituting equation 9.11 into equation 9.10:

$$C_{AA} - 2(C_{AA})^{1/2} (C_{BB})^{1/2} + C_{BB} = \left[(C_{AA})^{1/2} - (C_{BB})^{1/2} \right]^2 \qquad (9.12)$$

Using the assumption that the energy of interaction is equal to the sum of the interactions between all pairs of molecules, an expression for the enthalpy of mixing can be obtained and is given by:

$$\Delta H_m = V_m \, \Phi_A \Phi_B \, (n_A + n_B) \left[(C_{AA})^{1/2} - (C_{BB})^{1/2} \right]^2 \qquad (9.13)$$

where V_m is the molar volume of the mixture and the $V_m \Phi_A \Phi_B$ part of the expression, when multiplied by the total number of moles, $n_A + n_B$, is a measure of the change in the number of pair-wise contacts on going from the initially separate pure A and pure B liquids to the state where they are randomly mixed*.

Hildebrand defined a quantity called a *solubility parameter*, given for a liquid A by:

---

* To obtain equation 9.13 you must actually consider the energy of interaction between all pairs of molecules in pure A and pure B, then subtract the sum of these quantities from the energy of interaction between all pairs of molecules in a random mixture of the two.

$$\delta_A = (C_{AA})^{1/2} = \left(\frac{\Delta E_A^v}{V_A}\right)^{1/2} \qquad (9.14)$$

and using this quantity instead of cohesive energy densities, the free energy of mixing becomes:

$$\frac{\Delta G_m}{RT}\left[\frac{1}{n_A + n_B}\right] = x_A \ln x_A + x_B \ln x_B + \frac{V_m}{RT}\Phi_A \Phi_B (\delta_A - \delta_B)^2 \qquad (9.15)$$

where we have divided both sides of the equation by the total number of moles to give an expression for the free energy per mole, $\Delta G_m / (n_A + n_B)$.

In spite of some of the broad assumptions that have been made, the theory works reasonably well. We will briefly mention where it breaks down after considering the Flory-Huggins theory for polymer solutions. Before getting to this, however, it is essential that you grasp two important points from equation 9.15. The first is that the (combinatorial) entropy of mixing, the sum of the first two terms on the right hand side, is negative (the log of a fraction is a negative number) and therefore favorable to mixing. Remember, *one* condition for mixing is that ΔG_m should be negative. The second point is that if we are dealing solely with dispersion forces (and perhaps weakly polar forces), as in mixtures of simple hydrocarbons, then the enthalpy of mixing is positive and unfavorable to mixing, because it is proportional to a difference term *squared*, $(\delta_A - \delta_B)^2$. Accordingly, if the liquids we are mixing are similar (δ_A has a value that is close to δ_B so that the enthalpy of mixing is small), then A will mix with B, or as we usually say, the two usually form a *miscible mixture*. This is because the entropy of mixing small molecules is large and in this situation will dominate the free energy. However, if the two liquids are very dissimilar, like oil and water, then the positive enthalpy of mixing can become large enough to make the free energy of mixing positive. These liquids would not mix or would be *immiscible*. Note that in the form we have written the free energy the enthalpy term is divided by T, however, so that as we raise the temperature the enthalpy of mixing gets smaller and the two liquids are more likely to mix. This result reflects the intuitive knowledge of everyday experience; if one thing won't dissolve in another, heat it up and sometimes it will.

Finally, if you go further into mixing theory you will find some of the limitations of the regular solution approach, but it is still a powerful first approximation and provides the most essential component of any good theory—understanding.

The Flory-Huggins Theory

The equation for the free energy of polymer solutions obtained independently and almost simultaneously by Flory and Huggins uses many of the assumptions of regular solution theory and has a strikingly similar form to the regular solution result:

$$\frac{\Delta G_m}{RT} = n_s \ln \Phi_s + n_p \ln \Phi_p + n_s \Phi_p \chi \qquad (9.16)$$

where to begin with we will use the subscripts s and p to refer to solvent and polymer, respectively. The first two terms are again the combinatorial entropy and it can be seen that the symbols Φ_s and Φ_p, which represent the *volume fractions* of the solvent and polymer, replace the mole fractions used in regular solution theory. The third term uses the parameter χ to describe the interactions between components and, as in the regular solution model, it is related to the change in energy when solvent/solvent and polymer/polymer contacts and hence interactions are replaced by solvent/polymer interactions (i.e., in terms of cohesive energy densities it is related to $C_{ss} + C_{pp} - 2C_{sp}$). The same assumptions of pair-wise additivity and nearest neighbor interactions used in regular solution theory are also employed and one might immediately guess that the χ parameter can be related to solubility parameter differences. We will discuss this in the following section.

The derivation of the combinatorial entropy of mixing a polymer with a solvent is a much more difficult problem than the simple mixing of equal size spherical molecules that we considered in some detail in the preceding section, even if we again use the simplification allowed by a lattice model in order to count the number of configurations. To start with, how do we place a polymer on a lattice? Obviously it occupies a far greater volume than a solvent molecule. The essential first step made by Flory and Huggins was to assume that the polymer is a flexible chain of segments and that each of these segments is equal in size to a solvent molecule, as illustrated in figure 9.3, where it can be seen that the polymer occupies a set of adjacent lattice sites. Obviously, this assumption can be carried a step further and the solvent can also be considered to be a molecule of two or three linked segments. The extension to polymer blends, where both molecules are long chains of linked segments then follows directly. We won't worry about this complication now, however, because even the simple model illustrated in figure 9.3 presents a lot of problems when we try to

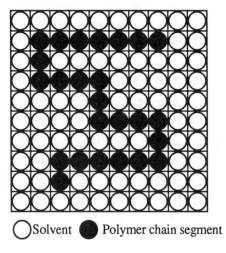

○ Solvent ● Polymer chain segment

Figure 9.3 Schematic representation of a polymer chain on a lattice.

count configurations. The answers obtained by Flory and Huggins actually represent different levels of approximation and the combinatorial entropic part of the equation:

$$-\frac{\Delta S}{R} = n_s \ln \Phi_s + n_p \ln \Phi_p \qquad (9.17)$$

is the simpler Flory result, but differs from that obtained by Huggins by an apparently negligible amount relative to the other approximations inherent in this approach. A discussion of the details of the derivations is beyond the scope of what we want to discuss in this introductory text[*], but there are three points concerning equation 9.17 that need to elaborated. First, the replacement of the mole fractions found in the logarithmic terms of regular solution theory by volume fractions accounts for the large volume of a polymer molecule relative to a solvent molecule. The mole fraction of polymer molecules is simply:

$$x_p = \frac{n_p}{n_p + n_s} \qquad (9.18)$$

and if this concentration variable were used it would be equivalent to assuming that both the polymer and solvent molecules each occupy a single lattice site! In contrast, the volume fraction of polymer is given by:

$$\Phi_p = \frac{n_p M V_s}{n_p M V_s + n_s V_s} = \frac{n_p M}{n_p M + n_s} \qquad (9.19)$$

while the volume fraction of solvent is:

$$\Phi_s = \frac{n_s V_s}{n_s V_s + n_p M V_s} = \frac{n_s}{n_s + n_p M} \qquad (9.20)$$

where $n_p M V_s$ is the number of polymer molecules multiplied by the number of segments in each polymer chain (M), multiplied by the volume of each segment (V_s) (i.e., each polymer occupies M lattice sites). Note that M is not (necessarily) the degree of polymerization of the polymer in terms of the number of chemical repeat units, but is equal to the volume occupied by a polymer chain divided by the volume of a solvent molecule, because for the purposes of mixing theory we have defined the polymer as being equivalent to a chain of segments, each of which has the same volume as a solvent molecule.

The second point we wish to make is more subtle. The number of polymer molecules per unit volume is smaller than the number of "small" molecules occupying an equivalent volume. In other words, if we could take the chain shown in figure 9.3 and break the linkages, thus forming M molecules, then the entropy of mixing these small molecules with the solvent would be much greater, because each of these segments could now be anywhere on the lattice, rather than

---

[*] We strongly urge that those of you who intend to pursue this part of the subject in greater depth work your way through the original papers, referenced in Flory's classic book *Principles of Polymer Chemistry*. In addition, the application of lattice models to polymers is discussed in great depth in Guggenheim's book *Mixtures* and although this book is older than most students, a battered and beaten up copy can usually be found in most good university libraries.

being required to take up the adjacent positions imposed when they are linked in a chain (i.e., they have more configurations available to them). The combinatorial entropy of mixing small molecules is thus larger than the entropy of mixing a polymer and a solvent, which in turn is larger than the entropy of mixing two polymers. We will rearrange our equation for the free energy so as to demonstrate this more explicitly later in this section.

The final point concerning equation 9.17 that we wish to make concerns the relationship between solubility and chain flexibility. It has been our experience that many students somehow catch on to the idea that if one polymer is more flexible than another, then it will somehow dissolve more readily in a particular solvent. In a way this makes sense, in that a flexible polymer can take on a broader range of conformations or shapes than a rigid one and therefore has more configurations. In determining the entropy of mixing, however, we are concerned with the *change* in entropy on going from the pure components to the mixture. If the conformations available to the chain are unaffected by intermolecular interactions, then there will be no entropy *change* (the polymer has as many configurations available to it in the pure state as it does in the mixture). If you make the effort to work through the Flory derivation of equation 9.17, then you will find that there is an explicit term for what Flory called the entropy of disorientation that is eliminated in the final expression. *In general*, the thermodynamic properties of polymer solutions are independent of chain flexibility. The exceptions to this rule are in those systems where the polymers are stiff rods and those where there are strong specific interactions such as hydrogen bonds.

The enthalpy of mixing in the original Flory-Huggins treatment is described by the χ parameter, which in a similar fashion to regular solution theory is related to the interactions between pairs, but here the pair is not two molecules, but a solvent and a *segment* of a polymer chain. If the energy parameters ω_{ss}, ω_{pp} and ω_{sp}^* describe the interactions between pairs of solvent molecules, pairs of polymer segments and solvent/polymer segment pairs, respectively, then the energy change describing the formation of a single sp contact, $\Delta\omega_{sp}$, is given by:

$$\Delta\omega_{sp} = \omega_{sp} - \frac{1}{2}\left(\omega_{ss} + \omega_{pp}\right) \tag{9.21}$$

(If you have forgotten why we have the factor of 1/2, go back and reexamine figure 9.2 and think about it!). Flory then argues that the number of contacts between segments and solvent molecules is approximately equal to $zn_s\Phi_p$, where z is the coordination number of the lattice (number of neighbors of a given site), and thus obtained:

$$\Delta H_m = z\,\Delta\omega_{sp}\,n_s\,\Phi_p \tag{9.22}$$

The χ parameter was then introduced by defining:

$$\chi = \frac{z\,\Delta\omega_{sp}}{kT} \tag{9.23}$$

---

* Note that we are using Flory's definition of pair-wise energies here rather than the cohesive energies used by Hildebrand. We will show how χ and solubility parameters are related later on.

so that:

$$\Delta H_m = kT \chi n_s \Phi_p \qquad (9.24)$$

Combining this expression with that for the combinatorial entropy (equation 9.17), the Flory-Huggins equation, given at the beginning of this section, is obtained (equation 9.16). You should be aware that in later work the meaning of the χ parameter was broadened and generalized to include entropic factors not accounted for in the combinatorial part of the free energy, but we won't go into that here. Instead, we now turn to the situation where instead of considering the mixing of a polymer with a solvent, we mix two polymers, A and B, to form a blend. This requires us to define a reference volume, V_r, which in the lattice model is equal to the lattice cell size.   However, instead of being equal to the molar volume of a solvent molecule, the reference volume can now be made equal to the molar volume of the chemical repeat unit of one of the polymers (or, in fact, some other arbitrarily defined volume).

The quantity V/V_r, the total (molar) volume of the system, V, divided by the reference (molar) volume, V_r (remember, for polymer solutions we use $V_r = V_s$, the molar volume of a solvent molecule), is equal to the number of moles of lattice sites. We can divide both sides of the free energy equation 9.16 by V/V_r to obtain (using the definition of a volume fraction):

$$\frac{\Delta G_m^{'}}{RT} = \left[\frac{\Delta G_m}{RT} \right] \left[\frac{V_r}{V} \right] = \frac{\Phi_A}{M_A} \ln\Phi_A + \frac{\Phi_B}{M_B} \ln\Phi_B + \Phi_A\Phi_B\chi \qquad (9.25)$$

where we have reverted to the more general subscripts A and B, rather than the s and p used to designate solvent and polymer, respectively. Equation 9.25 is the free energy of mixing per mole of lattice sites and this form of the free energy equation is informative because it immediately makes plain some of the points concerning the entropy of mixing that we made qualitatively above*. The quantities M_A and M_B can be thought of as the "degrees of polymerization" of A and B, as long as we keep in mind that this definition is in terms of a reference volume that needs to be defined.

Equation 9.25 is quite general, and if A and B are both small molecules of equal size, equal to the reference volume V_r, then M_A and M_B would both be equal to 1 and the entropy of mixing would be (relatively) large. If A is a solvent molecule and B a polymer, then it is most convenient to let V_r be equal to the molar volume of the solvent, so that M_A would be equal to 1, but M_B would be very large (>10,000) for a high molecular weight polymer. Accordingly, the entropy of mixing would be significantly less than for two small molecules (about half the value for a 50/50 mixture by volume). Finally, if A and B are both polymers, we could let V_r be equal to the molar volume of one of the chemical repeat units, then both M_A and M_B would both be large and the entropy of mixing would be very small. Because χ is positive for most hydrocarbon (non-polar) molecules, most polymers don't mix with one another. This is a big problem in recycling, for example, where polymers of different types are

---

* For the purposes of determining phase behavior it doesn't matter if we use the free energy, the free energy per unit volume, or the free energy per mole of lattice sites, etc.

collected together (polyethylene, polypropylene, polystyrene, PVC, nylons etc.). We cannot just physically mix them all up together and put them through an extruder and try to make something useful. The components of such a mixture phase separate into separate domains, like oil and water, and the object you are trying to make often has no cohesive strength and will just fall apart. (This problem is much more complex than we have just depicted and depends, amongst other things, upon the size of the phase separated domains. You will have to consider such complications if you intend to delve more deeply into this part of the subject.)

Solubility Parameters and the Flory-Huggins χ Parameter

If we are dealing with simple hydrocarbon mixtures, where we can make the geometric mean assumption, then one would anticipate that there should be a relationship between χ and solubility parameter differences. A direct comparison of equations 9.15 and 9.25 would suggest that:

$$\chi_{AB} = \frac{V_r}{RT} \left(\delta_A - \delta_B \right)^2 \qquad (9.26)$$

where the subscript AB indicates that we are considering the interaction between A and B segments (or molecules)[*]. As originally formulated, χ is an adjustable parameter that can be obtained from experimental measurements (e.g., from osmotic pressure measurements, which we discuss in Chapter 10.) Clearly, it would be advantageous if we could use solubility parameters to at least obtain an estimate of χ, and in fact solubility parameters are widely used in the surface coating industry. However, there are two problems with using equation 9.26. First, if one (or both) component of the mixture is polymers, how do we determine their cohesive energy densities? Polymers cannot be "evaporated" by heating to obtain their energy of vaporization, because they decompose well below the theoretical temperature necessary for vaporization. The second problem is that for polymer *solutions* equation 9.26 does not work very well. A comparison of theoretical calculations and experimental results suggests that a relationship of the form:

$$\chi_{sp} = 0.34 + \frac{V_r}{RT} \left(\delta_s - \delta_p \right)^2 \qquad (9.27)$$

is more appropriate (note that we have reverted to the subscripts s and p to emphasize that this equation is only applicable to polymer solutions). We will initially consider the problem of calculating solubility parameters for polymers, then consider the origin of the "fudge factor" introduced into equation 9.27 (i.e., the constant 0.34; incidentally, some texts have this equal to 0.30). This will then lead us to a general consideration of the limitations of the Flory-Huggins theory, which will be discussed in the context of phase behavior.

---

[*] Note that equations 9.15 and 9.25 have to be put on the same basis in order to obtain this result. This can be achieved by multiplying equation 9.15 by the total number of moles, $(n_A + n_B)$, and equation 9.25 by V/V_r.

There are essentially two ways of determining solubility parameters for polymers. One is experimental and usually involves the swelling of a lightly cross-linked polymer in a series of solvents. It is then assumed that the polymer has the same solubility parameter as the solvent that yields the maximum swelling*. The problem with this approach is first one of experimental accuracy (just check out the range of experimental values of solubility parameters listed in van Krevelen's useful compilation**); second, convenience, as it is not always a simple task to prepare lightly cross-linked networks and there are also complications introduced by factors such as crystallinity, etc.; finally, the solubility parameter of the polymer can be significantly different to that of the solvent that gives the maximum swelling because of factors and complications we have not yet considered, such as free volume differences***.

Accordingly, the second method for determining solubility parameters is the most common and this involves a calculation based on so-called group contributions. The essence of this approach is to assume that a molecule can be "broken down" into a set of functional groups. A simple paraffin:

$$CH_3 - (CH_2)_n - CH_3$$

is most conveniently assumed to consist of CH_2 and CH_3 groups, for example. Using the energy of vaporization of a series of such paraffins (i.e., with different values of n) so-called *molar attraction constants*, the contribution of CH_2 and CH_3 groups to solubility parameters, can be calculated. By including branched hydrocarbons, and molecules containing other functional groups (ether oxygens, esters, nitriles etc.), a table of constants can be obtained and subsequently used to calculate the solubility parameter of a given polymer. Such a compilation is shown in table 9.1, which is reproduced from some of our own work. To calculate the solubility parameter of a polymer one uses the simple relationship:

$$\delta = \frac{\sum_i F_i^*}{\sum_i V_i^*} \quad (cal.\ cm^{-3})^{0.5} \qquad (9.28)$$

where F_i^* is the molar attraction constant of the ith group (e.g., a CH_2 group), while V_i^* is the corresponding molar volume constant of this group.

For example, say we wish to calculate the solubility parameter of poly(methyl methacrylate) (PMMA):

---

* This follows from the theory of swelling polymer networks, where it can be shown that chains will swell more in "good" solvents, i.e., $(\delta_p - \delta_s) \rightarrow 0$.
** P. W. van Krevelen, *Properties of Polymers*, Elsevier, Amsterdam, 1972.
*** There is an excellent discussion of this problem in a review article by D. Patterson, *Rubber Chem. Technol.*, **40**, 1, (1967).

Table 9.1 *Molar Volume and Attraction Constants.*

Group	Molar Volume Constant V* (cm^3 mole^{-1})	Molar Attraction Constant F* ((cal. cm^3)$^{0.5}$ mole^{-1})
-CH$_3$	31.8	218
-CH$_2$-	16.5	132
>CH-	1.9	23
>C<	- 14.8	- 97
C$_6$H$_3$	41.4	562
C$_6$H$_4$	58.5	652
C$_6$H$_5$	75.5	735
CH$_2$=	29.7	203
-CH=	13.7	113
>C=	-2.4	18
-OCO-	19.6	298
-CO-	10.7	262
-O-	5.1	95
-Cl	23.9	264
-CN	23.6	426
-NH$_2$	18.6	275
>NH	8.5	143
>N-	- 5.0	- 3

$$\begin{array}{c} CH_3 \\ | \\ -CH_2-C- \\ | \\ \underset{O}{\overset{}{C}}\diagup\diagdown O-CH_3 \end{array}$$

then we simply need the sum of the molar attraction constants for one CH$_2$, two CH$_3$, one >C< and one –OCO– (132 + 436 - 97 + 298 = 769) divided by the sum of the molar volume constants for the same set of groups (16.5 + 63.6 - 14.8 + 19.6 = 84.9). Thus the solubility parameter value for PMMA is calculated to be 769/84.9 = 9.1 (cal. cm^{-3})$^{0.5}$.

This approach works reasonably well under the circumstances where the use of solubility parameters is most appropriate, namely in those systems where interactions are largely non-polar. It breaks down when there are strong polar forces or specific interactions such as hydrogen bonds. There is a whole literature concerning the use (and abuse) of solubility parameters, but we kind of

like a book entitled *Specific Interactions and the Miscibility of Polymer Blends* [*]
and the interested reader may want to start there.

To conclude this section we will now touch on the reasons why equation
9.26 does not work well for polymer *solutions*, requiring the addition of a
"fudge factor" of about 0.34 (equation 9.27). Originally, it was proposed that
this factor had its genesis in the level of approximation used in the Flory method
for counting configurations. In a series of papers published in the 1970's,
however, Patterson and coworkers showed that the most likely origin of this
correction term lay in so-called free volume effects that are neglected in the Flory-
Huggins treatment. As we discussed earlier in our treatment of the glass
transition, in the liquid state the motion and vibrations of the molecules leads to
density fluctuations and we called this free volume (one way to model this is to
put holes on the lattice). The free volume associated with a low molecular weight
liquid is usually larger than that of a polymer, so that in mixtures of the two there
is a mismatch of free volumes. This leads to the additional term in equation
9.27. In fact, a more general way of writing the Flory χ term would be to let it
have the form:

$$\chi = a + \frac{b}{T} \qquad (9.29)$$

where the quantity 'a' can be thought of as an entropic part of χ, accounting for
non-combinatorial entropy changes such as those associated with free volume,
while 'b' would be the enthalpic part (note that 'a' and 'b' need not be constants,
but could depend upon composition and temperature).

In general, however, for mixtures of small molecules, or for mixtures of two
polymers, there is a closer match of free volume parameters and equation 9.26
without the "fudge factor" (i.e., a = 0 in equation 9.29) is often a good
approximation. This brings us to our next topic, phase behavior, which will
include a brief discussion of the limitations of the Flory-Huggins treatment.

C. THE PHASE BEHAVIOR OF POLYMER SOLUTIONS AND BLENDS

The Chemical Potential and the Conditions for Phase Separation

We have so far only mentioned one condition for miscibility, which we have
defined as the formation of a single phase solution where the components are
intimately mixed at the molecular level, namely that the free energy should be
negative. This is not the only condition, however, as we can demonstrate by
considering a plot of the free energy of mixing as a function of composition, as
illustrated in figure 9.4. Here we show a curve that starts at zero in pure A (the
free energy of mixing molecules of type A with one another is zero) and finishes
at zero in pure B (for the same reason) and in between has a shape that is concave
"upwards" (i.e., towards the top of the page). The shape of the free energy curve

---

[*] M. M. Coleman, J. F. Graf and P. C. Painter, Technomic Publishing Co., Lancaster, PA
(1991).

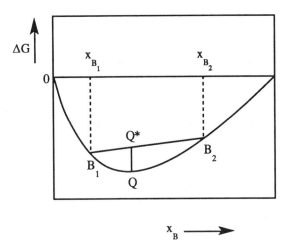

Figure 9.4 *Schematic diagram of free energy as a function of composition.*

is crucial, and if it is concave upwards, then the mixture is miscible (single phase) at all compositions. This is because any point on the curve, e.g., Q, has a lower free energy than two phase system of the same overall composition. To understand this, consider the points B_1 and B_2 on figure 9.4, representing mixtures of composition x_{B_1} and x_{B_2}, respectively, where x_B is the mole fraction of B units (we could have used volume fractions as our composition variables, but the use of mole fractions will allow us to introduce the chemical potential in a more straightforward manner). Now imagine that these two mixtures are contained in the same vessel, but separated by a barrier. What happens if the barrier is removed? If the two mixtures remain phase separated, then their free energy would be given by their composition weighted sum shown as point Q* (i.e., if the mixture is 50/50 by mole fraction, then the free energy of the phase separated mixture would be 0.5 times the free energy at B_1; 0.5 ΔG_{B_1}, plus 0.5 ΔG_{B_2}).

However, it can be seen from figure 9.4 that if the mixtures B_1 and B_2 formed a new single phase it could get to a lower free energy, given by the point Q on the free energy curve describing the mixture. Therefore, the two will mix, as will any compositions of A and B whose free energy of mixing is described by the type of concave-upward curve shown in figure 9.4.

Now consider the free energy curve shown in figure 9.5. The free energy is still negative across the entire composition range, but there are regions that are concave downwards instead of upwards. The shape of the curve has been exaggerated in this figure, to make it easier to see what's going on, but free energy curves of this type are obtained as a result of how the entropic and enthalpic parts of the free energy vary in a different manner with composition. It can be seen that the points B_1 and B_2 can be connected by a straight line that is a tangent at both points (i.e., a double tangent). This line actually represents the free energy of a phase separated mixture, where the compositions of the two phases are given by x_{B_1} and x_{B_2}, and each position on the line represents different

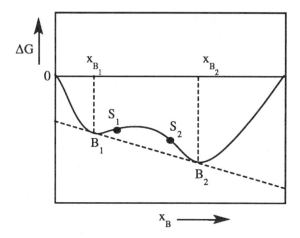

Figure 9.5 *Schematic diagram of free energy as a function of composition.*

proportions of the two phase separated mixtures (i.e., at B_1 all you have is a single phase mixture of composition x_{B_1}, then as you add more of component B you find more of the mixture of component x_{B_2} present in equilibrium with the phase separated mixture of composition x_{B_1}, until finally at point B_2 there is only a single phase mixture of composition x_{B_2}). The crucial point is that the free energy of any (hypothetical) single phase mixture between the compositions x_{B_1} and x_{B_2} is greater than the free energy of the phase separated mixtures, so that the system is *immiscible* over this composition range. The composition range where phase separation occurs is defined by the points of contact of the double tangent line and you will recall from basic calculus that the slope of this line is defined by the first derivative of the free energy with respect to composition. This gives us the relationship:

$$\left[\frac{\partial \Delta G}{\partial x_B}\right]_{x_{B_1}} = \left[\frac{\partial \Delta G}{\partial x_B}\right]_{x_{B_2}} \tag{9.30}$$

where the derivative evaluated at compositions x_{B_1} is equated to that at x_{B_2}. Rather than use mole fractions, it is more common to express the derivatives in terms of numbers of moles, n_B:

$$\left[\frac{\partial \Delta G}{\partial n_B}\right]_{n_{B_1}} = \left[\frac{\partial \Delta G}{\partial n_B}\right]_{n_{B_2}} \tag{9.31}$$

and the two results (equations 9.30 and 9.31) are, of course, identical (check the definition of a mole fraction). If we now use the symbols μ_B^1 and μ_B^2 to represent the derivatives evaluated at points B_1 and B_2 (with respect to n_B) we can write;

$$\mu_B^1 = \mu_B^2 \tag{9.32}$$

or if we choose to use the number of moles of A as our composition variable we have the equivalent result:

$$\mu_A^1 = \mu_A^2 \qquad (9.33)$$

You will hopefully recall from your basic physical chemistry classes that these derivatives are the important quantities known as the *chemical potentials*. Equations 9.32 and 9.33 essentially state that the chemical potential of the components A and B in one phase are equal to their chemical potential in the second phase. If you have forgotton what this means, examine figure 9.6, which shows a polymer solution phase separated into solvent rich and solvent poor domains. The interface between the two is not an impermeable barrier, so polymer and solvent molecules move back and forth between the two domains, but at equilibrium the number of solvent molecules moving from phase 1 to phase 2 is the same as the number of solvent molecules moving in the other direction (and the same is true for polymer molecules). We say that the chemical potential is the same when there is no net flow between the two phases. If the chemical potential between the two phases was different there would be a net flow until equilibrium is achieved. Equation 9.32 (or 9.33) can be used to define the composition limits, x_{B_1} and x_{B_2}, for phase separation (the phase boundary). If we use the Flory-Huggins equation for the free energy (equation 9.16), change our subscripts to s and p to designate solvent and polymer, respectively, we can obtain by simple differentiation:

$$\frac{\mu_s - \mu_s^o}{RT} = \ln(1 - \Phi_p) + \left(1 - \frac{1}{M}\right)\Phi_p + \Phi_p^2 \chi \qquad (9.34)$$

where we have used $\Phi_s = (1 - \Phi_p)$, μ_s^o is the chemical potential of the reference state (usually pure solvent) and μ_s is the chemical potential of the solvent in the mixture[*]. Equation 9.34 can be written for phase 1 (i.e. in terms of concentrations Φ_s^1 and Φ_p^1) and also for phase 2 (Φ_s^2 and Φ_p^2) and the equations can be equated and solved (but there is not an analytical solution; the equation has to be solved iteratively).

The points B_1 and B_2 represent the compositions of the two phases that would be present at equilibrium. However, if you examine the shape of the free energy curve around B_1 and B_2, you will notice that it is *locally* still concave upwards until the points of inflection S_1 and S_2 are reached, where the curve "turns over". What this means is that mixtures that have compositions between points B_1 and S_1 and (B_2 and S_2), are stable against separation into phases consisting of *local* compositions. The mixture still wants to phase separate into domains of composition x_{B_1} and x_{B_2}, but it is in a metastable region where it takes some effort (energy) to accomplish this (in these composition ranges phase separation proceeds by a process of nucleation and growth). The points of inflection, you will again hopefully recall from elementary calculus, are characterized by the condition that the second derivative of the free energy with respect to composition is equal to zero:

---

[*] In performing the differentiation you have to remember that both Φ_p and Φ_s depend on n_s, the number of solvent molecules; go back and check the definition of volume fraction.

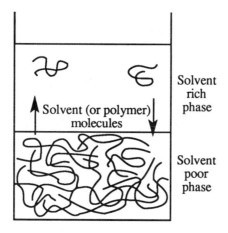

Figure 9.6 *Schematic diagram of a phase separated polymer solution.*

$$\frac{\partial^2 \Delta G}{\partial x_B^2} = 0 \qquad (9.35)$$

Here, however, we obtain an analytical solution that is much more easily solved, as we will show below.

At the point of inflection the second derivative of the free energy changes from positive to negative and we can now summarize the criteria for forming a stable single phase mixture or solution as:

a) The free energy should be negative.

b) The second derivative of the free energy with respect to composition should be positive (which means the free energy curve is concave upwards).

The compositions of the phase separated domains obtained from equating the chemical potentials (equation 9.32) and the compositions defining the metastable region are with respect to a specific temperature. The equations can be solved at different temperatures (remember that χ varies as $1/T$) to obtain a *phase diagram*, as illustrated in figure 9.7. The line obtained from solving the equations for the chemical potential is called the *binodal*, while the line defining the limit of the metastable region is called the *spinodal* (i.e., the metastable region is between the binodal and spinodal). It can be seen that the phase diagram appears like an upside down "U". The point at the very top of the inverted "U" is a critical point and is called the upper critical solution temperature (often abbreviated to UCST), because it is at the top of a two phase region. The binodal and spinodal both meet at this point, labelled C in figure 9.7, where it can be shown that the third derivative of the free energy with respect to composition is equal to zero; i.e.,

$$\frac{\partial^3 \Delta G}{\partial x_B^3} = 0 \qquad (9.36)$$

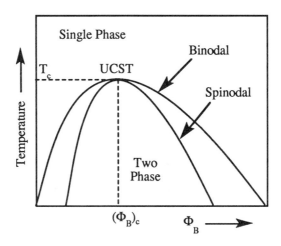

Figure 9.7 *Schematic representation of a phase diagram.*

We can use the criteria that both the second and third derivatives of the free energy are equal to zero at the critical point, together with the Flory-Huggins equation, to obtain a (theoretical) value of χ at the critical point, χ_c, and a predicted value of the composition at which the critical point occurs. In calculating the spinodal and the third derivative of the free energy we can use any composition variable we like. The algebra is a little easier if we use volume fractions and the Flory-Huggins free energy written in the form of equation 9.25, which we reproduce here for convenience[*]:

$$\frac{\Delta G'_m}{RT} = \left[\frac{\Delta G_m}{RT} \right]\left[\frac{V_r}{V} \right] = \frac{\Phi_A}{M_A}\ln\Phi_A + \frac{\Phi_B}{M_B}\ln\Phi_B + \Phi_A\Phi_B\chi$$

Differentiating both sides of this equation with respect to Φ_A (we could equally well use Φ_B) and remembering that $\Phi_B = (1 - \Phi_A)$, we obtain:

$$\frac{\partial^2(\Delta G'_m/RT)}{\partial\Phi_A^2} = \frac{1}{\Phi_A M_A} + \frac{1}{\Phi_B M_B} - 2\chi = 0 \tag{9.37}$$

$$\frac{\partial^3(\Delta G'_m/RT)}{\partial\Phi_A^3} = -\frac{1}{\Phi_A^2 M_A} + \frac{1}{\Phi_B^2 M_B} = 0 \tag{9.38}$$

which can be solved to show that at the critical point (subscript c):

$$(\Phi_A)_c = \frac{M_B^{1/2}}{M_A^{1/2} + M_B^{1/2}} \tag{9.39}$$

---

[*] Remember that this is a general form of the Flory-Huggins equation, applicable to both solutions and polymer blends.

and:

$$\chi_c = \frac{1}{2} \left(\frac{1}{M_A^{1/2}} + \frac{1}{M_B^{1/2}} \right)^2 \qquad (9.40)$$

For a polymer solution we can let the component A be the solvent, so that $M_A = 1$ and $M_B = M$, so that the critical value of χ is given by:

$$\chi_c = \frac{1}{2} \left(1 + \frac{1}{M^{1/2}} \right)^2 \qquad (9.41)$$

For a high molecular weight polymer ($M \gg 1$) χ_c is therefore about 0.5. In other words for values of χ greater than 0.5 phase separation can occur (depending upon composition), but at values of χ less than 0.5, the system is predicted to be single phase and the polymer should dissolve in the solvent. If we are dealing with two polymers, however, equation 9.40 tells us that χ_c is very small (to see this calculate χ_c for $M_A = M_B = 1000$). This means most polymers that interact through weak dispersion forces only won't mix, because the value of χ for most mixtures is larger than χ_c.

The critical value of χ can be translated into a critical value of solubility parameter differences using equation 9.28, so that for a polymer solution ($\chi_c \approx 0.5$).

$$\chi_c = 0.34 + \frac{V_r}{RT} \left(\delta_A - \delta_B \right)_c^2 = 0.5 \qquad (9.42)$$

and assuming RT at room temperature is about $2 \times 300 = 600$ cal./mole, and V_r is of the order of 100 cm^3/mole, the critical value of the solubility parameter difference is about 1. This provides a useful guide for deciding whether or not a polymer will dissolve in a particular solvent. Say, for example, the solubility parameter of the polymer of interest is equal to 9 (cal. cm^{-3})$^{0.5}$, we would expect that it would dissolve in solvents whose solubility parameters lie between 8 and 10 (9 ± 1) (cal. cm^{-3})$^{0.5}$. This "rule of thumb" is commonly used in the surface coating industry for predicting polymer solubility, but remember this is only applicable to polymer/solvent mixtures involving dispersion (and perhaps weak polar) forces. When strong forces, such as hydrogen bonds, are involved all bets are off!

A Comparison of Observed and Predicted Phase Behavior—The Limitations of the Flory-Huggins Model

The Flory-Huggins model does a good job of predicting some of the gross features of polymer solution behavior and the characteristics of polymer mixtures. For example, experimental measurements on various solutions have shown that χ_c is indeed close to 0.5. Also, most simple hydrocarbon polymers don't mix with one another. A more detailed examination of phase behavior shows several deviations between theory and experiment, however.

In studies of polymer solutions it is usual to perform some form of light scattering experiment to determine a cloud point. The solution will (most often) phase separate into domains of different refractive index, which therefore scatter an incident beam of light. The temperature (and hence the value of χ) at which this occurs is presumed to correspond to the binodal (this experiment is much more difficult to apply to polymer blends, because the rate at which the molecules diffuse past one another and hence phase separate is much slower). In addition, neutron scattering experiments can be used to construct an experimental spinodal curve. Here we will just consider the results of one or two classic experiments in order to illustrate some of the limitations of the Flory-Huggins approach.

In the early 1950's Schultz and Flory[*] compared calculated binodal curves to experimental measurements of cloud points for a set of polystyrene fractions of different molecular weights dissolved in cyclohexane. The results are illustrated in figure 9.8 and it can be seen that qualitatively the theory predicts the shape of the curve, but the calculated critical concentrations, $(\Phi_B)_c$, are too small and the theoretical and experimental curves deviate significantly at higher concentrations. In one way, however, this comparison does the Flory-Huggins theory an injustice. Implicit in the use of the equations as we have presented them is the assumption that the polystyrene samples are monodisperse (i.e., all the polymer chains have the same molecule weight). Schultz and Flory performed their experiments before "living" anionic polymerizations were introduced and therefore had to obtain their samples using fractional precipitation. This is imperfect and the samples still had a rather broad distribution of molecular weights. It has subsequently been demonstrated that the binodal depends upon the molecular weight distribution in a complex manner that has to be determined numerically. The general effect is to broaden and flatten the calculated binodals

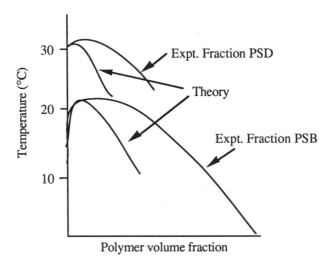

Figure 9.8 *Comparison of experimental cloud point measurements to theoretical predictions. Redrawn from the data of A. R. Schultz and P. J. Flory.* JACS, *74, 4760 (1952).*

---

[*] A. R. Schultz and P. J. Flory, *JACS*, **74**, 4760 (1952).

and thus give a better agreement between theory and experiment. This is still not enough, however, and in order to improve the fit still further it is necessary to assume that χ is concentration dependent; i.e. has the form:

$$\chi = \chi_1 + \chi_2 \Phi_p + \chi_3 \Phi_p^2 + \text{----------} \qquad (9.43)$$

Nevertheless, if the only discrepancy between observed and calculated phase behavior lay in the deviations of binodals from observed cloud points then the Flory-Huggins theory would be considered satisfactory. The observation of *lower critical solution temperatures* (LCSTs, which we will illustrate below) in polymer solutions and blends posed a much more dramatic difficulty, which could not be solved by some minor tinkering with Flory-Huggins theory (e.g., making χ composition dependent).

The observation by Freeman and Rowlinson[*] of LCST's in solutions of polyisobutylene in various non-polar solvents was a turning point in the study of polymer solutions. These authors did not present phase diagrams in their paper, so we reproduce here cloud points reported later by Siow et al. in their classic studies of the polystyrene/acetone system, shown in figure 9.9. For low molecular weight polystyrene samples (e.g., M = 4800) not only is there phase separation upon cooling, but also a phase separation upon heating the sample. The upper phase boundary looks like a highly flattened "U" and the critical point at the bottom of this "U" is the LCST[**]. As the molecular weight of the polysty-

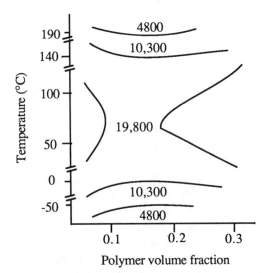

Figure 9.9 *Phase diagrams of polystyrene in acetone. Redrawn from the data of K. S. Siow, G. Delmas and D. Patterson,* Macromolecules, 5, 29 (1972).

[*] P. I. Freeman and J. S. Rowlinson, *Polymer*, 1, 20 (1959).
[**] This can be confusing because the *lower* critical solution temperature is at a higher temperature than the *upper* critical temperature, the top of the inverted "U", shown in the same figure. Just remember that the UCST is at the *top* of a two phase region, while the LCST is at the *bottom* of a two phase region.

rene is increased, it can be seen that the upper and lower phase boundaries shift towards one another and eventually coalesce to form the hourglass shape phase diagram characteristic of a phase separated system. The molecular origin of this counter-intuitive observation, that polymer solutions (and as it turns out, polymer blends) can phase separate when they are heated, lies in factors neglected in the Flory-Huggins theory. For non-polar polymer solutions and blends the most important factor is the free volume differences between the components of the mixture. As the temperature is raised, these differences increase and they are unfavorable to mixing, so that at some temperature phase separation occurs. In mixtures of polar molecules, particularly those that hydrogen bond, LCSTs can also have their origin in the entropic changes associated with forming a strong specific interaction. A full discussion of these effects is beyond the scope of this book, however, and is left for more advanced treatments. We will conclude this section by listing some of the limitations of the Flory-Huggins model (the list is not complete), one of which we haven't touched on at all, but that will bring us to the final topic in this chapter.

To summarize, the Flory-Huggins treatment:

a) Uses a lattice model to count configurations and there are various levels of approximation involved in this approach

b) Does not apply to systems where there are strong specific interactions.

c) Neglects so-called free volume effects

d) Does not apply to dilute or semi-dilute solutions

In spite of these limitations the theory has been very useful and can be considered one of the major achievements of polymer science

D. DILUTE SOLUTIONS, EXCLUDED VOLUME AND THE THETA TEMPERATURE

An implicit assumption of treating interactions in the Flory-Huggins model is that there is a random mixing of polymer segments with solution molecules (or, in the case of a polymer blend, with segments of a second polymer). Unlike low molecular weight materials, however, where in dilute solutions it can be assumed molecules are still randomly dispersed throughout the mixture, in a dilute polymer solution there are regions where the concentration of segments is very different. This is because the segments of each chain are connected and therefore constrained to be in a particular volume that can be defined approximately by the chain's end-to-end distance or, more accurately, its radius of gyration (which is the root mean square distance of the segments from the center of gravity of the coil; there is a simple relationship between end-to-end distance and radius of gyration). Between the chains, which on average are far apart in dilute solution, there are no segments at all, so that the system can be pictured as small but widely separated "clouds" of segments, as illustrated schematically in figure 9.10. The derivation of the Flory-Huggins equation rests on the assumption that there is a uniform distribution of segments and so does not apply to dilute solutions. We therefore have to use a different approach. Unfortunately, the

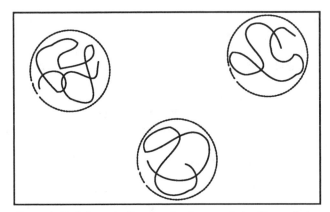

Figure 9.10 *Schematic diagram of polymer coils in dilute solution.*

most rigorous treatments require the use of some difficult and demanding mathematical tools. Flory (see *Principles of Polymer Chemistry*) devised an approximate but useful treatment that accounts for the principal features of these solutions. However, even this requires more work than we wish to impose on you in this introductory overview. Accordingly, we will present here a largely qualitative description, but include a watered down and simplistic treatment of one of the main results.

One principal result of the Flory treatment is the prediction of an *excluded volume* effect in a "good" solvent (usually $\chi \ll 0.5$). This means that the polymer chains can be thought of as "repelling" other chains, so that the clouds of segments pictured schematically in figure 9.10 exclude one another from the volume that they occupy (i.e., they behave like hard spheres). Of course, they don't actually "repel", but the segments of a particular chain can be thought of as having a stronger preference for solvent than for other polymer segments, so that there is a strong driving force for keeping chains far enough apart that they do not overlap*.

At first sight this type of argument can be a little confusing, for the following reason. We have seen in our discussion of phase behavior that the critical value of χ is about 0.5 (i.e., if χ is larger than this, phase separation occurs). If we have a system where χ is, say, 0.1, we call the solvent "good". It not only dissolves the polymer, but in dilute solutions the chains are swollen (see below). However, a positive value of χ means that the pair-wise interactions between polymer segments, and those between solvent molecules, are favored over those between polymer and solvent! In our discussion above and in other texts on this subject, it is implied that the "interactions" drive the excluded volume effect. The driving force is actually largely entropic (there are more configurations available

---

* You may recall that we also mentioned "excluded volume" when we discussed the conformations of individual chains. In that context, however, we meant that the segments of a chain cannot be placed so that they overlap other segments of the same chain. Here we are now talking about the overlap of the volume occupied by *different* polymer chains in dilute solution (see figure 9.10), an intermolecular rather than intramolecular excluded volume effect.

to the chains if they do not overlap) and many authors implicitly include this in their definition of "interactions".

This type of entropic driving force also comes into play within the volume occupied by a single coil. The individual segments of a chain would like to spread out and occupy as much volume as possible, but the degree they can accomplish this is limited by the fact that they are connected by covalent bonds. The net effect is that the chain swells relative to the dimensions that it would have when surrounded by other polymer chains or even relative to its dimensions in concentrated solutions (where the segments of different chains have to overlap).

It is possible to calculate the extent to which the chain swells by using an assumption due to Flory, that the chain extension is determined by the balance between the solvent/polymer "interactions" (i.e., both entropic and enthalpic) driving expansion, and the loss of entropy that results from the chain becoming more stretched out and therefore having fewer configurations available to it. Flory accomplished this by assuming that the polymer chain could be thought of as a "swarm" of segments that are distributed in a Gaussian fashion about the center of gravity of the molecule (i.e., the concentration of segments is at a maximum near the center of gravity and decreases according to an exponential relationship with distance from this center). We will consider a different approach due to Di Marzio[*], where in order to describe the interactions it is assumed that the polymer segments are uniformly distributed throughout a sphere, which we will define in terms of the end-to-end distance of the chain, as this is simpler, but still gives us the essential form of the main result.

An expression for the free energy of the system defined by the sphere "occupied" by a polymer chain can be written as the sum of contributions from the mixing of segments with solvent molecules and from elastic deformation of the chains. For the former we can simply use the Flory-Huggins theory (because we have assumed a uniform distribution of segments within the local volume defined by the dimensions of a chain), but we will not tackle rubber elasticity and chain deformation until Chapter 11. Accordingly, here we will simply write down an expression for this free energy and ask you to take it on trust.

The free energy of mixing a single polymer molecule with the large number of solvent molecules that are in the sphere defined by the chain end-to-end distance is:

$$\frac{\Delta F_m}{kT} = 1 \ln\Phi_p + n_s \ln\Phi_s + n_s \, \Phi_p \, \chi \qquad (9.44)$$

Because $n_s \gg 1$, the first term is usually omitted and we write:

$$\frac{\Delta F_m}{kT} = n_s \ln\Phi_s + n_s \, \Phi_p \, \chi \qquad (9.45)$$

The expression for the entropy of elastic deformation of a chain is simply (and approximately) given by:

$$\frac{\Delta S}{k} = -\frac{1}{2} \left(\lambda_1^2 + \lambda_2^2 + \lambda_3^2 - 3 \right) \qquad (9.46)$$

---

[*] E. A. Di Marzio, *Macromolecules*, **17**, 969 (1984).

where λ_1, λ_2 and λ_3 are extension ratios parallel to the axes of a Cartesian coordinate system (i.e., $\lambda_1 = x/x_0$, etc. where x is the projection onto the x-axis of the (root mean square) end-to-end distance of the unperturbed chain and x_0 is the unperturbed x-component of this distance). If there is no enthalpy change upon stretching and if the chain expansion is isotropic, so that we can let $\alpha = \lambda_1 = \lambda_2 = \lambda_3$, then[*] :

$$\frac{\Delta F_{el}}{kT} = -\frac{\Delta S}{k} = \frac{3}{2}\left(\alpha^2 - 1\right) \qquad (9.47)$$

The overall free energy of our system is assumed to be the sum of the mixing and elastic terms and is therefore given by:

$$\frac{\Delta F}{kT} = n_s \ln\Phi_s + n_s \Phi_p \chi + \frac{3}{2}\left(\alpha^2 - 1\right) \qquad (9.48)$$

Our task is to now determine the chain expansion α which minimizes the free energy (i.e., the value that balances the opposing mixing terms, which drive chain expansion, and elastic terms, which oppose it). This is, of course, achieved by differentiating the free energy with respect to α and setting the result equal to zero. In order to do this it has to be appreciated that the number of solvent molecules that must be considered changes with α. Remember that our system is defined in terms of a sphere whose diameter is defined by the root mean square end-to-end distance of a chain. If the end-to-end distance (on average) increases, then the sphere will be bigger and contain more solvent molecules. The number of polymer segments stays the same, but the volume fraction Φ_p decreases, because the volume of the sphere increases. Accordingly, we need an expression relating Φ_p (and hence n_s and Φ_s) to α. This can be obtained from simple geometry. Figure 9.11 shows the volume occupied by a single chain. The diameter of this sphere is defined by the root mean square end-to-end distance of the chain, which is equal to $M^{1/2}\ell$ in the unperturbed state (see Chapter 7) and $\alpha M^{1/2}\ell$ in the swollen state. If we consider the chain to be a collection of spherical segments of diameter ℓ, then the volume fraction of polymer segments is given by:

$$\Phi_p = \frac{\text{Volume occupied by polymer segments}}{\text{Volume of sphere}} = \frac{\pi/6 \; M\ell^3}{\pi/6 \; (\alpha \, M^{1/2}\ell)^3} \qquad (9.49)$$

or:

$$\Phi_p = \frac{1}{\alpha^3 \, M^{1/2}} \qquad (9.50)$$

---

[*] It should be noted that Flory obtains a logarithmic term in his expression for the free energy of deformation ($\ln \alpha^3$) which is not included here. This logarithmic term has been the source of considerable controversy over the years and should, in our opinion, be included in a more rigorous treatment of swelling. See P. C. Painter and S. L. Shenoy, *J. Chem. Phys.*, **99**, 1409 (1992). However, it does not affect the *form* of the final result, so we will continue with this "simplest possible treatment" approach.

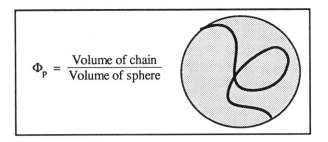

Figure 9.11 *Schematic diagram of the volume occupied by a single chain.*

It follows that:

$$\frac{\partial \Phi_p}{\partial \alpha} = -\frac{3}{\alpha^4 M^{1/2}} \tag{9.51}$$

Also, because $\Phi_s = (1 - \Phi_p)$:

$$\frac{\partial \Phi_s}{\partial \alpha} = \frac{3}{\alpha^4 M^{1/2}} \tag{9.52}$$

The other derivative we will need is $\partial n_s / \partial \alpha$ and using:

$$\Phi_p = \frac{M}{M + n_s} \tag{9.53}$$

(i.e., assuming a solvent molecule occupies the same volume as a polymer segment) we obtain:

$$\frac{\partial n_s}{\partial \alpha} = \frac{\partial n_s}{\partial \Phi_p} \cdot \frac{\partial \Phi_p}{\partial \alpha} = \frac{M}{\Phi_p^2} \cdot \frac{3}{\alpha^4 M^{1/2}} \tag{9.54}$$

We will leave the derivation of the result as an exercise and using:

$$\ln \Phi_s = \ln (1 - \Phi_p) = - \Phi_p - \frac{\Phi_p^2}{2} - \frac{\Phi_p^3}{3} - \text{.........} \tag{9.55}$$

together with:

$$\frac{n_s}{M} = \frac{\Phi_s}{\Phi_p} \tag{9.56}$$

we obtain:

$$\alpha^5 = M^{1/2} \left(\frac{1}{2} - \chi \right) + \frac{\Phi_p}{3} + \text{.......} \tag{9.57}$$

This result tells us two important things. First, because Φ_p is small $\alpha \sim M^{0.1}$ (remember that we are using the symbol $\sim$ to mean "varies as" or "is proportional to") Accordingly, in a dilute solution in a good solvent the root mean square end-to-end distance $<R^2>^{1/2}$ varies as $M^{0.6}$ (we used the symbol N instead of M in Chapter 7).

$$<R^2>^{1/2} \sim \alpha\, M^{0.5} \sim M^{0.6} \tag{9.58}$$

We have now obtained, using a simplified treatment and lots of unjustifiable assumptions, the result we presented earlier in our discussion of random walk statistics (Chapter 7, Section C). Fortunately, more rigorous treatments give the same answer and you should now go back and read these pages again as the arguments might now start to make more sense to you, if you were initially having trouble with some of these ideas. The crucial feature of this result is that in a "good" solvent ($\chi \ll 0.5$) the chain expands (e.g., if we have a chain of 10,000 units, $M^{0.5}$ is equal to 100, whereas $M^{0.6}$ is about 250; i.e., the root mean square end-to-end distance more than doubles for a chain of this size, if we assume the constants of proportionality are roughly the same).

The second aspect of the result given by equation 9.57 is that the chain expansion factor α clearly varies with χ, so that as χ approaches the value of 0.5, α tends to zero. In other words, the chain has its ideal or unperturbed dimension. Because χ varies as $1/T$, this allows for the definition of a so-called theta (θ)-temperature, the temperature at which deviations from ideality vanish. (Flory discussed this in a different manner, in order to show that the θ-temperature is the temperature at which the excess free energy vanishes. The interested reader should consult *Principles of Polymer Chemistry*.) To summarize, at high temperatures, when χ is small, a polymer chain in dilute solution is expanded, with its root mean square end-to-end distance varying as $M^{0.6}$. As the temperature is reduced, χ increases and at the θ-temperature where $\chi = 0.5$, the chain is nearly ideal. When $\chi > 0.5$, the solvent is poor and phase separation can occur, depending upon the temperature and composition of the mixture (the critical value of χ, $\chi_c = 1/2$ occurs at a specific composition and corresponds to the "highest point" on the phase boundary, as shown in figure 9.7.) In the polymer rich phase the chain has nearly ideal dimensions. In the very dilute, solvent rich phase "collapsed" chains may occur[*], but this is a topic for more advanced studies.

E. STUDY QUESTIONS

1. (a) By working through the algebra show explicitly how equation 9.8 can be obtained from equation 9.6.

   (b) Also show explicitly how equation 9.25 is obtained from equation 9.16.

2. (a) Using group contributions calculate the solubility parameters of poly(vinyl chloride) and poly(ε-caprolactone) [see table 9.1 for the appropriate parameters and the appendix of Chapter 1 for the chemical structures].

   (b) On the basis of your calculations would you expect mixtures of these two polymers to be miscible in the amorphous state?

   (c) Experimentally, it has been determined that these two polymers are actually miscible in the amorphous state. Can you rationalize why this is so?

---

[*] P. G. de Gennes, *Scaling Concepts in Polymer Physics*, Cornell University Press, 1979.

3. Obtain equation 9.34 by differentiating the Flory-Huggins equation.

4. Obtain expressions for the critical value of χ and the concentration of polymer, $(\Phi_p)_c$, at the critical point by differentiating equation 9.34 with respect to Φ_p (note that the two equations you obtain will not be the same as 9.37 and 9.38, but you will be able to solve them to obtain the values of χ_c and $(\Phi_p)_c$ given in the text).

5. The vapor pressure of the solvent in a polymer solution relative to the vapor pressure of the pure solvent may, to a first approximation, be equated to the activity of the solvent, which in turn is related to the chemical potential by:

$$\ln a_s = \frac{\mu_s - \mu_s^0}{RT}$$

For the system polystyrene/cyclohexane the following values of $\ln a_s$ were obtained at 34°C (Krigbaum and Geymer, *JACS*, **81**, 1859 (1959)):

Φ_p	$\ln a_s$	Φ_p	$\ln a_s$
0.343	− 0.004	0.637	− 0.063
0.388	− 0.0042	0.690	− 0.089
0.435	− 0.013	0.768	− 0.151
0.485	− 0.018	0.818	− 0.232
0.543	− 0.024	0.902	− 0.460

(a) Calculate χ for each value of Φ_p (assume M is large).
(b) Plot a graph of χ vs Φ_p and comment on the result.

6. Using the simple models presented in this chapter, calculate the free energy of mixing per mole of lattice sites divided by RT, $(\Delta G'_m/RT$, see equation 9.25), as a function of composition (Φ_p), for a hypothetical polymer, polycrudoline, dissolved in a poor solvent, solvnasty, at 27°C. Assume the solubility parameters of polycrudoline and solvnasty are $\delta_p = 9.2$ and $\delta_s = 11.6$ (cal. cm^{-3})$^{0.5}$ and $V_s/RT \approx 1/6$, in appropriate units. Also assume the degree of polymerization of polycrudoline relative to V_s is 1000.

(a) Plot your result graphically.
(b) Repeat the exercise for the same polymer dissolved in a good solvent, solvgood, $\delta_s = 9.4$ (cal. cm^{-3})$^{0.5}$ (assume same V_s).
(c) Comment on any apparent differences in these free energy curves and what this tells you about phase behavior.

7. Assume that you have a monomer and a solvent of almost equal size.
(a) Estimate, using a simple lattice model, the number of distinguishable arrangements for a solution of 1 mole of the monomer in 9 moles of solvent.
(b) What would be the entropy of mixing the monomer and solvent ?
(c) If the single mole of monomer is polymerized to chains of degree of polymerization 100, what would be the entropy of mixing the polymer with the same 9 moles of solvent ?

8. 1g of poly(N-isopropylacrylamide) (PNIPA) is dissolved in 100g of water at room temperature in a sealed glass vial and observed to be perfectly clear (transparent) solution.

$$\left[\text{CH}_2-\underset{\underset{O}{\overset{\displaystyle |}{\overset{\displaystyle C}{\diagdown}}}\;\underset{\underset{H}{\overset{|}{N}}-\underset{\diagdown}{\overset{\diagup CH_3}{CH}}}{}}{\text{CH}}\right]_n$$

PNIPA

The vial is then taken to the laboratory sink and placed under a stream of hot water. The polymer/solvent mixture now becomes turbid and looks like milk. Next the vial is placed under a stream of water from the cold water tap and the polymer/solvent mixture reverts to a perfectly clear solution. How would you explain this phenomenon ?

F. SUGGESTIONS FOR FURTHER READING

(1)   P. J. Flory, *Principles of Polymer Chemistry*,
        Cornell University Press, Ithaca, New York, 1953.

(2)   E. A. Guggenheim, *Mixtures*, Clarendon Press, Oxford, 1952.

(3)   J. H. Hildebrand and R. L. Scott, *The Solubility of Non-Electrolytes*,
        Third Edition, Reinhold Publishing Co., New York, 1950.

(4)   H. Yamakawa, *Modern Theory of Polymer Solutions*,
        Harper and Row, New York, 1971.

CHAPTER 10

Molecular Weight and Branching

*"Drop the idea of large molecules.
Organic molecules with a molecular
weight higher than 5000 do not exist."*
—Advice given to Hermann Staudinger[*]

A. INTRODUCTION

As the quotation given below the heading to this chapter might suggest, the struggle to establish the concept that polymers consist of covalently bonded very large molecules was a long and difficult one. The development of different physical methods of characterization and the weaving together of various threads of evidence was pivotal to the success of this endeavor. To paraphrase von Bayer, every triumphant theory passes through three stages; first it is dismissed as untrue; then it is rejected as contrary to religion; finally it is accepted as dogma and each scientist claims that he or she had long appreciated its truth. The ultimate test of the existence of macromolecules is, of course, provided by the measurement of size, but as we will see later in this chapter, the measurement of polymer molecular weights is by no means an easy task. Historically, the work of Staudinger on solution viscosity[**] and Svedberg's work on the development of the ultracentrifuge was crucial.

The macromolecular concept is now firmly established, but the determination of the average molecular weight and molecular weight distribution of polymer materials remains of fundamental importance, as all physical properties ultimately depend upon chain length to one degree or another.

In Chapter 1 we spent some time defining number and weight average molecular weight and also briefly mentioned the z-average (if you have forgotton or don't have a firm grasp of these concepts you should go back and review this material). As we will see, measurements of various solution properties can be used to obtain these averages directly (that's why we put this chapter after the one on polymer solutions, rather than earlier in the book). However, even though the determination of the number average ($\overline{M}_n$), weight average ($\overline{M}_w$) and z-average ($\overline{M}_z$) molecular weights gives a "feel" for the breadth of the distribution

---

[*]From H. Morawetz, *Polymers. The Origin and Growth of a Science*, Wiley, New York, 1985.
[**] Staudinger actually assumed that polymers behaved as rigid rods rather than flexible coils in solution and so the numbers he obtained were wrong. Nevertheless, this seminal work served to advance the development of polymer science.

(usually measured as the polydispersity, $\overline{M}_w / \overline{M}_n$, see Chapter 1), it is often insufficient to accurately reflect the properties of samples with highly asymmetric distributions, or those that are multimodal (i.e., those that appear to consist of a number of overlapping distributions, perhaps because the sample of interest is actually a mixture that is a result of two or more consecutive polymerizations). What we would really like to know is the entire molecular weight distribution and it so happens that there is one technique, size exclusion (SEC) or gel permeation (GPC) chromatography, that can provide this information. This does not mean we can immediately consign all other characterization techniques to the dustbin of history, because as we will see SEC/GPC is a *relative* method. Calibration, typically with monodisperse (or as close as you can get) standards of known molecular weight, is required and the molecular weight of these standards must be independently determined by some "absolute" method—examples of which are listed in table 10.1. In other words, absolute methods give a direct measure of an average molecular weight, without the necessity for calibration. Furthermore, modern SEC/GPC instruments are often coupled to absolute molecular weight devices, such as laser light scattering, to directly provide absolute molecular weights of the fractionated polymer molecules as they are eluted. Therefore, from this point of view alone, we must know something about these absolute methods. In addition to measuring molecular weight, however, these techniques also provide information on other solution properties, such as polymer/solvent interactions, the size of the chain in solution (radius of gyration or hydrodynamic volume) etc., so they are of general importance.

In this chapter we will focus on four methods. The first two will be the absolute methods osmotic pressure and light scattering. The former is a colligative property measurement, which therefore depends upon the number of molecules present and so provides a measure of $\overline{M}_n$. Light scattering, on the other hand, depends upon the size or mass of the molecule and provides a measure of $\overline{M}_w$. We will then turn our attention to viscosity measurements and

Table 10.1 Methods Used to Determine Absolute Molecular Weights.

Molecular Weight Average	Symbol	Technique
Number average	$\overline{M}_n$	End group analysis Lowering of vapor pressure Ebulliometry (elevation of boiling point) Cryoscopy (depression of freezing point) Osmometry (osmotic pressure)
Weight average	$\overline{M}_w$	Light scattering Neutron scattering Ultracentrifugation
z-average	$\overline{M}_z$	Ultracentrifugation

SEC/GPC, both of which provide a measure of hydrodynamic radius and can be combined to produce a universal calibration curve for GPC/SEC columns. We then conclude this chapter with a discussion and how we can use GPC/SEC in conjunction with intrinsic viscosity measurements to determine the amount of long-chain branching in polymer materials.

B. OSMOTIC PRESSURE AND THE DETERMINATION OF NUMBER AVERAGE MOLECULAR WEIGHT

Osmotic pressure experiments belong to that family of techniques that come under the heading of *colligative property measurements*. It is useful for the characterization of polymers because unlike other measurements of this type, such as freezing point depression and boiling point elevation, dilute solutions of high molecular weight materials provide a detectable osmotic pressure, while the change in many other colligative properties is usually too small to be accurately measured (e.g., a polymer of $\overline{M}_n$ of the order of 1×10^6 would elevate the boiling point of a solvent by about only 5×10^{-5} °K, at 1 atmosphere).

Osmotic pressure arises when you have two solutions of different concentration, or in our situation a solution and a pure solvent, separated by a semi-permeable membrane, as many of you no doubt recall from your high school biology classes. Ideally, the membrane should be completely selective, allowing the passage of solvent, for example, and nothing else. Biological membranes are beautifully adapted to such purposes and efficiently allow the transport of water from the soil into the roots of plants, or the separation of wastes from the blood, and so on. Synthetic membranes are nowhere near as good (but have improved over the years), and this puts a "floor" on the molecular weight that can be measured by osmometry, because small solute molecules can pass through the membrane almost as easily as solvent. The "ceiling" to osmotic pressure measurements is, of course, determined by how accurately the osmotic pressure can be measured. Modern osmometers can measure molecular weights between the limits of about 5×10^3 and 5×10^5 g/mole. The questions we must address now are what, exactly, do we measure and how is it related to molecular weight? We will start with a description of the experiment.

Osmometry—The Nature of the Experiment

The basic principles of membrane osmometry are illustrated schematically in figure 10.1. Figure 10.1(A) shows a polymer solution contained in a tube that is immersed in a reservoir of pure solvent. At the bottom of the tube is a cap which covers a membrane and prevents any exchange between solution and solvent reservoir. The tube is placed so that the solution and solvent liquids are level at the start of the experiment. If the cap is now removed solvent will flow through the membrane into the solution, as illustrated in figure 10.1(B). The driving force for this flow is thermodynamic. If you look back at our discussion of phase behavior (Chapter 9) you will recall that at equilibrium the chemical potential of each component in each phase is the same. Here, it only makes

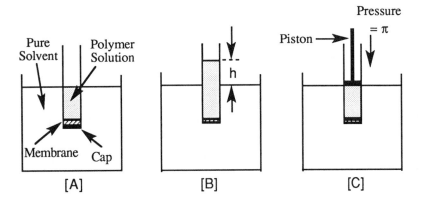

Figure 10.1 Schematic diagram of the osmotic pressure experiment.

sense to consider the chemical potential of the solvent (because the polymer solute is, in principle, physically prevented from passing through the membrane). For the chemical potential of the solvent on each side of the membrane to be equal *at the same temperature and pressure* requires that solvent should continue to flow until the solution becomes infinitely dilute (assuming the solvent is in contact with an infinite reservoir of solvent). In practice, however the pressure on each side of the membrane does not remain constant. The solution is pushed up the tube an amount h above the level of the reservoir until the excess pressure on the membrane provided by the weight of diluted solution exactly balances the pressure arising from the difference in chemical potential.

There are two problems here. First, the diffusion of solvent through the membrane is slow, so that it can take hours or even days to reach equilibrium in this type of "static" experiment. Because it is necessary to make measurements of osmotic pressure as a function of solution concentration, this experimental methodology is very time consuming. Much more crucially, however, this type of experiment can result in very large errors,[*] because of diffusion of low molecular weight polymer from the solution to the solvent side of the membrane. This changes the chemical potential gradient across the membrane and hence the measured osmotic pressure.

These problems have been largely alleviated by performing the experiments in a different manner. Figure 10.1(C) illustrates one method, applying a pressure, π, equal to the osmotic pressure, on the solution side of the membrane, thus preventing any *net* flow of solvent (molecules will still diffuse in both directions, however). Alternatively, the deflection of the membrane can be used to measure the initial osmotic pressure directly. Both methods provide reliable measurements of π in just a few minutes, before there has been any appreciable diffusion of solute or solvent.

---

[*] See H. P. Frank and H. F. Mark, "Report on Molecular Weight Measurements of Polystyrene Standards", *J. Polym. Sci.*, **16**, 129 (1953); ibid., **17**, 1 (1955).

The Relationship to Molecular Weight

Assuming that we can obtain accurate measurements of osmotic pressure, π, how do we relate this quantity to the molecular weight of the polymer solute? If you really mastered polymer solution thermodynamics (fat chance!) you might immediately recall that the expression for the chemical potential of the solvent contains a term in $1/M$, where M is the molecular weight of the polymer (assumed monodisperse in Chapter 9). But, rather than jumping right in with the Flory-Huggins equation, it is useful to start our discussion by initially considering an analogy to the properties of an ideal gas, which allows us to develop the idea of virial equations in what we believe is an easy to understand manner.

The thermodynamic properties of an ideal gas are described by the ideal gas law:

$$PV = NRT \qquad (10.1)$$

where P, V, R and T have their usual meaning and N is the number of moles of the gas. This can be rewritten:

$$\frac{PV}{NRT} = 1 \qquad (10.2)$$

If we now change the pressure and measure say V at constant temperature, we would obtain a straight horizontal line when plotting PV/NRT versus P, as illustrated schematically in figure 10.2.

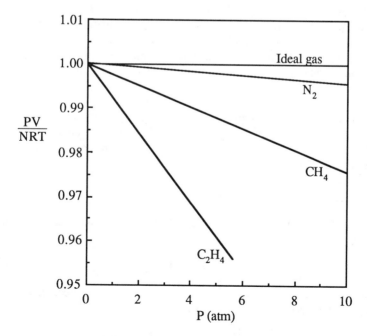

Figure 10.2 *Schematic plots of PV/NRT versus P.*

Real gases deviate from ideal behavior to various extents that depend upon the interactions of their molecules and typical curves for various gases are also shown in figure 10.2. Over the limited range of pressure shown in this figure the plots still appear to be straight, so that the behavior of these gases could be described approximately by an equation of the form:

$$\frac{PV}{NRT} = 1 + B'P \tag{10.3}$$

where B' is a constant at a given temperature. Over a wider range of pressures the lines become obviously curved, but their behavior can still be conveniently modeled, now using a polynomial that is called a *virial equation*:

$$\frac{PV}{NRT} = 1 + B'P + C'P^2 + D'P^3 + \text{--------} \tag{10.4}$$

where the coefficients B', C', D', etc., are the second, third, fourth, etc., *virial coefficients*. (We use a prime here to distinguish the virial coefficients of a gas from those of a solution.)

In much the same way that there is an equation describing the behavior of an ideal gas, we have an equation for an ideal solution. This is expressed in terms of the free energy, but before getting to that we believe understanding is enhanced if we just write by analogy an "equation of state" for an ideal solution in terms of osmotic pressure, π, as:

$$\frac{\pi V}{NRT} = 1 \tag{10.5}$$

Note that this equation depends on N, the number of moles (or molecules). Accordingly, when we consider polymers with a distribution of molecular weights we will calculate a number average (see below).

The quantity N/V, the number of moles of solute per unit volume, is just the molar concentration, m, so that this equation can be recast in the form:

$$\pi = mRT \tag{10.6}$$

known as the van't Hoff relationship (there is more than one relation due to van't Hoff, so don't get them mixed up). This is where the molecular weight comes in. If we measure the concentration of the solution in terms of the weight of solute per unit volume of the solvent, c, then using:

$$m = \frac{N}{V} = \frac{c}{M} \tag{10.7}$$

and recalling that the number of moles of solute is just the weight divided by molecular weight, M, we obtain:

$$\frac{\pi}{c} = \frac{RT}{M} \tag{10.8}$$

Of course, just like gases, solutions are not ideal, but we can again express deviations in terms of a polynomial in c:

$$\frac{\pi}{c} = \frac{RT}{M} + Bc + Cc^2 + Dc^3 + \text{------} \tag{10.9}$$

or in the alternative form:

$$\frac{\pi}{c} = \left(\frac{\pi}{c}\right)_{c \to 0} \left[1 + \Gamma_2 c + \Gamma_3 c^2 + \Gamma_4 c^3 + \ \text{------} \right] \tag{10.10}$$

where:

$$\left(\frac{\pi}{c}\right)_{c \to 0} = \frac{RT}{M} \tag{10.11}$$

and is the value obtained by extrapolating the plot of π/c versus c to zero concentration. In other words, we must measure π as a function of c and do this at low concentrations, so that higher order terms in the polynomial (i.e. those in c^2, c^3 etc.) can (hopefully) be neglected and we obtain a straight line of slope B and intercept RT/M.

This derivation has been by analogy, just to give you a feel for the equations in terms of something you should be comfortable with (the ideal gas law). Now we must consider a more formal derivation and what occurs when our solute consists of polymers with a distribution of molecular weights. For those of you who just want a feel for the subject, however, this may be enough and you can jump ahead, if you must, to the section on calculating $\overline{M}_n$.

An ideal solution is unlike an ideal gas in the sense that instead of assuming that there are no interactions between the molecules, it is assumed that the interactions between components (solute/solute, solute/solvent, etc.) are identical. In other words, there is no enthalpy change upon mixing. In your physical chemistry classes you should have learnt that the entropy of mixing ideal gases or an ideal solute with a solvent depends upon the mole fractions of the components. If you have forgotten this you can look back to our discussion of regular solution theory in Chapter 9 (Section B) which, in effect, assumes an ideal entropy of mixing (although we obtained an expression using a lattice model rather than the thermodynamic approach used in most physical chemistry texts). The free energy of mixing can therefore be written as:

$$\frac{\Delta G_m}{RT} = n_A \ln X_A + n_B \ln X_B \tag{10.12}$$

where X_A, X_B, are the mole fractions of components A and B.

If we let component A be the solvent, then the chemical potential:

$$\frac{\partial \Delta G_m}{\partial n_A} = \mu_A - \mu_A^0 \tag{10.13}$$

(where μ_A^0 is the chemical potential of some standard state) is given by:

$$\frac{\mu_A - \mu_A^0}{RT} = \ln X_A \tag{10.14}$$

The activity of the solvent is defined by:

$$\frac{\mu_A - \mu_A^0}{RT} = \ln a_A \tag{10.15}$$

so that for an ideal solution:

$$a_A = X_A \tag{10.16}$$

Now we are getting close to obtaining the expression for the osmotic pressure that previously we wrote purely by analogy (equation 10.4). We use a standard thermodynamic expression for the change in activity with pressure:

$$\left(\frac{\partial \ln a_A}{\partial P} \right)_{T,N_A} = \frac{\overline{V}_A}{RT} \tag{10.17}$$

where $\overline{V}_A$ is the partial molar volume of the solvent, which for a dilute solution can be equated to the molar volume of the pure solvent, V_A. We now integrate equation 10.17:

$$\int_{a_A}^{1} d \ln a_A = \int_{0}^{\pi} \left(\frac{V_A}{RT} \right) dP \tag{10.18}$$

where the integration limits describe the difference in activity across the membrane (from $a_A = X_A$ on the solution side to $a_A^0 = X_A = 1$ on the pure solvent side) and the difference in osmotic pressure across the membrane ($0 \rightarrow \pi$). If we assume the solution is incompressible (i.e. V_A independent of P), or practically so under the small pressures typical in osmometry experiments, then we obtain:

$$- \ln a_A = \frac{\pi V_A}{RT} \tag{10.19}$$

Substituting $a_A = X_A$ and noting that for dilute solutions $X_A \approx 1$, so that we can use the approximation*, $(- \ln X_A) \approx (1 - X_A) = X_B$ to obtain:

$$X_B = \frac{\pi V_A}{RT} \tag{10.20}$$

For a mole of solution the concentration of solute (weight per unit volume) c is related to X_B by:

$$c = \frac{\text{Weight}}{V_m} = \frac{M X_B}{V_m} \cong \frac{M X_B}{V_A} \tag{10.21}$$

where the molar volume of a dilute solution, V_m, can be approximated by the molar volume of the solvent V_A. Substituting in equation 10.20, for an ideal solution we obtain:

$$\frac{\pi}{c} \approx \frac{RT}{M} \tag{10.22}$$

A virial expansion can be used, as before, to account for deviations from ideal behavior to obtain equation 10.8, reproduced here for convenience:

$$\frac{\pi}{c} = \frac{RT}{M} + Bc + Cc^2 + Dc^3 + \text{------} \tag{10.23}$$

---

* From $\ln x = (x - 1) + 1/2 (x - 1)^2 + 1/3 (x - 1)^3 + \ldots$ and truncating after the first term for $x \approx 1$.

This equation assumes that the solute is a single species with a well-defined molecular weight, M. Now, we must consider the effect of a molecular weight distribution. This follows simply from our definition of the concentration of solute:

$$c = \frac{\text{Total weight of polymer molecules}}{\text{Volume of solution (V)}} \qquad (10.24)$$

The total weight of polymer is just the sum of the weight of all species (i.e., chain lengths) and equal to $\Sigma\, N_i\, M_i$. Recalling the definition of number average molecular weight:

$$\overline{M}_n = \frac{\sum N_i M_i}{\sum N_i} \qquad (10.25)$$

we have:

$$c = \frac{\sum N_i M_i}{V} = \frac{\overline{M}_n \sum N_i}{V} \qquad (10.26)$$

The term $\Sigma\, N_i$ is just the total number of moles of polymer (we could use numbers of molecules instead of moles, and change the terms appropriately). Converting to a per mole of solution basis (compare equation 10.21) we have:

$$c = \frac{\overline{M}_n X_B}{V_A} \qquad (10.27)$$

Substituting in equation 10.20 and performing a virial expansion we obtain:

$$\frac{\pi}{c} = \frac{RT}{\overline{M}_n} + Bc + Cc^2 + Dc^3 + \text{------} \qquad (10.28)$$

showing that osmotic pressure experiments provide a measure of number average molecular weight.

If experimental measurements of π as a function of concentration are made for dilute polymer solutions (where we can hopefully neglect terms in c^2 and higher powers) we should be able to obtain $\overline{M}_n$ from the intercept of a plot of π/c versus c. Before considering some real data, however, it is useful to look at a derivation from the Flory-Huggins equation, because this tells us something about the meaning of B and higher order coefficients and demonstrates that virial equations are not just a mathematical fudge used to fit the data, but have a proper basis in theory.

Osmotic Pressure Measurements and the Flory-Huggins Equation

We have seen that for an ideal solution the osmotic pressure is related to the activity of the solvent by equation 10.19:

$$- \ln a_A = \frac{\pi V_A}{RT}$$

which in turn is related to the solvent chemical potential by equation 10.15:

$$\frac{\mu_A - \mu_A^0}{RT} = \ln a_A$$

If you haven't really read the last part of Chapter 9 carefully you might think that all we have to do is use the Flory-Huggins theory to obtain the chemical potential and hence an expression for π. Following this path we have:

$$\frac{\mu_s - \mu_s^0}{RT} = \ln \Phi_s + \left(1 - \frac{1}{M_n}\right)\Phi_p + \Phi_p^2 \chi \qquad (10.29)$$

Note, to be consistent with the nomenclature used in Chapter 9 (see equation 9.34), we have replaced the subscripts A and B with s and p to denote solvent and polymer, respectively. Also, although we did not cover it in Chaper 9, Flory showed that for a polymer with a distribution of molecular weights, $\overline{M}_n$ replaces M in the expression for the solvent chemical potential*.

The osmotic pressure is then given by:

$$\pi = -\frac{RT}{V_s}\left[\ln \Phi_s + \Phi_p\left(1 - \frac{1}{M_n}\right) + \Phi_p^2 \chi\right] \qquad (10.30)$$

Using:

$$\ln \Phi_s = \ln\left(1 - \Phi_p\right) = -\Phi_p - \frac{\Phi_p^2}{2} - \frac{\Phi_p^3}{3} - \text{-------} \qquad (10.31)$$

We then obtain:

$$\pi = \frac{RT}{V_s}\left[\frac{\Phi_p}{\overline{M}_n} + \Phi_p^2\left(\frac{1}{2} - \chi\right) + \frac{\Phi_p^3}{3} + \text{------}\right] \qquad (10.32)$$

If you examine this equation carefully and compare it to the virial equation 10.28, you should recognize that it has the same form, but uses the concentration variable Φ_p instead of c (if you are a bit slow or just having a bad day, divide both sides of equation 10.32 by Φ_p). Naturally, we can substitute a term in c for Φ_p, as these two concentration variables are simply related, but we won't bother to do so. The important point that we want you to grasp is that a virial equation is not just an arbitrary construction, but arises from fundamental theory. The virial coefficients also have direct physical meaning, the second virial coefficient B would be related to $(1/2 - \chi)$, while the higher order coefficients would be constants, if equation 10.32 were correct. Unfortunately, it is not! First of all, pairwise interactions are assumed in deriving the Flory-Huggins equation and higher order coefficients (C, D etc. in equation 10.30) depend in a more complicated manner upon interactions. Much more crucially, however, the Flory-Huggins theory simply does not apply to dilute solutions, where osmotic pressure measurements are made. As we discussed in Chapter 9 (Section D), when we are considering *dilute solutions* we have to apply more complex theories that involve the configuration or size of the molecules, as well as

---

* P. J. Flory, *J. Chem. Phys.*, **12**, 425 (1944).

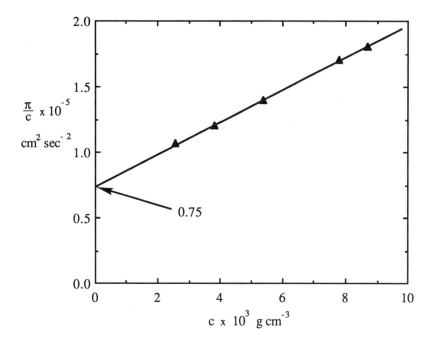

Figure 10.3 *Graph of π/c versus c for polystyrene in toluene. Redrawn from the data of Bawn et al.**

polymer/solvent interactions. (These interactions must also be described differently in order to properly account for chain connectivity and the variation in concentration of the segments as a function of the distance from the center of gravity of each polymer coil.) If you go further into polymer solution theory you find that you still obtain equations of the virial form, however, it just becomes more difficult to relate the coefficients to things you want to know (like χ or α, the chain expansion factor). We can abandon these problems here, because if all we want to know is $\overline{M_n}$, then we should just be able to use a virial equation in the form of equation 10.32, for example, and not worry about the meaning of the coefficients. We now examine how this is achieved.

Calculating the Number Average Molecular Weight from Osmotic Pressure Data

The calculation of $\overline{M_n}$ from osmotic pressure data would seem to be a very straightforward task. For example, figure 10.3 shows a plot of osmotic pressure, π, versus concentration (g/cm^3), c, for a sample of polystyrene in

---

* The experimental results were obtained by C. E. H. Bawn, R. F. J. Freeman and A. R. Kamaliddin, *Trans. Faraday Soc.*, 46, 862 (1950) and are also tabulated in a lovely little book by D. Margerison and G. C. East, *Introduction to Polymer Chemistry*, Pergamon Press, Oxford (1967).

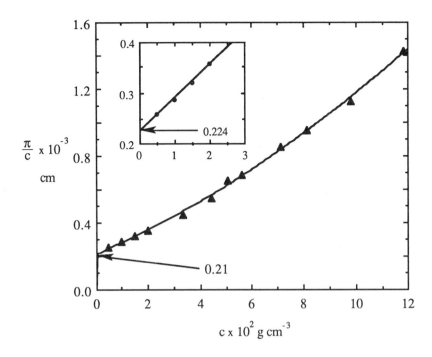

Figure 10.4 *Graph of π/c versus c for polyisobutylene in chlorobenzene. Redrawn from the data of Leonard and Doust[*]. (Note that the units of π are in g/cm².)*

toluene at 25°C. The osmotic pressures were originally reported in terms of the height of a column of toluene (see figure 10.1) in cm. This is converted to pressure (in dyne cm⁻²) by multiplying by the density of toluene (0.862 g/cm³) and g (981 cm sec⁻²). We tell you this so that if you look at other data (or are given a devious homework problem) you can figure out the correct units. The plot of π/c versus c gives a nice straight line with an intercept of 0.75 x 10⁵ cm² sec⁻² at c = 0 (the scale on the y-axis of figure 10.3 has been reduced by multiplying by 10⁻⁵). The number average molecular weight is then given by:

$$\overline{M}_n = \frac{RT}{0.75 \times 10^5 \text{ cm}^2 \text{ sec}^{-2}} = \frac{8.314 \times 10^7 \times 298}{0.75 \times 10^5} \left[\frac{\text{erg. mole}^{-1} \text{ °K °K}^{-1}}{\text{cm}^2 \text{ sec}^{-2}} \right]$$

$$= 330{,}000 \text{ g mole}^{-1}.$$

Not all data is as easy to handle as this, however. At what point, exactly, does a dilute solution become a moderately concentrated one? Consider the osmotic pressure data reported by Leonard and Doust[*] which is plotted in figure

---

[*] J. Leonard and H. Doust, *J. Polym. Sci.*, **57**, 53 (1962). You may have noticed that we are using data from "old" papers. That is because these have the almost forgotton virtue of tabulating their results. Many modern papers just give plots of data, which is OK, unless, like us, you want to recalculate everything!

10.4. This data extends to a concentration that is approximately an order of magnitude greater than that shown in figure 10.3 and the curvature is pronounced. (In fact, the data reported by Leonard and Doust extended to even higher concentrations, but we have arbitrarily cut these off.) We have fit this data to a third order polynomial which conforms to the first three terms of a virial equation;

$$\frac{\pi}{c} = \frac{RT}{M_n} + Bc + Cc^2 \qquad (10.33)$$

and obtained an apparently good fit with an intercept of 0.210×10^3 cm of chlorobenzene (which corresponds to an $\overline{M}_n$ of 1.08×10^5 g/mole, once you convert the units of π).

However, if we consider just the first few points, as shown in the insert, these data fit a perfectly acceptable linear equation (the first two terms of equation 10.33), but now the intercept is determined to be 0.224×10^3 cm^2 (corresponding to an $\overline{M}_n$ of 1.02×10^5 g/mole). This is a difference of roughly 5%, which may be good enough for some purposes, but if accurate results are required some decisions have to be made about the data to be included in a calculation and whether this data is appropriate for a first, second or higher order virial equation. In other words, at what concentration do the higher order virial coefficients contribute appreciably to values of π and have to be accounted for?

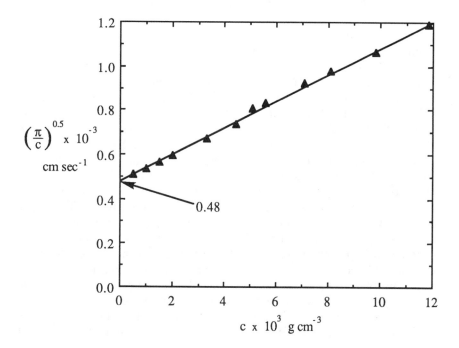

Figure 10.5 Graph of $(\pi/c)^{0.5}$ versus c for polyisobutylene in chlorobenzene (data as in figure 10.4, but with units of π converted to dynes/cm^2).

Stockmeyer and Casassa[*] suggested taking the virial equation in the form of equation 10.9:

$$\frac{\pi}{c} = \left(\frac{\pi}{c}\right)_{c \to 0} \left[1 + \Gamma_2 c + \Gamma_3 c^2 + \Gamma_4 c^3 + \text{------} \right]$$

Then by truncating after the Γ_3 term and assuming,

$$\Gamma_3 = 0.25\,\Gamma_2^2 \qquad\qquad (10.34)$$

(there is some theoretical justification for this assumption) the following equation can be obtained:

$$\left(\frac{\pi}{c}\right)^{0.5} = \left(\frac{\pi}{c}\right)^{0.5}_{c \to 0} \left[1 + \left(\frac{\Gamma_2}{2}\right) c \right] \qquad\qquad (10.35)$$

A so-called "square root plot" of $(\pi/c)^{0.5}$ versus c is shown in figure 10.5 and there is a good fit to a straight line with an intercept at zero concentration equivalent to 0.23 x 10^6 cm^2 sec^{-2} (corresponding to an $\overline{M}_n$ of 1.08 x 10^5 g/mole). Yamakawa[**] concluded that the square root plot is generally more reliable than the conventional π/c versus c plot, particularly for higher molecular weight samples.

C. LIGHT SCATTERING AND THE DETERMINATION OF WEIGHT AVERAGE MOLECULAR WEIGHT

The scattering of light, X-rays and neutrons are extremely important characterization methods in polymer science. In this introductory text we will only discuss light scattering, and then at an elementary level, leaving the bulk of this subject to more advanced and specialized treatments.

Like thermodynamics, the problem with discussing the scattering of light, even in a limited fashion, is knowing how much prior knowledge to assume. Most students should have some background in basic physics before tackling polymer science, but it is our experience that this knowledge has usually been imperfectly digested and as a result the apparent complexity of the equations for the scattering of light by dilute polymer solutions is a source of deep unease if not outright terror for students new to the subject. However, if examined in a calm manner, the final equation can be viewed as a virial equation of the form we discussed for osmotic pressure, in that it expresses the concentration dependence of scattering, superimposed upon which is a second expansion in terms of the dependence of the intensity of scattered light on the angle of observation. If this explanation is enough for your purposes, then we suppose you could jump to the final expression (equation 10.59). If you intend to have more than a superficial knowledge of this subject, however, you should read on and where necessary go back to your fundamental physics texts to remind yourself of concepts you've probably long forgotten. We will start with a description of the phenomenon of

[*] W. H. Stockmayer and E. M. Casassa, *J. Chem. Phys.*, **20**, 1560 (1952).
[**] H. Yamakawa, *Modern Theory of Polymer Solutions*, Harper and Row, New York (1971).

light scattering and leave a discussion of the experiments until we get to polymers.

Light Scattering from Gases

Starting almost at the beginning, with an extremely simple situation, consider a single, small (relative to the wavelength of light), spherical molecule located in a vacuum. We impinge a light beam onto this molecule and to keep the analysis as elementary as possible we imagine that this light is plane-polarized; in other words, the electric field of the light is confined to a plane. If we could sit at one point (say on the molecule, assumed stationary for the moment) and "watch" the field as a function of time, it would vary sinusoidally, as illustrated schematically in figure 10.6. The arrows in this figure indicate the magnitude of the electric field and the maximum value is E_0. Motion of this type is often described in terms of an angular frequency, ω (rads/sec); and can be written in the form:

$$E = E_0 \cos \omega t \qquad (10.36)$$

Thus, the electric field oscillates from E_0 to zero to $- E_0$. The oscillations of the electric field correspond to the projection of a vector traveling in a circular motion, as also illustrated in figure 10.6, and the angular frequency is, of course, just the angle swept out per second ($d\phi/dt$, see figure 10.6). The period of an oscillation, the time taken to go one full revolution is simply:

$$\text{Period} = \frac{2\pi}{\omega} \qquad (10.37)$$

In terms of the light wave itself a complete period is defined in terms of a wavelength λ, (see figure 10.6), and the period is also this distance divided by the velocity of light[*], $\tilde{c}$. Hence:

$$\frac{\lambda}{\tilde{c}} = \frac{2\pi}{\omega} \qquad (10.38)$$

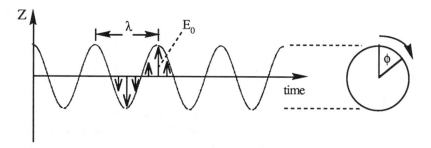

Figure 10.6 *Schematic diagram depicting the sinusoidal character of a light wave and its relationship to circular motion.*

---

[*] Unfortunately, c is the symbol used by convention to represent both the speed of light and concentration, so we have used a tilde to distinguish between the two.

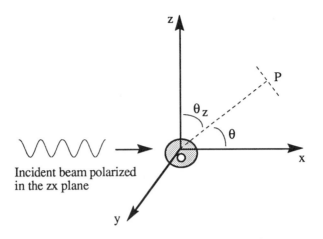

Figure 10.7 *Schematic diagram depicting the geometry of light scattering from a small molecule.*

or:

$$\omega = \frac{2\pi}{\lambda} \, \tilde{c} \tag{10.39}$$

so that the magnitude of the electric field can be written:

$$E = E_0 \cos\left(\frac{2\pi \tilde{c} t}{\lambda}\right) \tag{10.40}$$

Now we must consider how this electric field interacts with the molecule. Simply stated, the electric field drives or induces oscillatory motions of the electrons. The degree to which this occurs depends upon the details of atomic and molecular structure and is described in terms of a parameter called the *polarizability* of the molecule, α. The magnitude of the induced dipole moment, p, is given by:

$$p = \alpha E = \alpha E_0 \cos\left(\frac{2\pi \tilde{c} t}{\lambda}\right) \tag{10.41}$$

If the light is traveling in the x direction of a Cartesian system and is polarized in the xz-plane, then the dipole moment is a vector directed along the z-axis. Now you should recall from your courses in physics that an accelerating charge produces an electric field that varies inversely with the distance r from the oscillator and is proportional to the acceleration of the charge projected onto the plane of observation. In other words, the forced oscillations of the electrons by the incident beam of light result in the molecule becoming a new source of radiation which has the same frequency as the incident light. The light from the oscillator is emitted in all directions, including back towards the source and also "forward" in the direction of the incident light. This forward component combines with the incident beam and the resulting field has a "phase shift" (retardation) relative to the incident light, the origin of what we call the *refractive index*, n (the light appears to travel at a speed $\tilde{c}/n$). Our interest is the light "scattered" at some angle to the incident beam, for it is the light emitted by oscillations forced by the incident beam that is the scattered light.

As mentioned above, the electric field of the light scattered by the molecule we are considering is proportional to $1/r$ and the acceleration of the charge projected onto a plane of observation. The geometry of this is illustrated in figure 10.7. If the scattering molecule is placed at the origin of a Cartesian system and we observe the scattered light at P, where the line OP is at angle θ_z to the z-axis, then the projection of the acceleration of the charge perpendicular to P and in the PZ plane (see figure 10.7) is obtained by simply multiplying by $\sin \theta_z$. If you take our word for it (or go back to your old physics notes) that the amplitude of the wave scattered in the OP direction is given by the amplitude of:

$$\left(\frac{1}{\tilde{c}^2}\right)\left(\frac{d^2p}{dt^2}\right)$$

then by differentiating equation 10.41 and multiplying by $(\sin \theta_z)(1/r)$ we obtain:

$$E_{sc} = \left(\frac{\alpha E_0}{r}\right)\left(\frac{2\pi}{\lambda}\right)^2 \sin \theta_z \qquad (10.42)$$

Now the intensity of the scattered light is equal to the square of this amplitude, so we obtain our first result, an expression for the light scattered by a single particle in a beam of plane polarized light, i'_z:

$$i'_z = \frac{16\pi^4}{r^2 \lambda^4} I_{0,z} \alpha^2 \sin^2\theta_z \qquad (10.43)$$

where $I_{0,z} = E_0^2$ is the intensity of the incident beam.

Our next step is to make the problem slightly more complicated by considering what happens when the incident beam is unpolarized. This can be accomplished in a straightforward manner by resolving the incident intensity I_0 into two components polarized along the y and z axes. The total intensity of the incident beam is simply the sum of the components:

$$I_0 = I_{0,z} + I_{0,y} \qquad (10.44)$$

as is the intensity of the scattered light:

$$i' = i'_z + i'_y$$

$$= \left(\frac{8\pi^4 I_0 \alpha^2}{r^2 \lambda^4}\right)\left(\sin^2\theta_z + \sin^2\theta_y\right) \qquad (10.45)$$

It is a matter of simple geometry to show that the sin terms in brackets can be replaced by $(1 + \cos^2\theta)$, where θ is the angle that OP makes with the x-axis (see figure 10.7). Before doing this we generalize further by now assuming that if we have N molecules in a volume V, the scattering per unit volume is given by a simple superposition or sum of all the scatterers in this unit volume, N/V, to obtain*:

$$i'_\theta = \left(\frac{N}{V}\right)\left(\frac{I_0\, 8\pi^4}{\lambda^4}\right)(\alpha^2)\left(\frac{1+\cos^2\theta}{r^2}\right) \tag{10.46}$$

This looks complicated when you first see it, but by considering its construction piece by piece, we hope you now have a "feel" for the shape of the equation and the origin of the terms. The first term in brackets essentially accounts for the number of oscillators (molecules) per unit volume. The second term depends on the intensity and wavelength of the incident light. The third term is a molecular property (polarizability), while the final term depends upon the geometry of observation. All contribute to the intensity of scattered light. In fact, the only unknown is the polarizability, α, but since the change in refractive index of the assembly (here a gas, later a dilute solution) is small with respect to concentration, α is given by:

$$\alpha = \frac{1}{2\pi}\left(\frac{dn}{dc}\right)\frac{M}{A} \tag{10.47}$$

where M is the molecular weight and A is Avogadro's number (i.e., α depends upon the weight of a molecule, M/A). Substituting into equation 10.46 and rearranging we can obtain:

$$\frac{i'_\theta r^2}{I_0} = R_\theta = 2\pi^2\left(\frac{dn}{dc}\right)^2\frac{1}{\lambda^4}(1+\cos^2\theta)\frac{N}{V}\frac{M^2}{A^2} \tag{10.48}$$

where R_θ is the reduced intensity or the Rayleigh ratio and is the intensity of light scattered per unit volume, per unit incident radiation, per unit solid angle, per second (!) Note that the intensity of the scattered light is proportional to the square of the molecular weight. This will ultimately lead to a dependence of R_θ on $\overline{M}_w$ when we consider polymers.

Liquids and Solutions of Small Molecules

We now have our basic result. Next, we need to consider the modifications that are necessary to describe first, scattering from liquids and solutions of small molecules (i.e. much smaller than λ), then in the following section, polymer solutions. We are going to gloss over some fundamental physics and try to give you a feel for the important points without getting bogged down in a lengthy mathematical development.

The first point that should be made is that a gas is very different from a liquid in the scale of its discontinuities. In a gas there are a just a few molecules dispersed randomly through the volume of the container. A liquid is much more dense with the molecules in random close contact and if, in fact, one considered

---

* There are a lot of subtleties involved in this and we have glossed over them. If you are going to get more involved in scattering experiments you should master the underlying physics.

light incident on a theoretically "continuous" liquid (one whose density is equal at every point) you would end up with the prediction of no scattering at all! A crude way of looking at this is to reexamine equation 10.48 and note the dependence of scattering intensity on dn/dc. Adding molecules of a gas to a container (at constant volume) changes the refractive index considerably, as the number of oscillators per unit volume changes. If we consider an "incompressible" liquid, however, dn/dc would be zero. Adding more liquid molecules cannot change the density or concentration of "oscillators", just increase the volume of the sample.

Experimentally, of course, scattering is observed from liquids[*]. The origin of this scattering in pure liquids arises from fluctuations in density. Liquids are *not* incompressible (think back to our discussion of free volume). In fact we can think of scattering in general as being a consequence of discontinuities or fluctuations of various types. In a gas the molecules themselves are the "discontinuities", isolated scatterers in a vacuum. In a liquid, where all the molecules are identical, it is density fluctuations, the fact that at some instant of time there are a few more molecules per unit volume over here than over there. In a solution, where the solute will generally have different properties (polarizability, molecular weight, etc.) to that of the solvent, there will be fluctuations in the concentration of the solute from one position to another. One can visualize a *dilute* solution as being similar to a gas, except that we have to subtract out the "background" scattering from density fluctuations in the solvent (approximated to be equivalent to the pure solvent). Finally, there are also fluctuations in molecular orientation that can lead to scattering. In our treatment of a gas we assumed that the molecule was isotropic, so that light was emitted (scattered) equally in all directions. Real molecules usually have strange shapes with anisotropic distributions of electron densities. As the molecules rotate with thermal motion there are fluctuations in the distribution and arrangements of the "oscillators" relative to the incident beam. Because flexible polymers are irregularly coiled we can treat them as isotropic, however, so we don't have to worry about this type of fluctuation unless we have to deal with rigid rod or liquid crystalline polymers (which we won't; we have enough trouble in our lives!).

Although polymers simplify our treatment in that we can assume that they are isotropic, there is one complication that is introduced by the shear size of a polymer molecule. Implicit in our discussion of scattering so far has been the assumption that we can treat the molecules as point sources of radiation when oscillations of their electrons are forced by an incident beam of light. This assumption is reasonable when the molecules have dimensions less than about $\lambda'/20$, where λ' is the wavelength of light in the medium (λ/n, where n can be put equal to the refractive index of the solvent, as we will be dealing with dilute solutions). For light of wavelength 5145 Å (a common green colored laser line) and a solvent of refractive index 1.5, this means that the radius of gyration of the polymer chain can be no larger than about 170 Å in order to satisfy this

---

[*] Perhaps you have stood at right angles to a cell containing a liquid through which a laser beam is being passed. The fact that you see the beam means that there must be scattering.

condition. We will assume that this is so for now and then consider the general case of larger molecules in the following section.

As mentioned above, for a dilute solution scattering from the solute can be treated like a gas, in the sense that the intensity can be considered as the sum of contributions from the individual molecules. Superimposed upon this is the scattering from density fluctuations in the solution as a whole, but if we assume that these two contributions are independent and also assume that for a dilute solution density fluctuations are the same as in pure solvent, then we can write the total intensity of scattered light, $i_{\theta, t}^{0}$ per unit volume as:

$$i_{\theta, t}^{0} = i_{\theta, c}^{0} + i_{\theta, s}^{0} \tag{10.49}$$

where the subscripts c and s indicate the contribution from concentration fluctuations of the solute and the contribution from density fluctuations in the solvent, respectively. The superscript 0 indicates that we are considering molecules that are much smaller than λ'.

Our first modification to equation 10.48 is therefore very simple, we replace

$$i_{\theta}' \text{ with } i_{\theta, c}^{0} \ (= i_{\theta, t}^{0} - i_{\theta, s}^{0})$$

and for convenience we simply drop the subscript c; i.e., the measured intensity we use is relative to pure solvent.

There are a few more modifications that we need to make to equation 10.48 in order to apply it to dilute solutions. The polarizability α now becomes the excess polarizability of solute over the solvent and for dilute solutions we therefore replace equation 10.47 with:

$$\alpha = \frac{n_0}{2\pi} \left(\frac{dn}{dc} \right) \frac{M}{A} \tag{10.50}$$

where n_0 is the refractive index of the pure solvent and dn/dc is now the rate of change in the refractive index of the solution with concentration. The Rayleigh ratio (see equation 10.48) can then be written:

$$R_{\theta}^{0} = K \left(1 + \cos^2\theta \right) \left(\frac{M^2}{A} \frac{N}{V} \right) \tag{10.51}$$

where for simplicity parameters that are known or can be readily measured (e.g., dn/dc) are lumped together in the constant K:

$$K = \frac{2 \pi^2 n_0^2}{A \lambda^4} \left(\frac{dn}{dc} \right)^2 \tag{10.52}$$

We are almost through with our modifications to equation 10.48, there being two more factors that need to be taken into account. First the solutions we are considering are not ideal. There are interactions between the components of the solution that cannot be ignored. The treatment of these interactions and their effect on refractive index is acquired from an analysis of fluctuations. This is a difficult subject if you do not have a good background in statistical mechanics, so we will just give the result. You might intuitively expect that there should be some sort of dependence on the chemical potential, however, and that is indeed the case:

$$R_\theta^0 = K (1 + \cos^2\theta) \left(\frac{M}{A}\frac{N}{V}\right) \left(\frac{RTV_s}{(-\partial\mu_s/\partial c)}\right)$$ (10.53)

where as before V_s is the molar volume of the solvent and $(\partial\mu_s/\partial c)$ is the change in solvent chemical potential with concentration, c. From our discussion of osmotic pressure we know that we can make the substitution (see equation 10.9):

$$-\frac{\partial\mu_s}{\partial c} = V_s\frac{\partial\pi}{\partial c} = V_s\frac{RT}{M}\left(1 + 2\Gamma_2 c + \text{-------}\right)$$ (10.54)

to obtain:

$$R_\theta^0 = K (1 + \cos^2\theta) \left(\frac{M^2}{A}\frac{N}{V}\right) \left(\frac{1}{1 + 2\Gamma_2 c + \text{------}}\right)$$ (10.55)

Next, we now must consider the effect of molecular weight distribution, if the solute is a polymer. Because in dilute solution we are assuming that the scattering intensity at an angle θ is the sum of the separate contributions of each species, in equation 10.55 we can simply replace the term NM^2 with $\sum_i N_i M_i^2$. From the definition of weight average molecular weight (see Chapter 1) we have:

$$\sum_i N_i M_i^2 = \overline{M}_w \sum_i N_i M_i$$ (10.56)

The term $\sum_i N_i M_i$ is just the total weight of all the polymer molecules and using:

$$c = \frac{\sum_i N_i M_i}{V A}$$ (10.57)

we get:

$$R_\theta^0 = K (1 + \cos^2\theta)\left(\frac{\overline{M}_w c}{1 + 2\Gamma_2 c + \text{------}}\right)$$ (10.58)

our final result for scattering from solutions where the molecules are smaller than $\lambda'/20$. Note that the light scattering depends upon $\overline{M}_w$, and terms in the angle of observation θ and a virial expansion in the concentration.

So far we have not discussed what is actually measured in a light scattering experiment, though an examination of equation 10.58 should make this clear. The intensity of scattered light, relative to that of the solvent, is measured as a function of concentration and angle of observation. There are constants that depend upon the instrument (r, λ, etc.) or the solvent and solution (n_0, dn/dc; this latter quantity may have to be determined separately by additional experiments). In most texts a description of classical light scattering equipment is given, these instruments usually being of the type used when we were young graduate students (just after the fall of Constantinople). We won't do this because the technology has changed considerably, and modern instruments just look like a box attached to a computer. Inside the box is a laser and these days a number of detectors arranged at preset angles to the incident beam. The crucial point to grasp is that measurements of scattered light intensity as a function of θ and c must be made. When the size of the molecule is less than $\lambda'/20$, then the so-called excess scattering envelope (i.e., relative to the pure solvent), the distribution of the scattered intensity as a function of angle, looks like that shown in figure 10.8(B). This is a two dimensional representation and the three dimensional figure is obtained by rotating about the x-axis. If measurements are made at some angle (e.g., 90°) to the incident beam and equation 10.58 is rearranged to the following form:

$$\frac{K(1+\cos^2\theta)\,c}{R_\theta^0} = \frac{1}{\overline{M}_w}\left(1 + 2\Gamma_2 c + \cdots\cdots\right) \qquad (10.59)$$

then a plot of the parameters on the left hand side versus c should give a straight line with an intercept of $1/\overline{M}_w$. We won't give an example of such a plot, however, because we usually have to take into account what happens when the dimensions of the polymer coil approach the wavelength of the incident light.

Light Scattering from High Molecular Weight Dilute Polymer Solutions

When the polymer coil has dimensions larger than about $\lambda'/20$, then it can no longer be considered a point source of radiation. In other words, light scattered from different parts of the molecule is no longer in phase and we have to consider interference effects. If the experiment is arranged as in figure 10.9, then it can be seen that pathlength differences are larger in the "backward" direction than in the forward direction and destructive interference increases with scattering angle θ (it is zero when $\theta = 0°$). Accordingly the scattering envelope

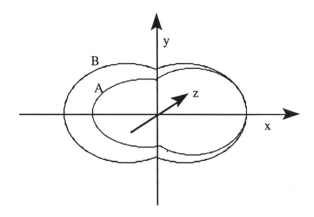

Figure 10.8 Schematic diagram depicting the scattering envelope from a high molecular weight polymer solution (A) relative to the scattering envelope calculated (B) for the same solution in the absence of interference.

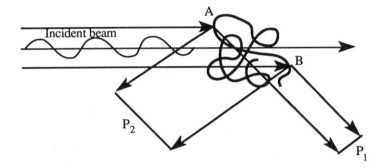

Figure 10.9 *Schematic diagram depicting the scattering of light from different parts of a polymer molecule.*

is now asymmetric, as illustrated earlier in figure 10.8. In order to account for this a term P(θ), called the particle scattering factor, is introduced into equation 10.59, which now takes the form:

$$\frac{K\,(1+\cos^2\theta)\,c}{R_\theta} = \frac{1}{M_w\,P(\theta)}\left(1+2\Gamma_2 c + \text{------}\right) \qquad (10.60)$$

The problem is that the factor P(θ) depends on the shape of the molecule, whether it is globular (like a protein), a rigid rod, a coil, etc. We are concerned here with the general situation of determining the molecular weight of a polymer that behaves like a coil in dilute solution and for this situation P(θ) can be expressed as a series in θ. It is usual to truncate this series after the second term such that:

$$\frac{K\,(1+\cos^2\theta)\,c}{R_\theta} = \frac{1}{M_w}\left(1+2\Gamma_2 c + \text{---}\right)\left(1+ S\,\sin^2\!\left(\frac{\theta}{2}\right)\right) \qquad (10.61)$$

where the term S depends upon various constants and also the radius of gyration of the molecule.

Equation 10.61 is our final result and it can be used in a number of ways, of which we will mention two. The first is conceptually simple, but experimentally demanding. At θ = 0, P(θ) = 1, (sin²(θ/2) = 0) and cos²θ = 1, so that :

$$\frac{2\,Kc}{R_0} = \frac{1}{M_w}\left(1+2\Gamma_2 c + \text{---}\right) \qquad (10.62)$$

Measurements of scattering as a function of concentration would then allow the determination of $\overline{M_w}$. Scattering measurements at zero angle cannot be made, but at small angles cosθ and P(θ) are both very close to unity (e.g., at 6°, cosθ = 0.994 and sin²(θ/2) = 0.003). Accordingly, weight average molecular weights of polymers have been routinely estimated by assuming that little error is introduced if the value of $2Kc/R_\theta$ at low angles is the same as that at zero angle.

Even so, measurements at these small angles are not easy and more often, a method attributed to Bruno Zimm[*] is employed, as follows.

A careful examination of equation 10.61 reveals that a plot of;

$$\frac{K(1 + \cos^2\theta)\, c}{R_\theta} \quad versus \quad \sin^2\left(\frac{\theta}{2}\right)$$

for a solution of a given concentration gives an intercept of $1/\overline{M}_w$ at $\sin^2(\theta/2) = 0$. (The slope of such a plot also gives S, which in turn is related to the radius of gyration through a few simple constants. However, determining the size of a polymer coil by this method is not very accurate for molecular weights less than about 80,000 g/mole.)

In a similar fashion, if instead of holding c constant, measurements of scattered intensity at constant θ on solutions of different compositions are made, an extrapolation to c = 0 again gives $1/\overline{M}_w$. In practice, measurements of scattering as a function of both c and θ are made and a double extrapolation to c = 0 and θ = 0, respectively, is performed on the same graph, as illustrated in figure 10.10. The quantity $K(1 + \cos^2\theta)c / R_\theta$ is plotted against $\sin^2(\theta/2) + 100c$

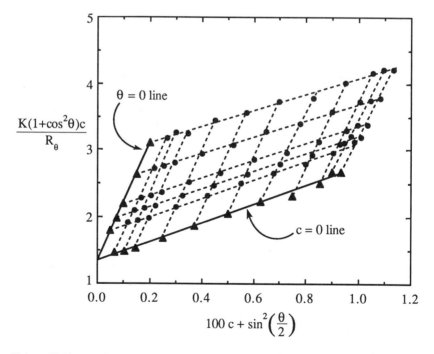

Figure 10.10 A Zimm plot showing both the angular and concentration dependence of scattering for polystyrene in benzene. Drawn from the data listed in D. Margerison and G. C. East, Introduction to Polymer Chemistry, Pergamon Press, Oxford, 1967.

---

[*] B. H. Zimm, *J. Chem. Phys.*, **16**, 1093 (1948).

(the factor 100, or sometimes 1000, is an arbitrary constant that serves to spread out the data and make it easier to see). There are two sets of lines, one corresponding to constant c and various values of θ, while the second set is the other way around. These two sets of lines trace out the characteristic grid pattern of a Zimm plot. (Question: how are the lines corresponding to c = 0 and θ = 0 data points determined? It is obvious, but we leave it as a homework problem as it is our experience that very few students really understand a Zimm plot unless they actually do one.) The c = 0 lines and $\sin^2(\theta/2) = 0$ lines should then intercept the y axis at a common point corresponding to $1/\overline{M}_w$, as shown in figure 10.10.

To summarize, light scattering provides (in principle) a wealth of data, not only $\overline{M}_w$, but also the radius of gyration and the second virial coefficient (Γ_2) and it is one of the most important characterization tools in polymer science.

D. SOLUTION VISCOMETRY AND THE VISCOSITY AVERAGE MOLECULAR WEIGHT

The viscosity of a fluid is simply a measure of its resistance to flow and as such reflects the frictional forces between the molecules*. In a solution one would therefore expect these frictional forces to increase with the size of the solute (simply because there will be more "contacts"), so that viscosity measurements should provide a measure of molecular weight. And indeed they do, as Staudinger realized in the very early days of polymer science. Unfortunately, there is not the clear theoretical relationship between measured or measurable parameters and molecular weight that we have with osmotic pressure and light scattering and the fundamental equation that we use is semi-empirical. As such, the parameters must be determined by experiments on polymer standards of known molecular weight, i.e., it is a *relative* not *absolute* method of molecular weight determination. Nevertheless, the theory that has been developed provides considerable insight and shows how viscosity measurements depend upon the effective size or hydrodynamic radius of a polymer coil. They are therefore useful for various purposes, as we will see later when we discuss GPC/SEC and branching. Accordingly, we will start our discussion of viscometry by first describing the nature of the experiments and the treatment of the data; then we will consider, in a largely qualitative fashion, the theory of the frictional properties of polymer molecules; finally, we will examine the empirical relationship mentioned above and a definition of a viscosity average molecular weight.

Measuring the Viscosity of Polymer Solutions

The most common method used to determine the viscosity of a polymer solution is to measure the time taken to flow between fixed marks in a capillary

---

* We will consider the macroscopic aspects of this in a little more detail in Chapter 11, when we discuss mechanical and rheological properties. It is useful to note here, however, that for a Newtonian fluid the viscosity is equal to the shear stress divided by the rate of strain and therefore has units of dyne sec. cm^{-2}, or *poise*.

tube under the draining effect of gravity. The (volume) rate of flow, υ, is then related to the viscosity by Poiseuille's equation*:

$$\upsilon = \frac{\pi P r^4}{8 \eta l} \tag{10.63}$$

where P is the pressure difference maintaining the flow, r and l are the radius and length of the capillary, respectively, and η is the viscosity of the liquid.

The viscosity of a polymer solution depends on frictional forces between solvent molecules, between polymer and solvent and also between polymer molecules. To obtain a simple relationship describing viscosity as a function of polymer molecular weight we need a parameter that relates just to the polymer/solvent frictional forces. As you might guess, the effect of polymer/polymer interactions can be eliminated by making measurements in dilute solution and extrapolating to infinite dilution (i.e., $c \rightarrow 0$, as in osmometry and light scattering experiments), but we will get to that shortly. The first thing that is measured is actually the *relative viscosity*, η_{rel}, which is the viscosity of a polymer solution (η) divided by the viscosity of the pure solvent (η_0). If the rate of flow between the two marks on a capillary is being measured, then by dividing Poiseuille's equation for the solution by that for the solvent and rearranging we obtain:

$$\eta_{rel} = \frac{\eta}{\eta_0} = \frac{t \rho}{t_0 \rho_0} \tag{10.64}$$

where t is the time taken for a volume V of solution (no subscript) or solvent (subscript 0) to flow between the marks. (An Ubbelohde suspended level type of viscometer is typically used for these measurements.) The pressure P maintaining the flow is proportional to the densities of the solution and solvent (ρ and ρ_0) for the same "head" of liquid used in each experiment. For dilute solutions we can assume that $\rho = \rho_0$ so that the relative viscosity is determined simply from the ratio of the time of flow of a solution relative to that of the pure solvent.

$$\eta_{rel} = \frac{\eta}{\eta_0} \approx \frac{t}{t_0} \tag{10.65}$$

If the relative viscosity of solutions of a polymer is plotted as a function of concentration, as in the example shown in figure 10.11, then it can be seen that there is a deviation from linearity. A power series, similar to that used in the treatment of osmotic pressure and light scattering data, is commonly used to fit relative viscosity data:

$$\eta_{rel} = \frac{\eta}{\eta_0} = 1 + [\eta] c + k c^2 + \ldots \tag{10.66}$$

---

* Poiseuille was a French physician whose interest in the measurement of blood led him to study the flow of fluids in capillaries. The units we use to descibe viscosity, poise, are named in his honor.

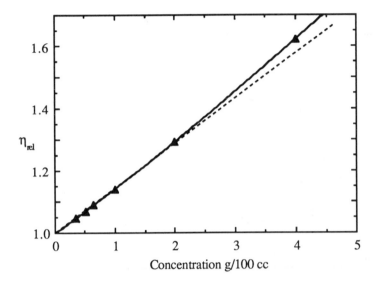

Figure 10.11 *Plot of η_{rel} versus c for poly(methyl methacrylate) in chloroform. Plotted from the data of G. V. Schultz and F. Blaschke, J. für* Prakt. Chemie, *158, 130 (1941).*

where both $[\eta]$ and k are constants. The former is called the *intrinsic viscosity* [*] and has special significance, as we will see in a moment (the square brackets are part of the representation of this constant). Note that this equation expresses the fact that at zero concentration of polymer η must equal η_0.

If viscosity measurements are confined to dilute solution, so that we can neglect terms in c^3 and higher, then equation 10.66 can be rearranged to give:

$$\left(\frac{\eta_{rel} - 1}{c}\right) = \frac{1}{c}\left(\frac{\eta - \eta_0}{\eta_0}\right) = [\eta] + k\,c \qquad (10.67)$$

The viscosity term on the left hand side:

$$\eta_{rel} - 1 = \left(\frac{\eta - \eta_0}{\eta_0}\right) = \eta_{sp} \qquad (10.68)$$

measures the increase in viscosity of the solvent that is a result of adding polymer relative to (i.e. divided by) the viscosity of the pure solvent. This is called the specific viscosity, η_{sp}. Note also from equation 10.67 that as c goes to zero (infinite dilution), then the intercept on the y-axis of a plot of (η_{sp}/c) against c is the *intrinsic viscosity*, $[\eta]$:

---

[*] The relative viscosity is dimensionless, so that the intrinsic viscosity must have dimensions of 1/c, or cm^3/g, to make all the units in this equation come out right. In many publications units of $100\ cm^3/g$ are used, however, so watch out!

Table 10.2 Definitions Employed in Solution Viscosity Measurements.

Name	Symbol and Definition
Relative viscosity Viscosity ratio (IUPAC)	$\eta_{rel} = \dfrac{\eta}{\eta_0} = \dfrac{t}{t_0}$
Specific viscosity	$\eta_{sp} = \eta_{rel} - 1 = \dfrac{\eta - \eta_0}{\eta_0} = \dfrac{t - t_0}{t_0}$
Reduced viscosity Viscosity number (IUPAC)	$\eta_{red} = \dfrac{\eta_{sp}}{c}$
Inherent viscosity Logarithmic viscosity number (IUPAC)	$\eta_{inh} = \dfrac{\ln \eta_{rel}}{c}$
Intrinsic viscosity (Staudinger Index) Limiting viscosity number (IUPAC)	$[\eta] = \left(\dfrac{\eta_{sp}}{c}\right)_{c=0} = \left(\dfrac{\ln \eta_{rel}}{c}\right)_{c=0}$

$$[\eta] = \left(\frac{\eta_{sp}}{c}\right)_{c \to 0} \tag{10.69}$$

and just to confuse you, η_{sp}/c, is also given a name, the *reduced viscosity*, η_{red}. The intrinsic viscosity is the parameter we want, however, as it describes polymer/solvent frictional forces only (i.e., with contributions from polymer/polymer and solvent/solvent interactions removed). It is this quantity that will be related to molecular weight.

At this point you may be starting to muddle up the various viscosity parameters that we have given you. It is not going to get better, because there's one more on the way, but to provide a handy guide we have collected the necessary definitions in table 10.2. There are two sets, the ones we have used that almost everybody else also uses, and a set devised by IUPAC, which they hoped everybody would use, but almost nobody does.

In practice, equation 10.67 is usually used in the semi-empirical form suggested by Huggins[*] , who pointed out that the slopes of the plots of η_{sp}/c versus c (i.e., k) for a given polymer solvent appear to be proportional to $[\eta]^2$, so that:

$$\frac{\eta_{sp}}{c} = [\eta] + k' [\eta]^2 c \tag{10.70}$$

There are other equations that have also been used to analyze viscosity data and the most common of these is one due to Kraemer, which is often used in conjunction with the Huggins equation. If instead of defining the intrinsic viscosity as:

$$[\eta] = \left(\frac{\eta_{sp}}{c}\right)_{c \to 0} \tag{10.71}$$

---

[*] M. L. Huggins, *J. Am. Chem. Soc.*, **64**, 2716 (1942).

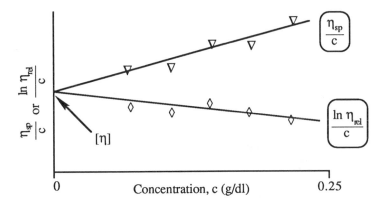

Figure 10.12 *Schematic diagram illustrating the graphical determination of the intrinsic viscosity.*

it is defined as:

$$[\eta] = \left(\frac{\ln \eta_{rel}}{c}\right)_{c \to 0} \tag{10.72}$$

where $\ln \eta_{rel}/c$ is known as the inherent viscosity, η_{inh}, then we would expect a power law relationship of the same form as the Huggins equation:

$$\frac{\ln \eta_{rel}}{c} = [\eta] + k''[\eta]^2 c \tag{10.73}$$

One would also anticipate that there should be a relationship between the constant k'' in the Kraemer equation and the constant k' in the Huggins equation, which can be obtained by expanding the logarithm:

$$\ln \eta_{rel} = \ln(1 + \eta_{sp}) = \eta_{sp} - \frac{\eta_{sp}^2}{2} - \text{------} \tag{10.74}$$

Comparing equations 10.70, 10.73 and 10.74 it follows that $k'' = k' - 1/2$. Because both the Huggins and Kraemer equations lead to the same extrapolated value of $[\eta]$, it is common practice to use both equations on the same plot, as shown schematically in figure 10.12. (Note that one does not require any further experimental data, only a few more calculations.)

Although the use of a combination of the Huggins and Kraemer equations increases confidence in the extrapolation procedure, we must caution that there are many cases, especially in systems with relatively strong intermolecular interactions between the polymer and solvent, where the errors involved are very large. Extreme examples include solutions of ionic polymers or polyelectrolytes. A typical, but not exceptional, result that might be encountered is illustrated schematically in figure 10.13. Neither equation is strictly valid at any of the experimental concentrations.

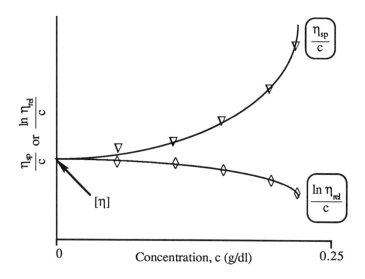

Figure 10.13 *Schematic diagram illustrating solution viscosity as a function of concentration for a polymer in a solvent where there are strong intermolecular interactions.*

This leads us to an important point. Unfortunately, most "extrapolation to zero concentration" procedures have a serious limitation. Where one would like to perform measurements is at the lowest concentrations possible, but this is generally where the greatest error in measurement occurs.

Frictional Properties of Polymers in Solution

We will briefly mention two models that have been used to describe the frictional properties of polymer molecules in solution. The first of these is the so-called freely draining model, which essentially assumes that the polymer chain is like a string of spherical beads and the frictional force on each bead can be described using Stokes law. The defining feature of this model is that the velocity of the medium is barely affected by the presence of the polymer, so that the solvent "streams" or "freely drains" through the polymer in a largely unperturbed fashion, as illustrated in figure 10.14. By considering the shear forces on the polymer molecule that are a result of viscous flow, then the viscosity can be calculated as a sum of contributions to energy dissipation from all of the beads*. This results in a prediction of the relationship between intrinsic viscosity and molecular weight that has the form:

$$[\eta] \approx M^{(1+\Delta)} \tag{10.75}$$

---

* This is described in Flory's book, if you are interested in working through the math.

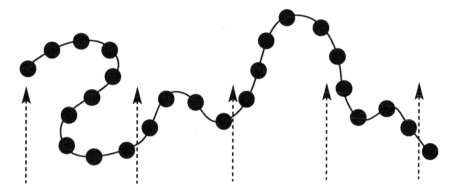

Figure 10.14 *Schematic diagram of the "freely draining" model. The arrows represent the largely unimpeded flow of solvent.*

where Δ is a small fraction. Experimentally, as we will see, $[\eta]$ varies with M^a, where the factor 'a' usually has a value between 0.5 and 0.8. Clearly the assumption that the polymer "beads" do not perturb the solvent is not a good one.

Going to the other extreme, it can be assumed that the solvent near the center of a polymer coil moves with the same velocity as the beads, so that the coil behaves like a sphere that is, in effect, impenetrable to solvent. Of course, this is not quite right, because the density of the polymer segments decreases with increasing distance from the center of gravity (see Chapter 9). But this is a useful and important start because it leads to the concept of an *equivalent hydrodynamic sphere*, which would display the same properties as the polymer coil. The viscosity of a solution of solid particles of various shapes has been treated by Einstein and Simha who obtained the relationship:

$$\eta_{rel} = 1 + \gamma \phi \tag{10.76}$$

where Φ is the volume fraction of the particles and γ is a constant which has the value 2.5 for spheres.

The volume fraction of the polymer coils or spheres is given by the number of polymer molecules, N, multiplied by the hydrodynamic volume, V_h, of each, divided by the total volume, V:

$$\phi = \frac{N V_h}{V} \tag{10.77}$$

Noting that the concentration (wt/vol) is given by:

$$c = \frac{N}{V}\left(\frac{M}{A}\right) \tag{10.78}$$

where A is Avogadro's number, then:

$$\phi = V_h \left(\frac{A}{M}\right) c \tag{10.79}$$

so that:

$$\eta_{rel} - 1 = \eta_{sp} = 2.5\,\phi = 2.5\,V_h\left(\frac{A}{M}\right)c \qquad (10.80)$$

hence, as $c \to 0$:

$$\left(\frac{\eta_{sp}}{c}\right)_{c \to 0} = [\eta] = \frac{2.5\,V_h\,A}{M} \qquad (10.81)$$

The volume occupied by a sphere whose diameter is equal to the root mean square end-to-end distance of a polymer coil is given by:

$$V' = \frac{4\,\pi}{3}\left(<R^2>^{0.5}\right)^3 \qquad (10.82)$$

For a polymer in dilute solution we have seen that $(<R^2>^{0.5})$ is proportional to $M^{0.5}\,\alpha$, where α is the chain expansion factor (see Chapter 9; note that previously we related $(<R^2>^{0.5})$ to the number of segments, but this is obviously equal to the molecular weight of the chain, M, divided by the molecular weight of a segment, M_0.) Accordingly, if we assume that V_h is proportional to V', then the intrinsic viscosity should be related to the molecular weight by:

$$[\eta] = K'\,M^{0.5}\,\alpha^3 \qquad (10.83)$$

where the various constants of proportionality have been lumped into K'. For a good solvent α varies as $M^{0.1}$, while in a theta solvent $\alpha = 1$, so that equation 10.83 can be rewritten as:

$$[\eta] = K\,M^a \qquad (10.84)$$

where a is a constant that varies between 0.5 and 0.8, depending upon solvent and temperature. Suprisingly, given the simplicity of this model, a relationship of this form provides a good description of the data and also provides a basis for using viscosity measurements to determine molecular dimensions (α) as well as molecular weight. We won't go into this any further, however. Clearly, this model is flawed as we would not expect the solvent molecules within the coil to move with exactly the same velocity as the coil. In other words, the coil should be permeable to some degree to solvent. If both K and a are treated as parameters that vary with polymer, solvent and temperature, however (i.e. the interactions between polymer and solvent), then a useful semi-empirical equation is obtained, as we will now discuss.

The Mark-Houwink-Sakurada Equation

If the logarithm of the intrinsic viscosities of a range of samples is plotted against the logarithm of their molecular weights, then linear plots that obey equation 10.84:

$$[\eta] = KM^a$$

are obtained, as illustrated in figure 10.15 using the data reported by Grubisic et al. This equation was originally established in an empirical fashion and is known

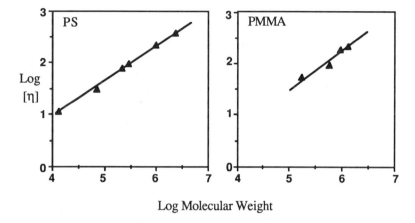

Log Molecular Weight

Figure 10.15 *Plots of the log [η] versus log M for polystyrene and poly(methyl methacrylate). Replotted from the data of Z. Grubisic, P. Rempp and H. Benoit,* J. Polym. Sci. Polym. Letters, *5, 753, (1967).*

as the Mark-Houwink-Sakurada equation*. As we have seen, this relationship does have some theoretical justification. It has been found experimentally that K is not a universal constant, but like the exponent "a" varies with the nature of the polymer, the solvent and the temperature.

The Mark-Houwink-Sakurada equation provides the basis for determining molecular weights from solution viscosity measurements. To reiterate, the values of M obtained are not absolute because the theoretical interpretation of the constants K and a is incomplete. Usually, K and a are determined graphically

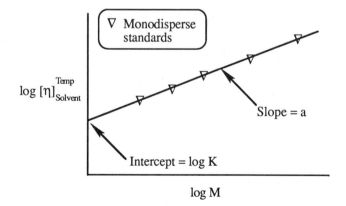

Figure 10.16 *Schematic diagram illustrating the graphical determination of the Mark-Houwink constants K and a.*

---

* In older texts this equation is often referred to simply as the Mark-Houwink equation.

(see figure 10.16) by measuring the intrinsic viscosity of a number of monodisperse samples (or fractions with narrow molecular weight distributions) of the polymer whose molecular weight has been determined independently from an absolute method such as osmometry or light scattering.

There are numerous values of K and as reported in various texts[*] for different polymers in different solvents at different temperatures. We caution, however, that many reported values were determined from fractions or whole polymer samples that were not monodisperse and where one particular average molecular weight was employed. One must expect errors when using these values.

The Viscosity Average Molecular Weight

For osmotic pressure and light scattering we saw that there is a clear relationship between experimental measurement and a particular molecular weight average ($\overline{M}_n$ or $\overline{M}_w$). Because viscosity measurements are related to molecular weight by a semi-empirical relationship, we have to consider a new average, the viscosity average, $\overline{M}_v$, for polydisperse polymer samples. This can be obtained by assuming that the solution is so dilute that the specific viscosity is simply the sum of the contributions from all the polymer chains of different length (i.e., they do not interact with one another)

$$\eta_{sp} = \sum_i \left(\eta_{sp}\right)_i \tag{10.85}$$

where $(\eta_{sp})_i$ is the contribution from all the chains of size i. Substituting:

$$\frac{\left(\eta_{sp}\right)_i}{c_i} = K\,M_i^a \tag{10.86}$$

into equation 10.85 we obtain:

$$\eta_{sp} = K \sum_i M_i^a c_i \tag{10.87}$$

Then because the solution is very dilute:

$$[\eta] = \frac{\eta_{sp}}{c} = \frac{K \sum_i M_i^a c_i}{c} \tag{10.88}$$

where:

$$c = \sum_i c_i \tag{10.89}$$

The term c_i/c is simply the weight fraction of i, w_i, in the whole polymer sample and therefore by definition is given by:

$$w_i = \frac{N_i M_i}{\sum_i N_i M_i} \tag{10.90}$$

---

[*] See for example, Chapter IV of the *Polymer Handbook*, J. Brandrup and E. H. Immergut, Editors, J. Wiley & Sons, New York, 1975.

Substituting into equation 10.88:

$$[\eta] = \frac{K \sum_i N_i M_i^{(a+1)}}{\sum_i N_i M_i} \tag{10.91}$$

Remember:

$$[\eta] = KM^a$$

so that it makes sense to define a *viscosity-average* molecular weight as:

$$\overline{M}_v = \left[\frac{\sum_i N_i M_i^{(a+1)}}{\sum_i N_i M_i} \right]^{\frac{1}{a}} \tag{10.92}$$

Thus $\overline{M}_v$ is a function of the solvent through the Mark-Houwink parameter a. In a theta-solvent, a = 0.5 and $\overline{M}_v$ lies between $\overline{M}_n$ and $\overline{M}_w$, i.e.:

$$\overline{M}_v = \left[\frac{\sum_i N_i M_i^{1.5}}{\sum_i N_i M_i} \right]^2_\theta \tag{10.93}$$

For the most probable distribution (see Chapter 4) in a θ-solvent it can be shown that:

$$\overline{M}_n : \overline{M}_v : \overline{M}_w = 1 : 1.67 : 2 \tag{10.94}$$

Finally, in good solvents, a approaches unity and $\overline{M}_v$ approaches $\overline{M}_w$. However, in such solvents the graph of (η_{sp}/c) against c is steeper and often curved so that extrapolation to c = 0 is less reliable. Sometimes you can't win for losing.

E. SIZE EXCLUSION (OR GEL PERMEATION) CHROMATOGRAPHY

Gel permeation chromatography (GPC), a technique which has been fundamental to the development of polymer science, is a traditional analytical method that is now included under the umbrella of the more general classification of separation techniques referred to as size exclusion chromatography (SEC). A schematic diagram of a typical size exclusion chromatograph is shown in figure 10.17.

In SEC, beads which contain pores of varying size and distributions, commonly made from different types of glass (or cross-linked polystyrene in the case of GPC—from which, incidently, the name gel in GPC originates), are packed into a column. This column, or more typically set of columns, is, in effect, a maze for the molecules. A solvent, or mixture of solvents, is pumped

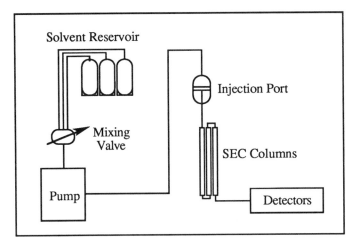

Figure 10.17 *Schematic diagram of an SEC instrument.*

through the column(s) and then through one or more detectors. A small sample of a solution of the polymer to be analyzed (in the same solvent) is introduced though an injection port. A fractionation of the polymer sample then occurs. Molecules of different size are eluted at different times and the detectors measure the amount of the different species as they pass by.

This is not the place to discuss the precise mechanisms postulated for the complex separation of molecules by SEC, or the many subtleties pertaining to problems such as instrumental broadening, resolution, detector response, etc.

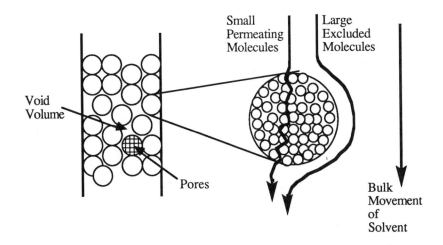

Figure 10.18 *Schematic diagram depicting the separation of molecules of different size by SEC.*

(the interested reader is referred to the article by Balke for further details*). We will keep it very simple and present only a qualitative picture of SEC, which is summarized schematically in figure 10.18. The essential principle is that small molecules can permeate and diffuse into the pores of the beads much more readily than large ones. The path of the small molecules may be likened to taking a walk with an inquisitive puppy, who sniffs and pokes his nose into every nook and cranny along the way. Very large molecules, on the other hand, cannot enter the pores and are physically excluded. In our analogy, this corresponds to an old dog, who is bored, seen it all, and can't wait to get home to take a nap! Intermediate-size molecules can permeate some but not all of the passages in the particles. For a given volume of solvent flow, molecules of different size travel different pathlengths within the column. The smaller ones travel greater distances than the larger molecules due to permeation into the molecular maze. Hence, *the large molecules are eluted first* from the column, followed by smaller and smaller molecules.

The Calculation of Molecular Weight Averages from SEC

SEC is not an absolute method for the determination of molecular weights and calibration of the instrument is required to provide a relationship between the time or amount of liquid eluted (called the elution volume) and molecular size. For example, essentially monodisperse polystyrene standards of known molecular weights are commercially available and solutions of these standards may be passed through the particular column or set of columns of an instrument to generate a calibration curve for polystyrene in a given solvent at a particular temperature. A calibration curve may be formed by plotting the log molecular weight against elution volume as depicted in figure 10.19. It should be emphasized that a particular column set has a restricted range where selective permeation occurs and any subsequent measurements made using this column set must be employed within this selective permeation region. Outside this range total exclusion of molecular sizes greater than a critical size occurs, so that these high molecular weight polymers are not separated. Conversely, total permeation of low molecular species occurs below a critical size, at the low molecular weight end.

Let us start with the simplest case and assume that SEC data has been obtained for a polydisperse sample of a linear amorphous polystyrene and we wish to determine the number and weight average molecular weight of this polymer. An experimental SEC curve might look like that shown in figure 10.20. This curve may be effectively cut into "slices" of equal increments of elution volume (ΔV) depicted also in figure 10.20. If sufficient "slices" are taken, each slice may be considered to be monodisperse, and the total area of the curve may be represented, without significant error, by the sum of the heights of the individual slices.

---

* S. T. Balke, "Characterization of Complex Polymers by Size Exclusion Chromatography and High Performance Liquid Chromatography", in *Modern Methods of Polymer Characterization*, Edited by H. G. Barth and J. J. Mays, Wiley, New York (1991).

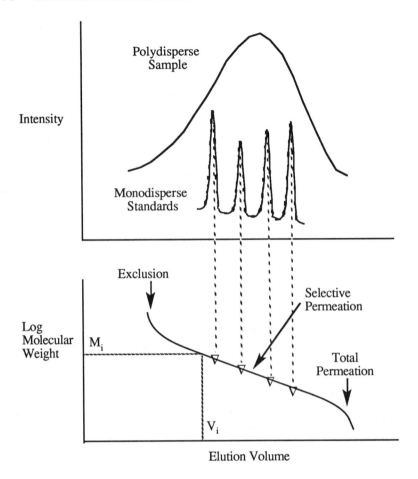

Figure 10.19 *Schematic diagram depicting the calibration of an SEC instrument.*

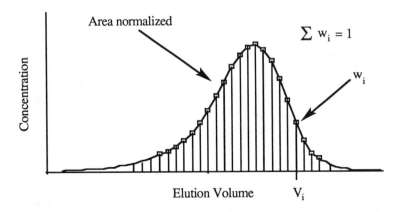

Figure 10.20 *Schematic diagram depicting a typical SEC curve.*

The weight fraction of any "slice", w_i, is then given by:

$$w_i = \frac{h_i}{\sum_i h_i} \qquad (10.95)$$

The molecular weight of the ith species, M_i, is obtained from the calibration curve at point V_i (see figure 10.19). The molecular weight averages are then calculated from:

$$\overline{M}_n = \frac{1}{\sum \frac{w_i}{M_i}} \qquad (10.96)$$

$$\overline{M}_w = \sum_i w_i M_i \qquad (10.97)$$

Note that in equation 10.96 the number average molecular weight is expressed in terms of w_i and M_i. This facilitates calculation since the same data is used to calculate the weight average through equation 10.97. We leave it to the student to verify that equation 10.96 is valid (and, by the way, it might interest the swift and attentive student to know that this makes a very nice little question at exam time!). Note also, that the intrinsic viscosity of the polydisperse polystyrene may be calculated from the relationship:

$$[\eta] = \sum_i w_i [\eta_i] = K \sum_i w_i M_i^a \qquad (10.98)$$

where K and the exponent 'a' are the Mark-Houwink-Sakurada constants for polystyrene in the same solvent as the SEC experiment and at the same temperature (this is more than just an idle aside, as you will soon see).

To generalize, the above methodology is fine for calculating molecular weight averages from an SEC plot if monodisperse standards of the *same* linear polymer are available. However, what do we do if such standards are not available or cannot be synthesized? Molecular weight averages *based upon polystyrene standards*, can be, and often are, reported, but these can be subject to significant error. SEC does not simply separate on the basis of molecular weight, but on the molecular size of the molecule in solution and the major assumption here is that the solution properties of the polymer in question are the same as polystyrene. We know from solution thermodynamics that this is unlikely to be the case and this leads us to the classic studies of Benoit and his coworkers.

The Universal Calibration Curve

If the molecular weight of well characterized monodisperse polystyrenes of different molecular architecture (e.g., linear, star-shaped, comb-like, etc.) are plotted against elution volume they do not fall on a single calibration curve. In other words, if we had three *monodisperse* polystyrenes, one linear, one star-shaped and one comb-like, *all with the same molecular weight*, they would not come off the column at the same time (elution volume). Similarly, *different* monodisperse polymers of *the same molecular weight* generally elute at different times. Thus, for example, monodisperse samples of polystyrene and poly(methyl methacrylate) having the *same molecular weight* might come off the column at different times. In effect, this means we would require different calibration curves for different polymers and even the same type of polymer if the architecture is different. This is illustrated schematically in figure 10.21.

To understand the reason for this consider a monodisperse linear polymer, polyA, in a good solvent and at a given temperature. For a particular molecular weight, M, the polymer would elute at a particular elution volume, V_L^A. If, on the other hand, polyA was a monodisperse star-shaped polymer with say six arms having an identical molecular weight to its linear analogue, V_S^A, it is intuititively obvious that this polymer would have a smaller size in the same solvent at the same temperature (more of the segments are bunched together in the middle of the molecule where the arms of the star are connected). It would, in fact, elute from the column sometime after the linear case. Finally, consider polyB, a monodisperse linear polymer of very different chemistry to polyA, for which the solvent is deemed poor. The size of this molecule in solution will be smaller than that of polyA for the *same molecular weight* and again will elute at some time later, V_L^B.

Benoit and his coworkers[*], in what is probably the most important discovery in this field, recognized that SEC separates not on the basis of molecular weight but rather on the basis of hydrodynamic volume of the polymer molecule in solution.

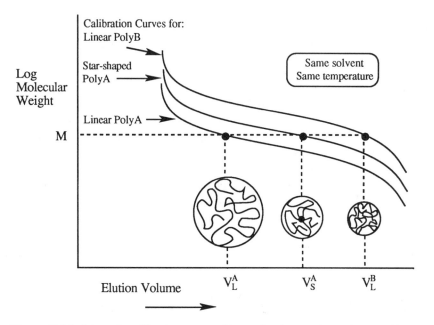

Figure 10.21 *Schematic calibration curves of log molecular weight vs elution volume for polymers of different chemistry and architecture.*

[*] H. Benoit, Z. Grubisic, P. Rempp, D. Decker and J. G. Zilliox, *J. Chim. Phys.*, **63**, 1507 (1966); Z. Grubisic, P. Rempp and H. Benoit, *J. Polym. Sci., Part B*, **5**, 753 (1967).

Recall that if we model the properties of the polymer coil in terms of an equivalent hydrodynamic sphere, then the intrinsic viscosity, $[\eta]$, is related to the hydrodynamic volume V_h via the equation:

$$[\eta] = \frac{2.5\ A\ V_h}{M} \tag{10.99}$$

where A is Avogadro's number and M is the molecular weight. Benoit and his coworkers recognized that the product of intrinsic viscosity and molecular weight was directly proportional to hydrodynamic volume. This being the case, these scientists reasoned that all polymers, regardless of chemical structure or architecture, should fit on the same plot of log $[\eta]$M vs. elution volume. This plot is referred to as the *universal calibration curve*.

As always, things are not perfect and there are exceptions to the rule (polymers with very rigid backbones, for example), but, in general, the universal calibration concept has withstood the test of time and has proven to be a very important and useful discovery. Figure 10.22 shows a plot of log $[\eta]$M vs. elution volume for various polymers.

How then can we use the concept of universal calibration to calculate the molecular weights of a polydisperse polymer for which we do not have monodisperse standards? First we define:

$$J_i = [\eta]_i\ M_i \tag{10.100}$$

and prepare a universal calibration curve using monodisperse polystyrene standards, as depicted in figure 10.23.

Note that we can experimentally determine the intrinsic viscosity of the linear polystyrene standards in the solvent and temperature of choice (see previous section), or we can simply calculate the intrinsic viscosity if we can obtain the Mark-Houwink-Sakurada constants, K_{PS} and a_{PS}, from the literature for polystyrene in the same solvent at the same temperature as the SEC experiment. Remember:

$$[\eta]_i = KM_i^a \tag{10.101}$$

Hence:

$$J_i = [\eta]_i\ M_i = K_{PS}\left[M_i\right]^{(1+a_{PS})} \tag{10.102}$$

Rearranging leads to:

$$M_i = \left[\frac{J_i}{K_{PS}}\right]^{1/(1+a_{PS})} \tag{10.103}$$

This is an important result because it relates the molecular weight of the ith species to the hydrodynamic volume of that species.

For the sake of discussion, let us assume that the SEC data shown in figure 10.20 was obtained from a polydisperse sample of poly(methyl methacrylate) (PMMA) on an SEC instrument using the same solvent and temperature that was

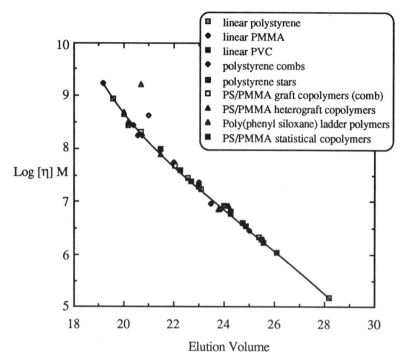

Figure 10.22 *A universal calibration plot of log [η]M vs elution volume for various polymers. Redrawn from the data of Z. Grubisic, P. Rempp and H. Benoit, J. Polym. Sci., Part B, 5, 753 (1967).*

used to prepare the universal calibration curve from PS standards (i.e., we measure V_i and obtain J_i from figure 10.23). Now, if we have the Mark-Houwink-Sakurada constants for PMMA in the same solvent and temperature, K_{PMMA} and a_{PMMA}, then the "true" molecular weights for the polydisperse PMMA may be readily calculated from:

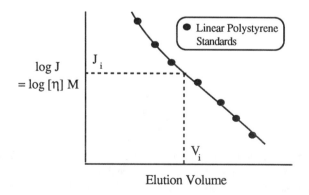

Figure 10.23 *Schematic representation of the universal calibration plot prepared from linear polystyrene standards.*

$$M_i = \left[\frac{J_i}{K_{PMMA}} \right]^{1/(1 + a_{PMMA})} \qquad (10.104)$$

again using:

$$\overline{M}_n = \frac{1}{\sum \frac{w_i}{M_i}} \qquad (10.105)$$

$$\overline{M}_w = \sum w_i M_i \qquad (10.106)$$

Similarly, the intrinsic viscosity of the polydisperse PMMA may be calculated from:

$$[\eta] = \sum w_i [\eta_i] = \sum w_i \left(\frac{J_i}{M_i} \right) \qquad (10.107)$$

F. SEC AND THE DETERMINATION OF LONG CHAIN BRANCHING

Nearly all polymers contain chain branches to a greater or lesser degree. The effect of branching upon the chemical, physical, mechanical and rheological properties of polymers depends upon the *number, type and distribution* of branches. As we mentioned in Chapter 1, a major distinction is made between *long* and *short* chain branches. Short chain branches, where the branches are small compared to the length of the polymer chain backbone, primarily affect crystallinity, as the branches disrupt chain packing, and we have mentioned various spectroscopic methods for measuring such branches in Chapter 6. In this section we will turn our attention to long chain branches, which are defined as branches whose length is commensurate with the length of the polymer chain backbone.

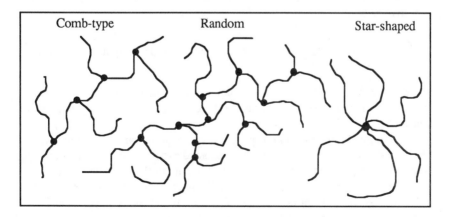

Figure 10.24 Schematic representation of different long chain branches.

Long chain branches can be in the form of random T-shaped molecules (trifunctional branches can arise from intermolecular chain transfer, for example) or X-shaped molecules (tetrafunctional branches can form by chain addition across an active double bond). In addition, long chain branches can occur in the form of comb-like molecules (usually synthesized by grafting reactions), or star-like molecules (synthesized either by initiators that generate several chains simultaneously or by termination of "living" anionic polymerization with a multifunctional reagent that couples chains together). These are depicted schematically in figure 10.24.

The presence of long chain branching can have a major effect upon the rheological and solution properties of polymers, but it is not easy to quantitatively determine the amount of long chain branching in a polymer sample using conventional analytical techniques, such as NMR or vibrational spectroscopy. This is primarily due to the very low concentration of any species that can be attributed to the presence of a long chain branch. For example, let us assume that we have a polyA chain with a degree of polymerization of 1000, onto which we graft a long chain of polyB which also has a DP of 1000.

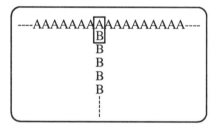

Only one unit, or one in one thousand, on PolyA or PolyB is affected and this is very difficult to detect. However, the effect of long chain branching on the solution properties is quite marked and we can take advantage of this.

The introduction of only one or two long chain branch points leads to a significant decrease in the mean-square dimensions of macromolecules compared to linear molecules of the *same molecular weight*. This statement may be expressed following Stockmayer and Fixman in terms of the ratio of the respective radii of gyration, g.

$$g = \frac{<\overline{S}^2>_b}{<\overline{S}^2>_l} \quad \text{(for the same molecular weight)} \quad (10.108)$$

where the subscripts b and l denote branched and linear molecules, respectively, and g is a function of the number and type of long chain branch points in the molecule.

If linear and branched monodisperse polymers of the same molecular weight are available, the parameter g may be determined experimentally using scattering techniques. In addition, the calculation of the unperturbed dimensions (i.e., the dimensions in a theta-solvent) of branched molecules of various types has been performed by many authors, most notably Zimm and Stockmayer[*] and Zimm

---

[*] B. H. Zimm and W. H. Stockmayer, *J. Chem. Phys.*, **17**, 1301 (1949).

and Kilb*. Calculations of the ratio of the mean-square radii of gyration for *lightly* branched and linear polymers of equal molecular weight have been made on purely geometric grounds. When the number of branch points becomes large statistical methods have to be employed. For star-shaped molecules with *arms of equal length*, for example, it was shown that if the branch point has a functionality, f, then:

$$g = \frac{3}{f} - \frac{2}{f^2} \quad \text{(for star shaped polymers)} \quad (10.109)$$

Thus for trifunctional stars g = 7/9 and tetrafunctional stars g = 5/8.

In this text we will be more interested in *random* branching. It was shown in the seminal paper of Zimm and Stockmayer that g is related to the number average number of branches per molecule, $\overline{m}_n$, for *randomly branched monodispersed* polymers, with tri- or tetrafunctional branch points, *in theta-solvents* by:

$$g_3 = \left[\left(1 + \frac{\overline{m}_n}{7}\right)^{1/2} + \frac{4\,\overline{m}_n}{9\,\pi} \right]^{-1/2} \quad \text{(trifunctional)} \quad (10.110)$$

$$g_4 = \left[\left(1 + \frac{\overline{m}_n}{6}\right)^{1/2} + \frac{4\,\overline{m}_n}{3\,\pi} \right]^{-1/2} \quad \text{(tetrafunctional)} \quad (10.111)$$

SEC chromatography has a major advantage over other techniques in that it separates polydisperse polymers into components that may be considered, albeit simplistically, as monodisperse fractions and thus we should be able to employ the above branching functions to determine the degree of branching using the universal calibration concept. We should caution, however, that given the number of assumptions and approximations in what we are about to do, the methodology can only be used to give a relative number of long chain branches. Nonetheless, this can still be very useful for industrial laboratories.

We start by relating g to the intrinsic viscosity. Zimm and Kilb introduced another branching parameter, g', which is defined as:

$$g' = \frac{[\eta]_b}{[\eta]_l} \quad \text{(for the same molecular weight)} \quad (10.112)$$

Naturally g' < 1 and the value of g' is a function of the type and number of long chain branches. It should also be mentioned that equation 10.112 was defined for branched and linear polymer chains under theta conditions and one assumption usually made is that the equation is also valid for good solvents (which are, of course, employed in the SEC experiment). For star type polymers Zimm and Kilb calculated that:

$$g' = g^{0.5} \quad \text{(star shaped)} \quad (10.113)$$

---

* B. H. Zimm and R. W. Kilb, *J. Polym. Sci.*, **37**, 19 (1959).

For most other polymer chain architectures simple relationships between g' and g have not been theoretically developed and experimentalists have relied upon empirical relationships based on an equation of the form:

$$g' = g^x \qquad (10.114)$$

where the superscript x takes values typically in the range of 0.5 to 1.5. For example, for low density polyethylene x has been reported as 1.2 ± 0.2. We shall rely on the studies of Kurata et al.[*] who determined for randomly branched polymers that:

$$g' = g^{0.6} \qquad \text{(random branching)} \qquad (10.115)$$

In a second paper, Kurata et al.[**] describe a rather ingenious and practical computer method for determining the *relative* extent of long chain branching in polydisperse polymers. At the heart of this work is the fact that for a branched polymer the value of the experimentally determined intrinsic viscosity $[\eta]_{obs}$, will be less than that calculated, $[\eta]_{cal}$, from the SEC data using the universal calibration curve and equation 10.107 (which implicitly assumes that the polymer chains are perfectly linear). These authors then introduced an appropriate branching function, g'(λ, M) that contains a branching parameter, λ, such that:

$$[\eta]_b = g'(\lambda, M) [\eta]_l = g'(\lambda, M) KM^a \qquad (10.116)$$

For example, in the case of randomly branched polymers with tetrafunctional branch points, Kurata et al. used the relationship:

$$g'(\lambda, M) = \left[\left(1 + \frac{\lambda M}{6}\right)^{0.5} + \frac{4 \lambda M}{3 \pi} \right]^{-0.3} \qquad (10.117)$$

which is a combination of equations 10.111 and 10.115, with the additional assumption that:

$$\lambda = \frac{\overline{m}_n}{M} \qquad (10.118)$$

where λ is defined as the number of branch points per unit molecular weight. Note that when $\lambda = 0$ in equation 10.118, g'(λ, M) = 1 and $[\eta]_b = [\eta]_l$; the linear case. Thus we have an approximate relationship that describes the intrinsic viscosity of a randomly branched polymer containing tetrafunctional branch points as a function of the molecular weight, M, and branching parameter, λ, with two constants, K and 'a', the Mark-Houwink-Sakurada constants for the *linear* polymer in the same solvent and temperature.

For a polymer with an unknown degree of branching, the methodology employed by Kurata and his coworkers was to use the following basic equations:

$$\log J_i = f(V_i) \qquad (10.119)$$

---

[*] M. Kurata, H. Okamoto, M. Iwama, M. Abe and T. Homma, *Polym. J.*, **3**, 729 (1972).
[**] Ibid, **3**, 739 (1972).

$$J_i = [\eta_i] M_i = K M_i^{(1+a)} \left[\left(1 + \frac{\lambda M_i}{6}\right)^{0.5} + \frac{4 \lambda M_i}{3 \pi} \right]^{-0.3} \tag{10.120}$$

and:

$$[\eta]_b = \sum_i w_i [\eta_i] = K \sum_i w_i M_i^a \left[\left(1 + \frac{\lambda M_i}{6}\right)^{0.5} + \frac{4 \lambda M_i}{3 \pi} \right]^{-0.3} \tag{10.121}$$

To see how this works, let us again return to our polydisperse polyA polymer and assume that we have experimentally determined its intrinsic viscosity in a particular solvent at a particular temperature, $[\eta]_{obs}$, and also have obtained SEC data, similar to that shown previously in figure 10.20, from a particular column set in the same solvent at the same temperature. Again, we can normalize this curve and "slice" it into small increments of elution volumes, V_i, of weight fractions, w_i. As before, from the universal calibration curve (figure 10.23) we can obtain J_i for a given V_i (equation 10.119). Starting with a value of $\lambda = 0$, which corresponds to the linear case, we can use equation 10.120 to find M_i as a function J_i, and simply calculate $[\eta]_{\lambda = 0}$ as before from equation 10.121. Note that we require the Mark-Houwink-Sakurada constants for the linear analogue of polyA in the same solvent and at the same temperature used in obtaining the SEC data. We now compare the calculated intrinsic viscosity $[\eta]_{\lambda=0}$ to the experimental value $[\eta]_{obs}$. If the two values are identical within experimental error, then it is safe to say that polyA is essentially linear.

However, if the value of $[\eta]_{\lambda=0}$ is significantly greater than that of $[\eta]_{obs}$, then, everything else being equal, we might suspect that polyA is branched. Starting again, we select a particular value of λ (something like 1×10^{-6} might be appropriate) and substitute it in equation 10.120. Unlike the simple linear case where we simply rearranged equation 10.103 to give M_i as a function J_i, this is not possible for the expression that involves random branching (equation 10.120) and this equation has to be solved using an iterative program where M_i is determined for values of J_i assuming a particular value of λ. It is now straightforward to calculate the intrinsic viscosity $[\eta]_{\lambda=\lambda}$ from equation 10.121 (where the symbolism $\lambda = \lambda$ denotes that a particular value of λ has been employed in the calculation). Again this calculated intrinsic viscosity $[\eta]_{\lambda=\lambda}$ is compared to the experimental value $[\eta]_{obs}$. If $[\eta]_{\lambda=\lambda} = [\eta]_{obs}$, within experimental error, then we have found the appropriate value of λ that describes the extent of branching in polyA. However, if $[\eta]_{\lambda=\lambda} \neq [\eta]_{obs}$ then another value of λ, $(\lambda + \Delta\lambda)$, is selected and the process repeated. Naturally, this is all done automatically using a second iterative loop in the computer program until $[\eta]_{\lambda=\lambda} = [\eta]_{obs}$. The values of M_i, corresponding to the finally determined value of λ are stored in a matrix in the computer program and the molecular weight averages of the branched polymer can be determined as before from:

$$\overline{M}_n = \frac{1}{\sum \frac{w_i}{M_i}} \tag{10.122}$$

$$\overline{M}_w = \sum_i w_i M_i \qquad (10.123)$$

Long Chain Branching in Polychloroprene

To illustrate the SEC–[η] method for the determination of long chain branching described above we will describe some of the results that one of us obtained[*] on polychloroprene samples. During the free radical emulsion polymerization of chloroprene, branching occurs via addition of a growing chain to the double bond contained in the minor amounts of 1,2- and 3,4- placements incorporated in an already formed polymer chain. For example:

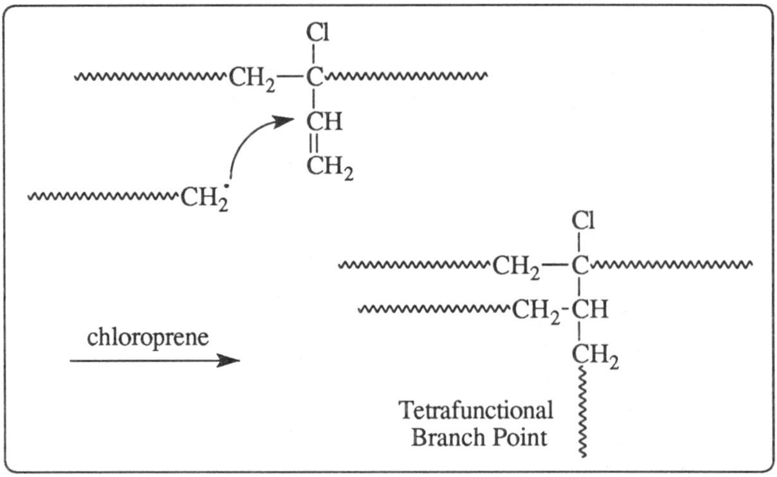

Naturally this branching reaction is favored when the monomer concentration is being depleted in a batch process; i.e., at high degrees of conversion. In fact, in commercial processes the polymer is isolated before about 70% of the monomer is converted to polymer, otherwise gelation occurs. The fact that the extent of long chain branching is increasing as a function of conversion permits us to qualitatively test the SEC–[η] method by isolating samples at various stages of the polymerization from the reaction vessel.

The experimental methodology was not trivial and involved:

1) Synthesis of *linear* polydisperse polychloroprenes. This was achieved by polymerizing chloroprene at low temperatures (≤ -20°C) and isolating the polymer after < 5% conversion.

2) Fractionation of the linear polydisperse polychloroprenes into a set of molecular weight *standards* of relatively narrow polydispersities. This was achieved using an automatic belt fractionator developed at DuPont's experimental station.

3) Characterization of the narrow polydisperse linear polychloroprene standards by light scattering to determine molecular weights and by viscosity measurements to determine the intrinsic viscosity (in THF at 30°C, the solvent

---

[*] M. M. Coleman and R. E. Fuller, *J Macromol. Sci. - Phys.*, **B11**, 419 (1975).

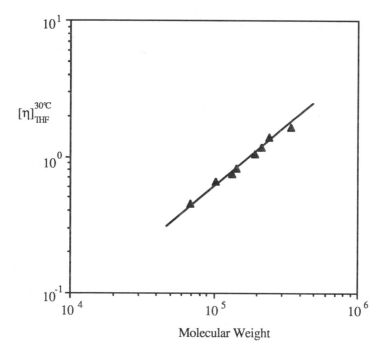

Figure 10.25 *Determination of the Mark-Houwink constants for linear polychloroprene in THF at 30 °C.*

and temperature that were employed in the subsequent SEC experiments).

4) Determination of the Mark-Houwink parameters, K and "a" for the linear polychloroprene in THF at 30°C from a graph of log [η] vs log molecular weight (see figure 10.25). Values of 4.18 x 10^{-5} dl/g and 0.83 were determined for linear polychloroprene from the intercept and slope, respectively.

5) Confirmation that the polychloroprene standards satisfactorily fit on the universal calibration curve prepared from polystyrene standards, which in fact they do (see figure 10.26).

6) Writing a computer program to calculate λ (the branching parameter) and the various molecular weight averages in a manner similar to that of Kurata et al. The SEC curve was digitized to yield the weight fraction, w_i, as a function of the elution volume, V_i. Values of J_i were determined from the universal calibration curve by interpolation from a non-linear calibration file. One subroutine calculates the molecular weight, M_i, for a given value of λ and J_i (equations 10.119 and 10.120). Another subroutine compares the observed [η]$_{obs}$ with that calculated (initially with λ = 0) and then iterates until a value of λ is found such that [η]$_{cal}$ = [η]$_{obs}$. The computer then calculates the various branched molecular weights.

7) Collecting several samples of polydisperse polychloroprenes obtained as a function of conversion (synthesized from conventional free radical emulsion polymerization at 40°C), from which was determined the extent of long chain branching. Results are summarized below in table 10.3.

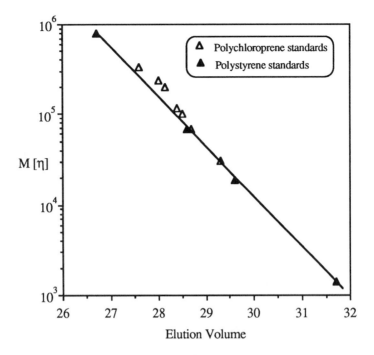

Figure 10.26 *SEC universal calibration curve.*

Six polychloroprene samples were isolated at conversions between 11.7 and 82.3%. Note that the experimental values of the intrinsic viscosity of these samples exhibit a minimum as a function of conversion, illustrating the two competing factors contributing to $[\eta]$; the increase in molecular weight, which increases $[\eta]$, counterbalanced by the increase in long chain branching, which tends to decrease $[\eta]$.

Table 10.3 *Characterization of Polychloroprene Samples.*

% Conversion	11.7	33.6	55.9	62.8	71.9	82.3
$[\eta]_{THF}^{30°C}$	1.50_0	1.49_2	1.47_8	1.49_0	1.52_8	1.54_1
Calculated $[\eta]_{\lambda=0}$	1.49	1.48	1.60	1.67	1.82	1.92
$\lambda \times 10^5$	-	-	0.15	0.23	0.36	0.52
$\overline{M}_w \times 10^{-5}$	3.25	3.25	4.05	4.44	5.49	6.15
$\overline{M}_n \times 10^{-5}$	1.44	1.44	1.19	1.07	1.05	1.26
Polydispersity	2.3	2.3	3.4	4.2	5.2	4.9
γ	0	0	0.38	0.50	0.66	0.76

The third row of table 10.3 shows $[\eta]_{\lambda=0}$ values which were calculated from the SEC curves of the polychloroprene samples *assuming that the polymer is totally linear*; i.e., when $\lambda = 0$. Note that the samples removed after 11.7 and 33.6% conversion have $[\eta]_{\lambda=0}$ values that are, within error, the same as the experimental values, which implies that these polychloroprene samples are essentially linear. At higher conversions, however, the $[\eta]_{\lambda=0}$ values are significantly greater than those experimentally determined. This suggests the presence of long chain branching. Incorporating the branching function described above, values of λ were determined for each sample, such that the calculated and experimental values of $[\eta]$ were the same. The results are displayed graphically in figure 10.27. As expected, we note that at extents of conversions greater than approximately 40%, the branching parameter, λ, increases rapidly towards the point of incipient gelation.

Calculated molecular weight averages are displayed graphically in figure 10.28. It is gratifying to see the anticipated rapid increase in the weight average molecular weight after $\approx 40\%$ conversion, while the corresponding number average molecular weight remains essentially constant. Naturally, this also dictates that the polydispersity increases with conversion (from about 2 to 5 from 40 to 80% conversion).

Finally, there is another branching parameter described independently by Stockmayer and Kilb that is particularly useful for diene type polymerizations,

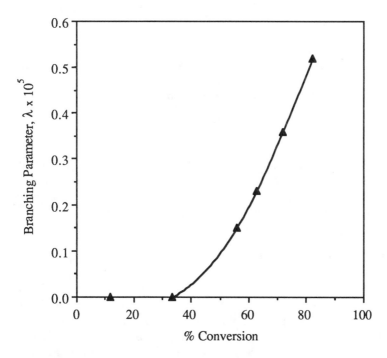

Figure 10.27 *The branching parameter, λ, for polychloroprene samples isolated as a function of conversion.*

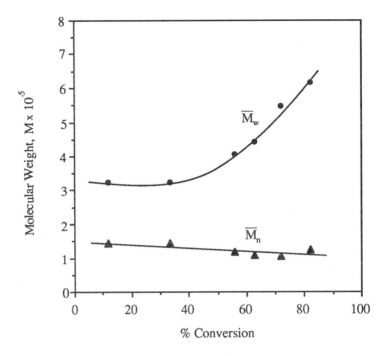

Figure 10.28 *Calculated molecular weight averages for polychloroprene samples isolated as a function of conversion.*

where the onset of gelation can have important practical ramifications. This branching parameter, denoted γ, has limits of zero for the completely linear polymer and unity at the incipient gel point, and is defined as:

$$\overline{m}_w = \lambda \, \overline{M}_w = \frac{\gamma}{1 - \gamma} \qquad (10.124)$$

Since we have the values of λ and $\overline{M}_w$ we can readily calculate γ and these results are shown in the final row of table 10.3. Note that γ is a useful indicator of how close we are to approaching the incipient gel point and it can be seen that we are rapidly approaching gelation in this particular system.

G. STUDY QUESTIONS

1. Seven monodisperse poly(ethylene oxide) (PEO) standards having the general formula: $CH_3-(CH_2-CH_2-O)_n-CH_2-CH_2-OH$ were carefully synthesized. A plot of molecular weight versus refractive index of these standards is shown below. A commercial polydisperse sample of PEO has a refractive index of 1.4525. This corresponds to an average molecular weight of about 1850 g mole⁻¹. What molecular weight average do you think has been determined ? Careful, this is not an easy question. Explain your answer.

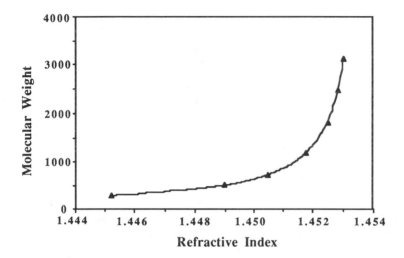

2. The following osmotic pressure data was obtained for polystyrene in toluene at 25°C.

Concentration (g/cm^3 x 10^3)	π (cm toluene)
2.56	0.325
3.80	0.545
5.38	0.893
7.80	1.578
8.86	1.856

The density of toluene is 0.8618 g cm^{-3} and g = 981 cm sec^{-2}. Calculate the number average molecular weight.

3. Below are values of $K(1+\cos^2\theta)c / R_\theta$ (mole g^{-1} x 10^6) tabulated in the book of Margerison and East (see Suggestions for Further Reading). These data were obtained from light scattering experiments performed on a series of polystyrene solutions in benzene. Construct a Zimm plot and calculate the weight average molecular weight.

c (g cm^{-3})	Angle θ (degrees)										
	30	37.5	45	60	75	90	105	120	135	142	150
0.00200	3.18	3.26	3.25	3.45	3.56	3.72	3.78	4.01	4.16	4.21	4.22
0.00150	2.73	2.76	2.81	2.94	3.08	3.27	3.40	3.57	3.72	3.75	3.78
0.00100	2.29	2.33	2.37	2.53	2.66	2.85	2.96	3.12	3.29	3.38	3.37
0.00075	2.10	2.14	2.17	2.32	2.47	2.64	2.79	2.93	3.10	3.21	3.20
0.00050	1.92	1.95	1.98	2.16	2.33	2.51	2.66	2.79	2.96	3.11	3.12

4. Below are two sets of experimental data recorded using an identical Ubbelohde viscometer. In the first set (denoted I) the data were obtained from a solution of a polyelectrolyte in deionized water, while in the second set (II) a small amount of sodium chloride (10^{-3} M) was added to the water solution.

Concentration (g dl^{-1})	Time (secs)	
	(I)	(II)
0.40	743.3	515.9
0.35	711.3	475.1
0.30	673.3	440.2
0.25	620.0	402.5
0.20	583.0	373.1
0.15	542.2	346.7
0.10	466.4	323.6
0.05	367.3	306.1
Deionized water	291.5	291.5

What is the intrinsic viscosity of the polymer and why are there such marked differences between these two sets of data ? [Note: this is a "real" situation—the results may be equivocal.] You will be expected to comment on your conclusions and be critical in your answer.

5. Six poly(methyl methacrylate) (PMMA) standards of varying molecular weights and with polydispersities of <1.1 were synthesized. Below are intrinsic viscosity data determined from solutions of these polymers in benzene at 25°C.

Molecular weight (g mole^{-1})	$[\eta]$ (dl g^{-1})
46000	0.258
82000	0.408
152000	0.663
253000	0.893
348000	1.226
856000	2.210

Determine the Mark-Howink-Sakurada constants, K and a.

6. A linear polydisperse PMMA is synthesized. From solution viscosity measurements in benzene at 25°C the intrinsic viscosity is determined to be 1.04 dl g^{-1}.

(a) Using the K and a values determined in question 5 above, calculate the molecular weight of the PMMA.

(b) What molecular weight average have you determined ?

(c) Assuming that the PMMA polymer has a most probable molecular weight distribution, what are the corresponding values of number and weight average molecular weights ?

7. Answer the following questions pertaining to the section on size exclusion chromatography:

(a) On what basis does size exclusion chromatography (SEC) separate polymers ?

(b) Two polymer molecules of similar chemistry and the same molecular weight, one linear and the other star-shaped, are injected together into a SEC. Which one is eluted first ? Why ?

(c) How is a universal calibration curve prepared ?

8. The number average molecular weight is defined as:

$$\overline{M}_n = \frac{\sum_i N_i M_i}{\sum_i N_i}$$

Show that if w_i is the weight fraction of the ith species:

$$\overline{M}_n = \frac{1}{\sum \frac{w_i}{M_i}}$$

9. Below is a GPC curve together with intensity versus elution volume data that was recorded from a sample of Polymer A in THF at 30°C.

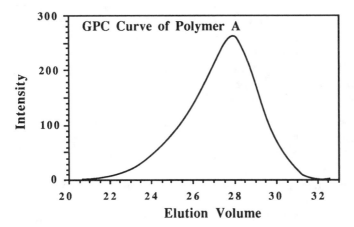

Elution Volume	Intensity or Height	Elution Volume	Intensity or Height
32.5	0.0	26.5	181.5
32.0	0.8	26.0	147.1
31.5	4.5	25.5	120.0
31.0	15.0	25.0	88.0
30.5	30.2	24.5	67.8
30.0	60.0	24.0	50.1
29.5	105.5	23.5	36.2
29.0	165.3	23.0	22.1
28.5	225.1	22.5	12.3
28.0	256.0	22.0	6.9
27.5	250.3	21.5	2.0
27.0	220.6	21.0	0.1
		20.5	0.0

Under identical experimental conditions six monodisperse samples of polystyrene having molecular weights of 2000000, 850000, 470000, 105000, 52000 and 11000 g mole^{-1} eluted at elution volumes of 23.3, 25.4, 26.7, 29.3, 30.3 and 32.2, respectively. The Mark-Houwink-Sakurada constants for polystyrene in THF at 30°C are K = 3.71 x 10^{-4} dl/g and a = 0.64.

(a) Assume that the GPC curve of Polymer A is that of a polydisperse linear polystyrene. Calculate the weight and number molecular weight averages, the polydispersity and the intrinsic viscosity of the polystyrene.

(b) Assume that the GPC curve of Polymer A is that of a polydisperse linear polychloroprene with Mark-Houwink-Sakurada constants in THF at 30°C of $K = 4.18 \times 10^{-5}$ dl/g and $a = 0.83$. Calculate the weight and number molecular weight averages, the polydispersity and the intrinsic viscosity of this polychloroprene.

(c) Assume that the GPC curve of Polymer A is that of a polydisperse tetrafunctionally star branched polychloroprene (i.e. a polydisperse polymer made up of different molecular weight star-shaped polymers having four arms of equal length). Calculate the weight and number molecular weight averages, the polydispersity and the intrinsic viscosity of this star-branched polychloroprene.

(d) To calculate the molecular weight averages and intrinsic viscosity you will have to extrapolate the calibration curves outside the range of the standards in order to obtain appropriate values of M_i or J_i at elution volumes that are less than 23.3. Would you expect this to introduce significant errors ? Test your answer by calculations using different assumptions.

10. A chemist polymerizes styrene by a free radical polymerization process in the presence of the chain transfer agent bromotrichloromethane, CCl_3Br. This yields a relatively low molecular weight product where essentially all of the chains have one Br end group. The polymer is washed and isolated. Subsequently, the polymer is dissolved in a suitable solvent and a stoichiometric amount, (based upon the concentration of Br groups), of a tetrafunctional amine is added. The chemist believes that this reaction should lead to four branched star-shaped polymers. How would you attempt to determine whether or not the chemist had been successful ?

H. SUGGESTIONS FOR FURTHER READING

(1)   D. Margerison and G. C. East, *Introduction to Polymer Chemistry*, Pergamon Press, Oxford, 1967.

(2)   H. G. Barth and J. W. Mays, *Modern Methods of Polymer Characterization*, Wiley-Interscience, New York, 1991.

CHAPTER 11

Mechanical and Rheological Properties

"Man's mind, stretched to a new idea,
never goes back to its original dimension."
—Justice O. W. Holmes
"When all else fails, use bloody great nails."
—J. E. Gordon

A. INTRODUCTION AND OVERVIEW

In treating the mechanical and rheological properties of materials there are a variety of approaches. Perhaps the most common is what can be called the "engineering" or "continuum approach". This essentially ignores the atomic or molecular nature of matter, treating the mechanical behavior of solids in terms of the laws of elasticity and the rheological properties of liquids in terms of the laws of fluid dynamics and viscous flow. The properties of solids and liquids are often considered as separate, non-overlapping topics and this separation is entirely reasonable and useful for many types of materials. Polymers, however, are much more awkward, displaying both elastic and viscous types of behavior at ordinary temperatures and loading rates. The extent of each depends upon temperature, the time frame of the experiment, the structure and morphology of the polymer, and the rest of the usual suspects. Accordingly, we will be describing what are called *viscoelastic properties*. We will *initially* still subdivide the subject into two parts, the first focusing on the more solid-like properties that make polymers useful in their various applications and the second focusing on the more liquid-like properties characteristic of the melt, the state in which most polymers are processed. But after this introduction these categories are going to overlap and what you should fix in your mind now is that while a particular polymer may appear to the naked eye to be a solid material, it can have some liquid-like aspects to its behavior. Similarly, a polymer melt has some elastic as well as viscous aspects to its behavior.

We will begin our discussion of mechanical and rheological properties by reminding you of some basic definitions and laws. We will then see how well these laws apply to polymers and introduce various manifestations of viscoelasticity. We will consider both mechanical models and microscopic or molecular theories, but the latter will only be treated qualitatively. Most of this discussion will deal with so-called linear models, but non-linear effects are important and reveal themselves in the properties of rubber elasticity and melt flow at high strain rates. Finally, we will proceed from non-linear to ultimate properties and describe the conditions under which polymers yield and fail.

B. A BRIEF REVIEW OF SOME FUNDAMENTALS

Stresses and Strains

Many students whose primary interest is synthetic chemistry somehow manage to avoid, until they get to the study of polymers, any courses on mechanical properties, or immediately forget even those simple parts of the subject that are taught in elementary physics classes. This section is aimed at those of you who are in such sorry straights. So first a test; if you know what compliance is, skip this section; if not, read on!

We will start by first defining stress and strain. This might seem to be elementary, but as Gordon* has pointed out, these concepts are sometimes muddled in the minds of those students who have not previously been exposed to basic aspects of mechanical behavior, because the words have been stolen by psychologists and such-like to describe certain mental states. Not only have these words been stolen, but they have been indiscriminately used to describe the same thing! We talk about being "under strain" or "under stress" (or, these days, "stressed out") interchangeably. We can have no truck with such lack of precision. Stress and strain have distinct, clear meanings and are the device by which we can distinguish between the intrinsic properties of a material and those properties that are a function of its shape and size. For example, we would expect that a half inch diameter steel rod would sustain a larger load (before breaking) than one that is only a tenth of an inch in diameter. Similarly, we would anticipate that a long piece of wire would deform or stretch more under a given load than a shorter piece of the same wire. We would also expect a rubber fiber having the same diameter and length as this steel wire to deform more under the same load. If we wish to compare the properties of rubber as a material to those of steel, then we must find a way to take out the effect of shape and size. We start by considering simple elongation, as illustrated in figure 11.1 and define tensile stress (σ) to be:

$$\sigma = \frac{F}{A} \tag{11.1}$$

or force (F) per unit area (A). Strain (ε), you will hopefully recall from other courses or books, is the change in length (Δl) divided by the original length (l_0):

$$\varepsilon = \frac{\Delta l}{l_0} \tag{11.2}$$

Hooke was the first to notice (or at least write down) that the extension of a material is apparently proportional to the load applied to it (i.e., if a wire stretches a certain amount under a load of 100 lbs, it will stretch twice as much under a load of 200 lbs). As we will see, this is only an approximation that is good for small deformations, but we will ignore this for now. Of more concern to us here is that the actual deflection is not only proportional to the load, but depends upon the material being stressed.

---

* J. E. Gordon, *The New Science of Strong Materials or Why You Don't Fall Through the Floor, Second Edition*, Penguin Books, (1976).

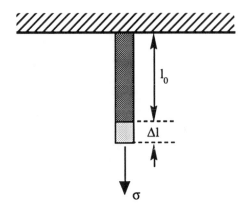

Figure 11.1 Uniaxial elongation.

This was made more explicit by Young, who realized that Hooke's law could be written:

$$\frac{\text{Stress}}{\text{Strain}} = \text{constant} \qquad (11.3)$$

where the constant is a characteristic of the *material*, not the size and shape of the object being stressed. For the simple stretching experiment shown in figure 11.1 we can therefore write Hooke's law as:

$$\sigma = E\varepsilon \qquad (11.4)$$

where the constant E is called Young's modulus. We can also define a quantity called the tensile compliance, D, which for this type of time independent experiment* is equal to 1/E or:

$$D = \frac{\varepsilon}{\sigma} \qquad (11.5)$$

This definition might seem trivial and possibly even useless at this point, but when we turn our attention to time dependent behavior (phenomena such as creep and stress relaxation) these separate definitions of E and D will have more meaning and utility.

When we stretch a material it does not just deform in the direction of the applied load, but also contracts in directions perpendicular to the stretch, as illustrated in figure 11.2. The amount of this contraction is proportional to the extension in the direction of stretch, so that if we let Δw be the change in the width of the material and w_0 be the original width, then:

---

* By this we mean the material deforms practically instantaneously as the stress σ is applied, then the strain ε stays constant over time until the load is removed.

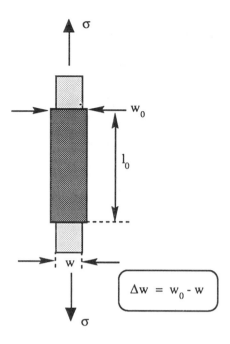

Figure 11.2 *Contraction in the perpendicular direction upon stretching.*

$$\frac{\Delta w}{w_0} = - \upsilon \frac{\Delta l}{l_0} \tag{11.6}$$

where υ is called *Poisson's ratio*.

If we are considering homogeneous, isotropic materials, Young's modulus E and Poisson's ratio υ are the only constants we need in order to completely specify the elastic properties. For example, if we now wish to examine what happens to a material subjected to a uniform hydrostatic pressure P, as illustrated in figure 11.3, then we can define a *bulk modulus* B:

$$P = - B \frac{\Delta V}{V_0} \tag{11.7}$$

where ΔV is the change in volume. As you might intuitively expect, it can be shown that there is a simple relationship between E, B and υ, which turns out to be:

$$B = \frac{E}{3(1 - 2\upsilon)} \tag{11.8}$$

Similarly, if we need to consider what happens to a material subjected to a shear force, illustrated in figure 11.4, then we can define a *shear modulus*, G, which is related to E and υ by:

$$G = \frac{E}{2(1 + \upsilon)} \tag{11.9}$$

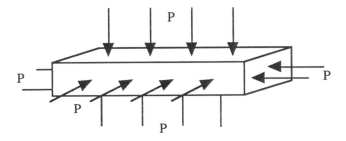

Figure 11.3 *A material subjected to uniform hydrostatic pressure.*

The jump from simple elongation to compression by a uniform pressure is straightforward, but in order to define a shear modulus we have to start paying much more attention to the direction of our forces and displacements, and this is where many students start to get upset with the subject, because matrix algebra and tensors now rear their ugly heads. The shear force we have illustrated in figure 11.4 is parallel to a plane and we will arbitrarily let this be the x, y plane in a Cartesian system. This raises the questions of how such deformations are defined, because both the "vertical" and "horizontal" dimensions of the cubic object shown in the figure are changing. The most convenient way is to express the deformation in terms of the angle by which the materials is twisted:

$$\tan \theta = \frac{\delta}{l_0} = \gamma_{xy} \tag{11.10}$$

Notice that our strain, γ_{xy}, is now defined as a deformation *parallel* to the direction of the applied force, divided by a length *perpendicular* to this direction. Also, the shear stress is the force applied *parallel* to a face or plane in the materials divided by the area of this face:

$$\tau_{xy} = \frac{F}{A_{xy}} \tag{11.11}$$

(contrast this to simple elongation, where the stress is equal to the force *perpendicular* to this face, divided by the area of this face).

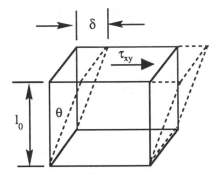

Figure 11.4 *Material subjected to a shear stress.*

Hooke's law for shear stress and shear strain is then:

$$\tau_{xy} = G\gamma_{xy} \tag{11.12}$$

and we can define a shear compliance as:

$$J = \frac{\gamma_{xy}}{\tau_{xy}} \tag{11.13}$$

where J=1/G.

The equations describing the normal and shear stresses can now be combined to describe the deformation of a small cubic element of a material under a load. This is most conveniently done in terms of matrices, hence the tensor representations that some of you may have stumbled across in other books. These can get very involved indeed when non-isotropic materials are considered, as the elastic moduli of such materials are different in different directions. In hypothetically perfect crystalline materials this is a consequence of structure, because the forces of attraction between the atoms of molecules can depend upon the mode of packing and arrangements in the crystal (i.e., they may differ along different crystallographic planes). Defects in the crystal obviously introduce all sorts of additional problems. In polymers, if we align the chains in a particular direction, then clearly the elastic properties parallel to the direction of the chains will differ to those perpendicular to this direction, as covalent bonds are far stronger than weak intermolecular forces. The basic principle governing all these situations is just Hooke's law, however, and the rest of the subject is the application of various mathematical tools to deal with these more difficult situations[*]. In this introductory treatment we do not have to consider most of these difficulties, but it is important that you appreciate one subtlety; even under a simple uniaxial tensile load shear forces are induced within a material.

Figure 11.5 shows a rectangular specimen (cross sectional area $A = l^2$) under a load F. The tensile stress is simply:

$$\sigma_t = \frac{F}{l^2} = \frac{F}{A} \tag{11.14}$$

Now consider an arbitrary plane defined within the material at an angle ϕ to the original cross section of the material (see figure 11.5). Clearly there are non-zero components of the load or force parallel and perpendicular to this plane that result in a shear stress, σ_s, and a normal stress, σ_n, respectively.

It is a matter of simple trigonometry to show that:

$$\sigma_n = \frac{F \cos\phi}{l^2/\cos\phi} = \frac{F}{A} \cos^2\phi \tag{11.15}$$

Similarly:

---

[*] This is a bit unfair, as it is like saying quantum mechanics is just the Schroedinger equation, or statistical mechanics is just the Boltzmann relationship, $S = k\ln\Omega$!

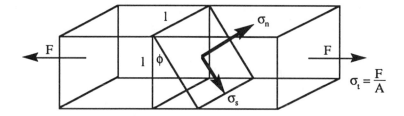

Figure 11.5 *Shear stresses in a material under a uniaxial tensile load (F).*

$$\sigma_s = \frac{F}{A} \sin\phi\cos\phi = \frac{F}{2A} \sin^2\phi \qquad (11.16)$$

When $\phi = 45°$

$$\sigma_s = \sigma_n = \frac{F}{2A} \qquad (11.17)$$

and σ_s is a maximum. As we will see later, when certain polymers are stretched beyond something called the yield point they exhibit shear bands that are oriented at $\pm45°$ to the draw direction. This is illustrated in figure 11.6, just to give you a taste of things to come. We will discuss this phenomenon in a little more detail later on, but it is clearly related to the fact that shear stress is a maximum at angles of $\pm45°$ in a sample subjected to uniaxial tension.

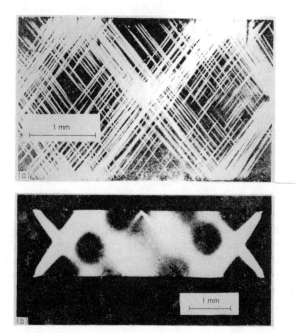

Figure 11.6 *Shear yielding in glassy polymers (a) polystyrene and (b) poly(methyl methacrylate). Reproduced with permission from P. B. Bowdon, Philos. Mag., 22, 455 (1970).*

Hooke's law only applies to small deformations of ideal solids and in real life there is no such beast. But, it works reasonably well for many materials, up to a point. Then there are deviations from (apparent) linear behavior, and the material breaks or irreversibly deforms. Nevertheless, because we are often most interested in using materials such as steel in the range in which they do not break or deform permanently, Hooke's law is extraordinarily useful and a pillar of the engineering sciences. Polymers, however, are much more complex in their mechanical response, and time dependent behavior often has to be considered. More on this later.

Viscosity

Just as in the elastic properties of solids, there is a simple law describing the flow of a liquid that is also an approximation, but provides an adequate description of the properties of many fluids. This law is due to Newton and connects the shear stress and the strain rate:

$$\tau_{xy} = \eta \, \dot{\gamma}_{xy} \tag{11.18}$$

where the constant of proportionality η is called the viscosity of the fluid.

We have defined what we mean by shear stress [force divided by the area parallel to which this force acts], but not strain rate. This is also fairly simple. Consider our fluid to be placed between two plates set a distance y apart, as illustrated in figure 11.7. We hold the bottom plate stationary and let the top one move with a velocity v_0. The velocity of the fluid *relative to each plate* is zero right at the surface (this is an experimental observation, not a hypothesis). This means that there is a velocity profile set up in the fluid between the plates, because the fluid particles right at the bottom surface are stationary, while those right next to the top plate must be moving with a velocity v_0. We are going to assume that our fluid is ideal and the velocity profile is therefore linear, as also shown in figure 11.7. Now, recall that in defining a shear modulus we put a cubic object under a load and deformed the top surface an amount δ horizontally. The shear strain γ_{xy} was equal to the quantity δ/l_0, where l_0 was the distance between the top and bottom surfaces of our cube. Here we are moving the top surface an amount v units per second (x/t, if the top surface moves x in t seconds) so it follows that our definition of the strain *rate* is simply v/y, where y

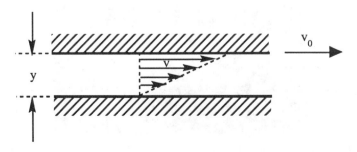

Figure 11.7 Velocity profile in an ideal fluid subjected to a shear stress.

is the distance between the plates. More generally, we can consider the velocity difference dv between two layers of a fluid a distance dy apart in laminar flow, so that the strain rate is given by:

$$\frac{dv}{dy} = \frac{d}{dt}\left(\frac{dx}{dy}\right) = \dot{\gamma}_{xy} \qquad (11.19)$$

Now we are more or less set. We have defined all the quantities we need to present a broad picture of the mechanical and rheological properties of polymers. We will briefly consider units and then go on to introduce the elastic and rheological properties of polymer materials.

An Aside on Units

First a confession; we are old-fashioned in many of our ways and are most comfortable with units like pounds per square inch (psi), rather than modern abominations such as Pascals (newtons/m²). This prejudice is no doubt a residue of our early education, where as insular English schoolboys we were at some point inflicted with what we then considered to be the bane of all history classes, the French Revolution. We really didn't care what they did across the channel and so throughout our youth remained blissfully unaware that this gave birth to three significant plans for the reform of measurement, one of which was the metric system (the other two involved the calendar and units of time). The modern SI, or "Système International des Unités", is the logical extension of the metric system, but we won't be consistent in our use of units, switching from lbs and inches to the cgs system and to SI units according to how the data we will use was reported in the original literature. Accordingly, we think it useful to briefly discuss units.

Stress was defined above as force per unit area and this gives us a problem right off the bat, because we have to distinguish between units of mass and weight. Originally, a kg was a unit of weight (force), corresponding to a prototype cylinder of a platinum-iridium alloy stored under triple bell jars in Sèvres, France (and which in spite of all the precautions is reported to be getting heavier as traces of gunk build up on its surface). What we call *weight* is the force exerted by gravity on the *mass* of a material, so we should differentiate between a kg of mass and a kg of force. Of course, the bodies that decide such things have indeed devised a system based on gravitational units (in the foot-pound-second system this is a *slug* and in the metric system a *metric slug*), but we have never seen these units used. As might be expected, the SI system, based on superior Gaulic logic, uses mass, so that a Newton (N) is a kg m/sec² and the corresponding units of stress are just newtons/m² = 1 Pa (pascal). The antiquated British and U.S. systems use lbs as a unit of *weight*, so stress is in terms of lbs/square inch (psi).

You will sometimes see a prefix like M or G standing for mega or giga (e.g., MPa, GPa; megapascals and gigapascals, equal to 10^6 and 10^9 Pa, respectively). To convert between these various units you can simply apply the following conversion factors;

$$1 \text{ Pa} = 10 \text{ dynes/cm}^2$$
$$1 \text{ MPa} = 145 \text{ psi}$$

Units of strain are easy, because there aren't any, being a length divided by a length. Young's modulus must therefore be in units of stress, and so on.

Turning to viscosity, we noted in our discussions of molecular weight measurement (previous chapter) that this is usually expressed in terms of poise, P, where:

$$1 \text{ P (poise)} = 1 \text{ dyne sec/cm}^2$$

There is an equivalent SI unit, N sec/m², the poiseuille, Pl and:

$$1 \text{ Pl (poiseuille)} = 10 \text{ P (poise)}$$

C. DEVIATIONS FROM IDEAL BEHAVIOR

For many engineering applications we usually wish to know just a few simple things about a polymer. How stiff is it, or, in other words, how much does it deflect under a given load? How strong is this material, or what level of stress can it support before it breaks or irreversibly deforms? Even if the material appears strong under static loads, is it brittle? Does it break easily on impact (i.e., under a high *rate* of loading)? In addition to these, there are mechanical properties such as hardness, the ability to withstand cyclic loading (fatigue), the resistance to loading in the presence of an organic solvent, etc., that are also important. Finally, what are the rheological properties of the polymer melt, will it be easy to process? If polymers were ideal, by which we mean they obeyed Hooke's law and Newton's law in the solid state and melt, respectively, then a few simple experiments would give us practically all the information we would need. The behavior of real polymers is far more complex and interesting,

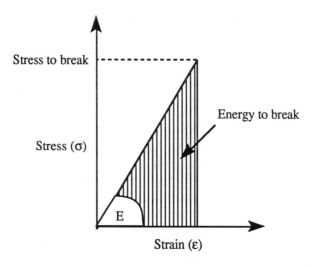

Figure 11.8 Stress-strain curve for a hypothetical material.

however, and particular attention has to be paid to time dependent properties. [In studying other materials, such as metals, these types of properties are usually (but not always) important only at high temperatures].

We will approach this subject by first considering gross features of the elastic and viscous properties of real polymer systems and the extent to which their behavior departs from ideality. We can then proceed to consider various aspects of these deviations in more detail, specifically viscoelasticity, non-linear behavior (particularly rubber elasticity) and ultimate properties (yielding and failure).

Stress-Strain Behavior

We will start this section by considering a classical mechanical experiment, which essentially involves placing a sample in a machine (often an Instron® tester), where it is stretched at a controllable rate until it breaks. The machine records strain as a function of applied stress and the resulting stress-strain plots provide information concerning some of the quantities we mentioned at the beginning of this section, stiffness, strength and, to some degree, toughness.

For a hypothetical ideal material we might expect the type of plot shown in figure 11.8; a straight line up to the point of failure, where the material breaks without yielding or plastic deformation, which causes the stress/strain curves to have weird shapes, as we will soon see. The curve, or in this case straight line, provides other pieces of information; the stress to break, which we can think of as the strength of the material; its stiffness, as measured by Young's modulus and given by the slope of the line (i.e., stress divided by strain); and finally the energy to break, proportional to the area under the curve, which we can think of as a measure of toughness.

Real materials of course, do not behave like this. Some come close (certain ceramics, for example), displaying a line that curves somewhat, but is nearly straight up until failure. In fact, even in a hypothetically perfect crystalline material we should expect some deviations from linearity (Hooke's law) and it is useful and interesting to first consider why this should be so. To illustrate this point we will use a very simple model, a one dimensional array of atoms, shown schematically in figure 11.9. We can imagine this array as being part of an ideal material that would have identical rows laid adjacent to one another to give a cubic crystal structure. Stretching this crystal in directions perpendicular to the array shown in figure 11.9 would then be governed by the same arguments as those given by our simplified model.

When a uniaxial stretching force is applied to this ideal crystal the atoms are displaced from their equilibrium positions and this generates a restoring force that is equal and opposite to the applied load. For this type of material the mechanical response is therefore governed by the internal energy, which is the sum of the potential energy associated with interatomic or molecular forces and the kinetic

Figure 11.9 One dimensional array of atoms.

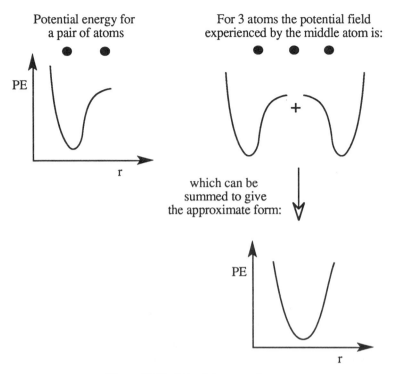

Potential energy for
a pair of atoms

For 3 atoms the potential field
experienced by the middle atom is:

PE

r

which can be
summed to give
the approximate form:

PE

r

Figure 11.10 Potential energy diagrams.

energy of motion (later, when we consider rubber elasticity, we will encounter mechanical responses that are governed by entropic rather than energetic forces, so in order to get you ready for some more thermodynamics we bring up the internal energy here). We can neglect the kinetic energy associated with the vibrations of the atoms around their mean positions, so that the only contribution to the internal energy we will consider is the potential energy associated with the chemical bonds that bind the atoms to one another in this lattice. For a pair of atoms this energy has a form that looks like that shown in figure 11.10. At small distances apart the atoms strongly repel, at large distances apart the forces of attraction are small, and there is an equilibrium position corresponding to a minimum energy. Because each atom is attached to two neighbors, each atom is acted on by a potential that is a sum of two such potentials, as also illustrated in figure 11.10. At small displacements this energy curve can be approximated by a quadratic function of the form:

$$PE = k'x^2 \qquad (11.20)$$

where we let x be the displacement from the minimum energy position, which we define to the origin of our one dimensional coordinate system. The factor k' is a constant and in order to get our answer in a simple (and more usual) form we will replace it with a factor k/2, so that:

$$PE = \frac{k}{2} x^2 \qquad (11.21)$$

[Those of you who are familiar with vibrational spectroscopy will recognize equation 11.21 as the form of the potential used to calculate the normal modes of vibration in the simple harmonic approximation].

At larger displacements there are some deviations, which we can approximate using a power series:

$$PE = \frac{1}{2} kx^2 + \frac{1}{3} k''x^3 + \text{-------} \qquad (11.22)$$

Because:

$$f = \frac{d(PE)}{dx} \qquad (11.23)$$

It follows that:

$$f = kx + k''x^2 + \text{-------} \qquad (11.24)$$

or, reverting to our former nomenclature and using Δl to represent the displacements:

$$f = k \Delta l + k''(\Delta l)^2 + \text{-------} \qquad (11.25)$$

For small values of Δl we can neglect higher terms [e.g., if $\Delta l = 0.02$ (2% strain), $(\Delta l)^2 = 0.0004$ etc.]. But, at higher strains we would expect to see deviations from Hooke's law. And indeed we do, but there are also deviations due to imperfections in the material we are studying and it is difficult to separate these two effects. One approach is to look at materials that are as perfect as we can make them, which usually means trying to grow single crystals. (Most crystalline solids are not "single crystals", but consist of collections of small crystals that grow from separate nucleation events until their boundaries impinge on one another, like the spherulites shown earlier in figure 7.29.) For most polymers this may never be possible, but there is a particular class of polymers

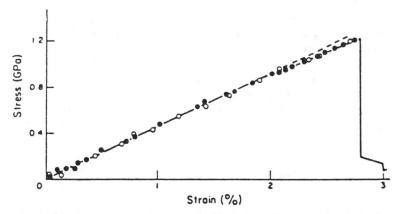

Figure 11.11 Stress-strain curve of a single crystal of a polydiacetylene. Reproduced with permission from C. Galiotis and R. J. Young, Polymer, 24, 1023 (1983).

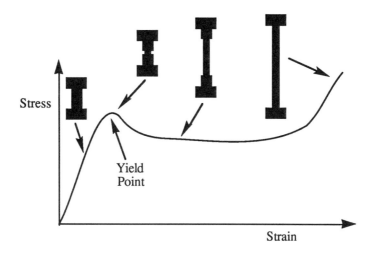

Figure 11.12 *Schematic representation of the stress-strain curve for a semi-crystalline polymer such as polyethylene, showing the yield point and the development of necking.*

called the polydiacetylenes which can be prepared by polymerizing the monomers as they sit in a crystalline lattice and these monomers can be grown in the form of single crystals. The stress-strain results obtained from a single crystal fiber of one of these polymers are shown in figure 11.11 and it can be seen that Hooke's law is indeed obeyed at small strains, up to about 2%, but beyond that deviations occur. Clearly, Hooke's law is only valid for small strains and beyond this, even in a "perfect" crystal, we must expect deviations.

Most materials are, naturally, imperfect in their structure. Even highly crystalline materials such as metals display *yielding and plastic deformation*. At the yield point a "neck" is formed in the sample (see below) and it deforms irreversibly. In highly crystalline materials like metals this process is associated with the motion of defects (dislocations) in the crystalline lattice. Amorphous materials can also display plastic deformation, but here the process involves a viscous flow mechanism. Polymer materials are at best semi-crystalline, their morphology is complex and involves chain folding (which can be thought of as a defect), while many do not crystallize at all. Accordingly, they deform irreversibly far more readily than other materials and this propensity to deform increases dramatically as the temperature is raised even relatively small amounts[*] (150°C or so above room temperature for many polymers). Consider polyethylene, as an initial example. A schematic representation of a typical room temperature stress-strain curve is shown in figure 11.12.

Flexible semi-crystalline polymers such as this often display a *yield point* (as long as the rate of loading is not too fast, a very important factor to keep in

---

[*] The word *plastic* is often defined as "capable of being formed" and the ability of most polymeric materials to be molded at relatively low temperatures led, we believe, to their being called "plastics".

mind). The materials deforms more-or-less elastically up to this point, but then a neck forms (also shown schematically in figure 11.12) and the material suddenly stretches out considerably. It subsequently appears to regain some resistance to stress and ultimately breaks at a higher load. (There is a simple molecular reason for this behavior that we will come back and describe later when we discuss yielding and ultimate properties in more detail.) Note also a strange little hump at the yield point. Plastic deformation is initiated, a neck is formed and there is an actual decrease in stress. (Remember that the experiment applies a varying load in order to deform the sample at a given rate). Deformation then continues at an almost constant stress. (A further complication is that there is a reduction in cross-sectional area at the neck, increasing the stress locally at this point. The stress/strain diagram shown in figure 11.12 actually plots "nominal" stress, which is with respect to the original cross-sectional area of the sample.)

The appearance of the stress/strain plot of a polymer will obviously depend upon its structure and morphology, whether it is an amorphous glass or rubber, whether it is semi-crystalline or some other type of multiphase system (e.g., a block copolymer). Typical curves are shown in figure 11.13. This is purely a schematic figure whose purpose is to illustrate types of behavior. The exact form of the curve and the value of the stress-to-break, etc. will depend upon the polymer. As you would no doubt anticipate, polymers that are glassy or that have amorphous regions that are glassy, have a higher modulus (slope of the stress-strain curve) than polymers that are above their T_g at the temperature we are considering (i.e., those with rubbery amorphous regions).

Because most of the stress-strain plots are not linear, but curve to greater or lesser extents, we define the modulus as the initial slope of the plot, the tangent to the curve at the origin. Glassy, stiff, polymers often have slightly curving

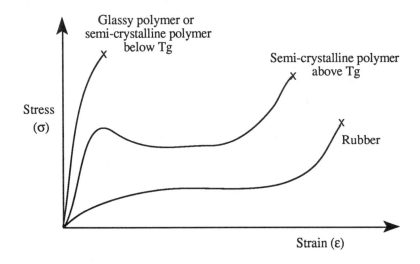

Figure 11.13 *Typical stress-strain curves for polymeric materials.*

stress-stain plots and only yield a relatively small amount before they break. This is in contrast to amorphous rubbery polymers or semi-crystalline polymers whose amorphous regions are above their T_g. We have already considered an example of the latter (polyethylene), so now let us look at elastomers and consider one that is lightly cross-linked, so we do not have to worry too much about viscous flow in the time frame of our stress-strain experiment. Because most people consider rubbers to be almost perfectly elastic materials, you would no doubt at first think that the stress-strain curve should be perfectly straight or at least almost so, but with a small value of the slope (i.e., low modulus). The observed stress-strain plot is actually an unusual curve, as illustrated in more detail in figure 11.14. This is not because of yielding or other failures, but is an inherent non-linear elastic property of these materials.

It turns out that we must distinguish between highly *elastic materials*, by which we mean those that deform reversibly a large amount under a load, and *elastic behavior*, by which we usually mean a material that (more or less) obeys Hooke's law, even though it may deform just one or two percent. The crucial point is that for elastomers we are no longer looking at the small deformations characteristic of polymer glasses, but now considering deformations over a much, much, wider range, sometimes up to 1000%, and the relationship between stress and strain is highly non-linear. We will deal with the molecular basis of rubber elasticity later, but you should know that this type of behavior can also be modeled in the context of continuum mechanics by eliminating the restrictions imposed by Hooke's law (that stress is a linear function of strain) and small deformations.

We will not go into the derivation, it involves something called strain invariants and the derivation proceeds through equations for the strain energy, but you should be aware of the result, the Mooney-Rivlin equation, which has been widely used in analyzing the stress-strain behavior of elastomers:

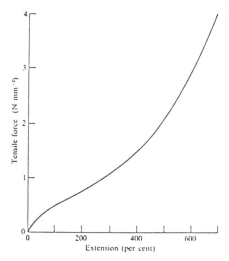

Figure 11.14 *Typical stress-strain curve for an elastomer. Reproduced with permission from L. R. G. Treloar,* The Physics of Rubber Elasticity, *Clarendon Press, Oxford, (1975).*

$$\sigma = 2C_1\left[\lambda - \frac{1}{\lambda^2}\right] + 2C_2\left[\lambda - \frac{1}{\lambda^3}\right] \tag{11.26}$$

or, alternatively

$$\sigma = 2\left[C_1 + \frac{C_2}{\lambda}\right]\left[\lambda - \frac{1}{\lambda^2}\right] \tag{11.27}$$

where λ is the extension ratio, l/l_0, not the strain, which would be $(1 - l_0)/l_0$. We will see later that the molecular theory of rubber elasticity gives the first term in equation 11.26, but not the second. As might be expected from a model that contains an additional fitting parameter (C_2), equations 11.26 or 11.27 provide a better correlation to the experimental data. Accordingly, there have been numerous attempts to elicit a theoretical basis for the Mooney-Rivlin equation, but so far the results have been unsatisfactory.

The stress-strain behavior of glassy polymers and elastomers illustrated in figure 11.13 obviously represents two ends of a range. We can obtain in-between behavior if we take a glassy polymer and measure its stress-strain properties as a function of temperature, as illustrated in figure 11.15. The same types of plots can be obtained by adding a plasticizer to the glassy polymer (where we now lower the T_g of the material rather than raise the temperature of the sample).

To summarize, we have seen in this section that Hooke's law is only obeyed at small deformations, even in hypothetically perfect materials. In addition to the curvature of the stress/strain plot at strains above (at most) just one or two percent, we also observe various types of yielding phenomena. Finally, the elastic properties of rubbers, which can reversibly deform by large amounts

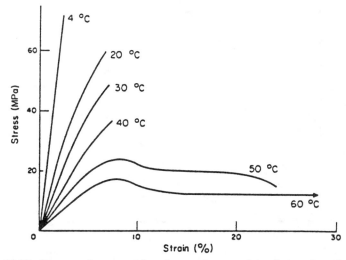

Figure 11.15 *Stress-strain curves of a glassy polymer, poly(methyl methacrylate), as a function of temperature. Reproduced with permission from T. S. Carswell and H. K. Nason, Symposium on Plastics, American Society for Testing Materials, Philadelphia, (1944).*

(>100%), is inherently non-linear. We will go on to consider all these phenomena individually, but there is one type of experiment we have not yet considered. What if we were to apply a small load, within the apparent Hooke's law range, but now measure deformation as a function of time? Would the measured strain stay constant? In other words, what is the time dependence of mechanical properties? We will come back to this important question after we consider the rheology of polymer melts.

The Viscosity of Polymer Melts

In the preceding paragraphs we considered elastic behavior by starting with Hooke's law and then discussed the deviations observed in real systems. We will use the same approach with melt viscosity and start by reminding you of the linear nature of Newton's law, where the shear stress is directly proportional to the rate of strain

$$\tau = \eta \, \dot{\gamma} \tag{11.28}$$

and the constant of proportionality η is called the viscosity. The viscosity is a measure of the frictional forces between the molecules in a liquid and depends upon factors such as intermolecular forces and free volume. Just as with our treatment of Hooke's law, it is possible to formulate theories for simplified model systems (e.g., describing hypothetical spherical molecules), but even these are more complex than we wish to consider here. Suffice it to say that (just as in Hooke's law) at low rates of strain a linear relationship is obtained, but at higher strain rates deviations are predicted to occur.

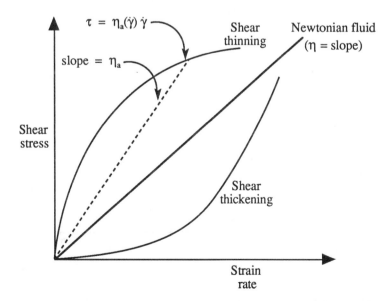

Figure 11.16 Schematic plot of shear stress versus strain rate for Newtonian and non-Newtonian fluids.

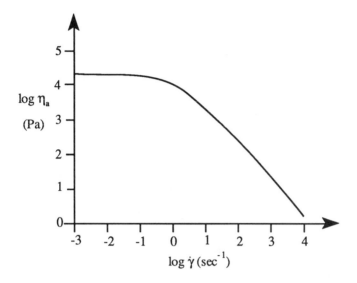

Figure 11.17 *Schematic plot of η_a against $\dot{\gamma}$ for a polymer melt.*

In real systems two types of deviations from Newton's law are observed, characteristic of *shear thinning* and (more rarely) *shear thickening* fluids. Polymer melts (and solutions) are usually shear thinning (some polymers that crystallize under stress can be shear thickening). As the name suggests and figure 11.16 indicates, this means that the viscosity decreases with increasing shear rate (which is of obvious importance in processing). Although we will mention some molecular theories of polymer rheology later, it is useful to point out here that this type of behavior is due to the variation of the rate of chain "disentanglement" with increasing strain rate. Accordingly we would expect the melt viscosity to be a strong function of molecular weight (and also temperature, as this affects molecular motion), which it is, as we will see later. Before getting to this, however, we need to define a measure of viscosity that can be used as a basis to compare behavior.

For non-Newtonian fluids the viscosity depends upon strain rate, $\dot{\gamma}$, which is also referred to as the shear rate, just to confuse you. We can therefore define an apparent viscosity, $\eta_a(\dot{\gamma})$, according to:

$$\tau = \eta_a(\dot{\gamma})\,\dot{\gamma} \qquad\qquad (11.29)$$

The apparent viscosity at a particular point, η_a, is given by the slope of the secant drawn from the origin to that point, as also illustrated in figure 11.16. If we now take values of η_a obtained from such plots (often measured on an instrument called a capillary rheometer), then we can in turn obtain plots of η_a against $\dot{\gamma}$, which typically appear as illustrated in figure 11.17. It can be seen that a linear proportionality between shear stress and strain rate is obtained at low values of $\dot{\gamma}$ (i.e., the viscosity is constant) and this leads to the definition of the *zero shear-rate viscosity* (i.e., value of η_a as $\dot{\gamma} \to 0$), which we will label η_m,

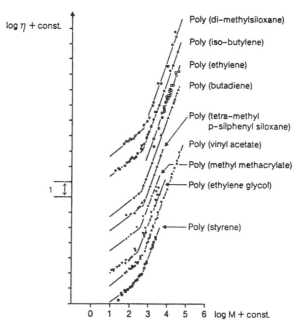

Figure 11.18 *Plot of log melt viscosity versus log molecular weight for various polymers. Reproduced with permission from G. C. Berry and T. G. Fox,* Adv. Polym. Sci., **5**, 261 *(1968).*

and used as a characteristic parameter for describing the behavior of polymer melts.

The variation of η_m with molecular weight is intriguing and, to some degree, still incompletely understood. If the log of the melt viscosity is plotted against the log of the number of carbon atoms in the backbone, then very similar curves are obtained for a wide range of polymers, as illustrated in figure 11.18. This suggests that there must be some underlying general theoretical relationship between these quantities.

For low molecular weight samples the relationship has the simple linear form

$$\eta_m = K_L (DP)_w^{1.0} \tag{11.30}$$

where $(DP)_w$ is the weight average number of carbon atoms in the main chain and K_L is just a constant. High molecular weight polymers obey a different relationship;

$$\eta_m = K_H (DP)_w^{3.4} \tag{11.31}$$

The change in the dependence of η_m on molecular weight occurs at chain lengths of the order of 300-500 main chain atoms. Although the plots presented in figure 11.18 make the transition appear sharp, a result of fitting the data points to straight lines, the change in the dependence on $(DP)_w$ occurs smoothly, but over a relatively narrow range of molecular weight. It is useful to assume (see

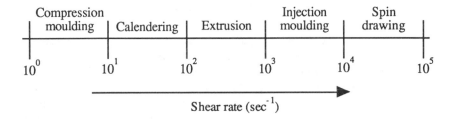

Figure 11.19 *Approximate range of shear rates encountered in typical commercial processing equipment.*

Graessley[*]) that the melt viscosity is related to two factors, the first of which depends upon local features such as free volume, which governs the viscosity of liquids of small molecules and results in a practically linear dependence of viscosity on molecular weight. The second factor depends upon the entanglements of the chains with one another and becomes important once a critical value of molecular weight (i.e., chain length) is reached. We will come back to discuss the molecular basis for this a little later (only briefly, however, because the mathematics gets out of hand for an introductory treatment), but will finish this section with some comments on the practical implications of these relationships.

If you examine equation 11.31 and consider the effect of doubling the molecular weight, it should be obvious that the polymer melt viscosity will increase by a factor of $2^{3.4}$, or about 10! This would seem to make the melt processing of high molecular weight polymers an imposing if not impossible task. We are saved by two things, however. First, recall that we are dealing with the zero shear-rate viscosity in the power law relationship. At higher shear rates the melt viscosity can be considerably lower. The approximate range of shear rates found in various types of processes are illustrated in figure 11.19.

Second, the melt viscosity also decreases with increasing temperature. Nevertheless, there is clearly an upper limit to the processing temperature we can use for a particular polymer that depends upon the degradation temperature. Accordingly, polymers become increasingly difficult to process as their molecular weight increases and at some point cannot be processed from the melt at all.

To summarize, in this and the preceding sections we have introduced the elastic properties of polymers as if they were unaccompanied by flow (apart from a brief mention of yielding) and the viscous properties of melts as if they were unaccompanied by elasticity. This is a good approximation of some materials, but for polymers we must consider how these properties usually occur together, which brings us to the topic of viscoelasticity.

[*] W. W. Graessley, *Adv. Polym Sci.*, **16**, 1 (1974).

D. INTRODUCTION TO VISCOELASTICITY

We have noted in the preceding paragraphs that when we stretch or otherwise deform a crystalline solid, elastic energy is stored in the stretched (or compressed) bonds and this provides the means for the material to return to its original shape upon removal of the load (assuming this was not a load large enough to break bonds, or induce slip along crystalline planes, etc.) In contrast, upon application of a shear stress to a Newtonian fluid, the energy is immediately dissipated in flow. Upon removal of the stress there is no elastic recovery. Some simple liquids and highly crystalline solids under small loads and deformations may approach these poles of ideal behavior, but most display characteristics that are between and their response is called *viscoelastic*. Viscoelastic behavior has actually been observed throughout recorded history. Archers in medieval times (and before), for example, knew never to leave their bows strung when not in use, because the tension in the bowstring would decrease over time as the wood in the bow "relaxed" and accommodated itself to the continuous load. Some of the first systematic studies of this phenomena were conducted by Weber (1835)[*], who noted that silk fibers displayed an immediate elastic deformation upon loading that was followed by an extension that gradually increased with time. Similarly, upon removing the load he observed both an immediate and delayed contraction. In many of the materials studied by Weber, such as glass and silver, "primary creep", which is the reversible component of the delayed deformation (there is often an irreversible component referred to as "secondary creep"), is usually a small proportion of the instantaneous elastic deformation (about 3%). Conversely, in the polymer materials studied, silk and natural rubber, the bulk of the deformation was observed to be time dependent.

In general, viscoelastic properties are particularly apparent in polymers and the extent and type of response depends upon chemical structure and morphology, the size of the applied load and, crucially, temperature. We will commence our discussion of this subject by first considering how viscoelasticity manifests itself in terms of some important mechanical methods of measurement[**]. This will lead to a consideration of "regions" of viscoelastic behavior and time-temperature correspondence. Finally (in this section), we will briefly discuss the molecular basis of what we will call relaxation behavior. This will serve as a foundation for the more detailed consideration of models and theories of viscoelastic behavior which will follow.

Experimental Measurements of Viscoelastic Behavior

We will introduce the subject of viscoelasticity by first considering the experiment that is the easiest to understand, *creep*. Here, we subject a sample to *constant stress* and measure deformation as a function of time. A perfectly elastic

---

[*] See the discussion in the classic book on creep by H. Leaderman, *Elastic and Creep Properties of Filamentous Materials and other High Polymers*, The Textile Foundation (1943).
[**] There are other important methods, such as dielectric spectroscopy, but these are topics you should study in more advanced treatments.

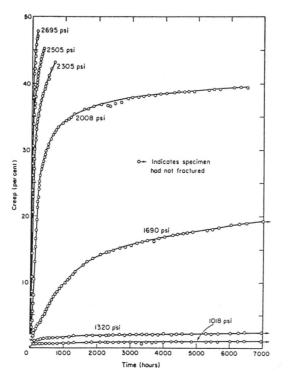

Figure 11.20 *Creep of cellulose acetate at 25 °C as a function of applied stress. Reproduced with permission from W. N. Findley,* Modern Plastics, *19, 71 (August 1942).*

solid would instantaneously deform an amount given by Hooke's law and there would be no further deformation with time. For a viscoelastic solid, however, strain will change with time. Consider, as an example, some classic experiments performed on cellulose acetate more than fifty years ago. Figure 11.20 shows strain, measured as percent elongation, plotted as a function of time. For loads of the order of 1,000 psi, creep is not extensive, but increases dramatically with increasing loads. Note the time frame of the experiments; 7000 hrs, the better part of a year.

It is important that you grasp that the type of behavior shown in figure 11.20 is not just a simple superposition of linear elastic and viscous responses. Strain/time plots for hypothetically ideal systems are shown schematically in figure 11.21, as is a strain/time plot characteristic of creep. As we just mentioned, elastic materials would deform (almost) instantaneously as a load is applied, the strain would then stay constant until the load is removed and the sample would then return to its original dimensions. In purely Newtonian viscous fluids strain increases continuously and linearly with time as energy is dissipated in flow, giving permanent deformation. Both of these types of behavior contribute to the schematic creep curve shown in figure 11.21, but it is clear that this curve is not a linear sum of ideal elastic and viscous behavior. The defining feature of the creep curve is a retarded elastic or anelastic response,

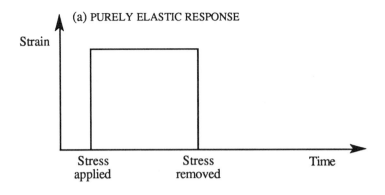

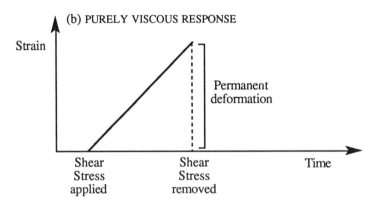

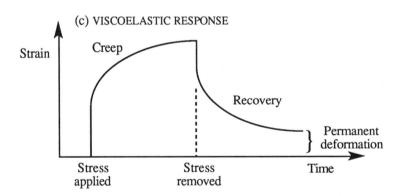

Figure 11.21 *Plots of strain versus time for purely elastic, purely viscous and viscoelastic materials.*

where viscous and elastic responses are coupled so that the material deforms gradually and in a non-linear fashion with time. In addition, there is a gradual recovery process when the load is removed. We will find that most polymers show various combinations of elastic, anelastic and viscous flow behavior, the extent of each varying with polymer structure and morphology, temperature, etc.

The second type of experimental measure of viscoelasticity is *stress-relaxation*. Here, instead of applying a constant stress to a sample and measuring strain as a function of time, the polymer is deformed (usually stretched) *instantaneously* (or, more accurately, as quickly as possible) to a given value of the strain and the stress necessary to maintain that strain is measured as a function of time. As the sample relaxes (i.e., as the chains change their conformations, disentangle or slide over one another, and so on), this stress decreases. Stress-relaxation experiments are somewhat easier to perform than creep experiments and typical stress relaxation curves are shown in figure 11.22. The results are reported as the relaxation modulus, E(t) (equal to the stress, which is changing with time, divided by strain, which is constant) and values obtained at different temperatures are then plotted as a function of time. We will show later how curves on log/log plots such as this can be strung together and superimposed to produce a single time/temperature master curve.

Finally, perhaps one of the most useful methods for determining viscoelastic behavior is the measurement of *dynamic mechanical properties*. In this type of experiment an oscillatory stress is applied to the sample. The frequency of the oscillation can be varied over an enormous range, but, as might be expected, different instruments are used for different frequency ranges. If the sample happens to be perfectly elastic and if the applied stress varies sinusoidally, then the strain would be completely in-phase with the applied stress and would vary as:

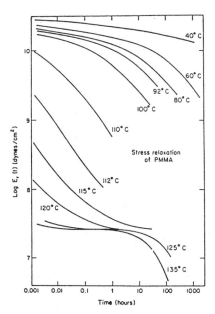

Figure 11.22 *Stress relaxation plots of log $E_r(t)$ versus log time (t) for poly(methyl methacrylate). Reproduced with permission from J. R. McLoughlin and A. V. Tobolsky, J. Colloid Sci., 7, 555 (1952).*

$$\gamma = \gamma_0 \sin (\omega t) \qquad (11.32)$$

where ω is the angular frequency of the applied stress in radians/sec (equal to $2\pi f$, where f is the frequency in cycles/sec). If we are considering experiments involving shear stresses and strains and small amplitudes of vibration then we can write Hooke's law as:

$$\tau (t) = G \gamma_0 \sin(\omega t) \qquad (11.33)$$

where we use the symbol $\tau(t)$ to indicate that the applied shear stress varies with time.

For a Newtonian viscous liquid, the resulting strain will be exactly 90° out-of-phase with the stress, because:

$$\tau (t) = \eta \dot{\gamma}(\tau) = \eta \frac{d}{dt}\left[\gamma_0 \sin (\omega t) \right] \qquad (11.34)$$

or:

$$\tau (t) = \eta \gamma_0 \omega \cos (\omega t) \qquad (11.35)$$

It follows that for a viscoelastic material we would expect the response to be characterized by a phase angle somewhere between 0° and 90° and this phase angle should be related to the dissipation of energy (viscous type response) relative to its storage (elastic type response). If we therefore apply a sinusoidal stress to a viscoelastic material such as a polymer:

$$\tau(t) = \tau_0 \sin(\omega t) \qquad (11.36)$$

We should now expect the resulting measured strain to lag behind by some degree. You may recall that when dealing with oscillating phenomena (e.g., light) it is usual to define this lag in terms of a phase angle δ, so we can write the resulting strain as:

$$\gamma = \gamma_0 \sin(\omega t - \delta) \qquad (11.37)$$

It is actually more convenient (and usual) to "reset the zero" and consider the stress "leading" the strain by an amount δ, so that

$$\tau(t) = \tau_0 \sin(\omega t + \delta) \qquad (11.38)$$

and:

$$\gamma = \gamma_0 \sin \omega t \qquad (11.39)$$

We can now rewrite equation 11.38 as:

$$\tau(t) = (\tau_0 \cos\delta) \sin\omega t + (\tau_0 \sin\delta)\cos\omega t \qquad (11.40)$$

which expresses the stress in terms of an in-phase component and an out-of-phase component with respect to the strain (the $\sin\omega t$ and $\cos\omega t$ terms, respectively). The relationship between stress and strain can now be *defined* in terms of these in-phase and out-of-phase components by

$$\tau(t) = \gamma_0\left[G'(\omega) \sin\omega t + G''(\omega) \cos\omega t \right] \qquad (11.41)$$

where:

$$G'(\omega) = \frac{\tau_0}{\gamma_0} \cos \delta \qquad (11.42)$$

and:

$$G''(\omega) = \frac{\tau_0}{\gamma_0} \sin \delta \qquad (11.43)$$

(If you are a bit under the weather today and don't see where equation 11.41 comes from, just substitute 11.42 and 11.43 into 11.41, note that equation 11.40 is obtained and recall that 11.41 only serves to *define* the quantities $G'(\omega)$ and $G''(\omega)$).

The in-phase component, $G'(\omega)$, is called the *storage modulus*, while the out-of-phase component, $G''(w)$, is called the *loss modulus*. We also have:

$$\tan \delta = \frac{G''(\omega)}{G'(\omega)} \qquad (11.44)$$

Although the use of a trigonometric representation (i.e., sines and cosines) is easiest to picture when you start this part of the subject, in terms of the mathematical manipulations required in subsequent developments it is actually more convenient to use a complex notation (i.e. $e^{i\omega t}$; terms like this are much easier to differentiate and integrate than combinations of sines and cosines). For our introductory treatment this is unnecessary, but to give you a flavor of what you can do we introduce this approach as one of the homework problems at the end of this chapter.

A schematic representation of the dependence of $G'(\omega)$, $G''(\omega)$ and $\tan \delta$ upon the frequency of oscillation (ω) is shown in figure 11.23 for an amorphous

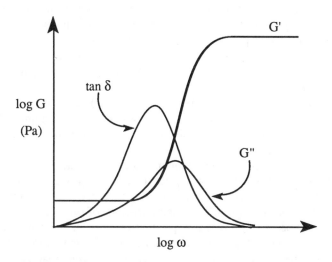

Figure 11.23 *Schematic diagram of the frequency dependence of the moduli G', G" and their ratio tan δ.*

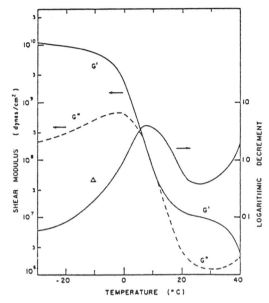

Figure 11.24 *The temperature dependence of G', G" and the log decrement for a styrene-co-butadiene. Reproduced with permission from L. E. Nielsen,* Mechanical Properties of Polymers, *Reinhold, New York, 1962.*

polymer. You should immediately note that this representation is idealized and real data can often appear more complex. Nevertheless, the key features correspond to those observed for all amorphous polymers. First, at low frequencies the storage modulus $G'(\omega)$ is characteristic of that found in rubbers. As the frequency increases this modulus increases several orders of magnitude and levels out at a value of the modulus characteristic of the glassy state. Keep in mind that this is a constant temperature experiment and let us assume for the sake of argument that it was performed at room temperature. What may surprise you is that both a glassy polymer and a rubber at this temperature would give this type of curve. It's just that these curves would be shifted along the frequency axis relative to one another.

It can also be seen that the loss modulus, $G''(\omega)$, starts at low values and also increases with frequency, while tan δ goes through a maximum and displays a peak in the range where $G'(\omega)$ is changing its value from one characteristic of rubbers to one characteristic of the glassy state.

Clearly, tan δ is showing a maximum at a position that is characteristic of a T_g, except that instead of measuring the modulus of a sample as we change temperature, we have measured modulus as a function of a time (in terms of the frequency of an oscillation, which has dimensions of time^{-1}). If a sample was subjected to an oscillating stress at a specific frequency and the temperature were allowed to vary, then a typical result would be that shown in figure 11.24. Here a logarithmic decrement $\Delta (\approx \pi \tan \delta)$ has been plotted rather than tanδ, but again there is a maximum at the T_g. We will discuss the molecular origin of this later, first we wish to explore this time/temperature correspondence in a little more detail and describe various "regions" of viscoelastic behavior. Then we can explore the type of molecular relaxation behavior that governs these various regions.

The Regions of Viscoelastic Behavior and Time-Temperature Correspondence

To start our discussion we will consider a stress relaxation experiment performed on an amorphous polymer as a function of temperature. We have already noted that the time frame of the experiment is important, so let us imagine that we have measured the (Young's) modulus over some arbitrary short time period, say ten seconds*. The sample is therefore stretched instantaneously to give a chosen value of the strain and after 10 seconds the stress is measured and the modulus calculated. A plot of the log of this modulus vs. temperature is shown in figure 11.25. A value of the modulus characteristic of the glassy state is found at "low" temperatures (i.e. well below the T_g of whatever polymer we are considering) and this is characterized by a modulus that is usually larger than 10^9 Pa. As the temperature increases we come to the glass transition region, where the value of the modulus decreases by several orders of magnitude. Within this range the

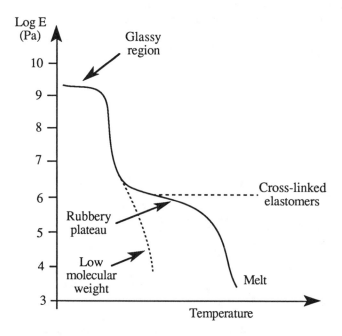

Figure 11.25 *Schematic plot of log modulus vs. temperature for a hypothetical amorphous polymer.*

---

* We followed J. J. Aklonis, and W. J. MacKnight, *Introduction to Polymer Viscoelasticity*, Second Edition, Wiley Interscience, (1983), in picking this arbitrary value.

polymer has properties that are similar to leather and shows a noticeable retarded elastic or anelastic response. As the temperature goes beyond the glass transition region, three things can happen, depending upon the molecular weight of the sample and whether or not it is cross-linked. Low molecular weight polymers, those whose chain lengths are less than that necessary for entanglements to be significant, quickly become liquid-like in their behavior. This is shown as the dotted line in figure 11.25. High molecular weight polymers exhibit a rubbery plateau (i.e., there is a fairly constant modulus over a range of temperature that depends upon the molecular weight of the polymer). Here the material is largely solid-like in its properties and displays (over our 10 second time range) the properties associated with rubber elasticity. If the sample is lightly cross-linked, this behavior extends to the degradation point, as shown by the dashed line in figure 11.25. Finally, if the sample is not cross-linked, the modulus starts to decrease more sharply with increasing temperature and the polymer can no longer hold its own shape, becoming more liquid-like in its properties. To summarize, for a non cross-linked high molecular weight amorphous polymer four regions of viscoelastic behavior are observed; the glassy region; the glass transition region, a rubbery plateau and a terminal flow region.

As might be expected from our description of dynamic mechanical properties in the preceding section, similar results are obtained if we make time our variable and hold the temperature constant. To be explicit, the experiment now involves stretching the sample a given arbitrary amount and measuring the stress required to maintain this strain as a function of time. A schematic representation of the modulus vs time plot is shown in figure 11.26.

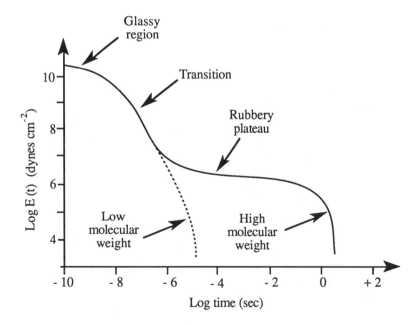

Figure 11.26 *Schematic representation of the modulus as a function of the time frame of the experiment.*

At short time periods [how short depends on the nature of the polymer (rubbery or glassy) at the temperature of the experiment] the measured modulus (stress/strain) is high, characteristic of the glassy state. As time goes on the stress required to maintain the given strain starts to decrease sharply, hence so does the calculated modulus, and this decrease is over a fairly well-defined range, corresponding to the T_g in the modulus vs temperature experiment. A rubbery plateau is then encountered. It should be noted that if we started with a glassy polymer at room temperature, then the time required for the stress to relax to the point that we calculate a modulus characteristic of the rubbery state could be very long indeed, of the order of hundreds of years. At even longer times the stress required to maintain the strain would be extremely small. In other words, the calculated modulus would be characteristic of a polymer melt, where in ordinary time frames (i.e., ~10 secs) a small stress will result in a significant deformation. For a glass in stress relaxation experiments this level of deformation is only reached after thousands of years.

So far we have only discussed amorphous polymers. The viscoelastic behavior of semi-crystalline polymers is far more complex, because of the superposition of the behavior of the crystalline and amorphous domains. In some polymers, where the degree of crystallinity is not too high, transitions characteristic of the amorphous state can be observed superimposed upon those due to the crystalline domains. This superposition is not necessarily linear, however, and coupling of the responses can occur, particularly as the degree of crystallinity of a sample is increased and the amorphous regions become constrained by the crystalline domains. Because of these factors the behavior of semi-crystalline polymers is much less uniform than those that are purely amorphous, displaying individual idiosyncrasies that often have to be described separately. There is one generalization that we can make, however, and that concerns the large change in modulus that is observed at the crystalline melting point, T_m.

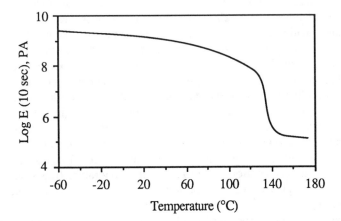

Figure 11.27 *Schematic plot of the log modulus vs. temperature for polyethylene.*

Figure 11.27 displays schematically the type of (ten second) modulus vs temperature behavior that would be obtained from a highly crystalline sample of polyethylene. Here there are small regions of amorphous material that are constrained by hard crystalline domains, so at low temperatures the sample displays a high modulus. As we have seen in our discussion of thermal properties, melting of smaller, less perfect, crystals can occur at temperatures below the melting of the bulk of the crystalline domains, so the modulus decreases somewhat as the temperature is raised, but the largest change in modulus occurs at what we usually define as T_m. At this temperature the polymer becomes entirely amorphous and subsequently displays a rubbery plateau and at still higher temperatures enters the terminal flow region. (The T_m is always at a higher temperature than T_g, so a semi-crystalline polymer will melt to give a rubbery type polymer.)

To conclude, we wish to emphasize that the most crucial point that you should take from this section is that time and temperature are inextricably intertwined when it comes to viscoelastic behavior and for *linear* viscoelastic properties this relationship can be expressed in terms of the *time-temperature superposition principle*, which we will discuss later on. First we wish to consider why and how polymer molecules respond to loads in the fashion we have just described, which brings us to a qualitative discussion of relaxation processes.

Relaxation in Polymers

It is useful to start this section by restating some fundamental principles and then use these to get into the heart of our discussion. First, you should recall the meaning of *temperature*; kinetic theory tells us that this is simply a measure of the mean kinetic energy of the molecules. If we add heat to a material and therefore raise its temperature we are simply making the molecules move faster (on average). Second, in molecules such as polymers there are energy barriers (due to steric repulsion and so on) inhibiting bond rotation, so that at low temperatures the atoms largely vibrate around their mean positions, but at higher temperatures they might have sufficient kinetic energy to rotate over these barriers and adopt new conformations. Third, it is a useful approximation to simply focus on a few local bond configurations or conformations defined in terms of local energy minima, such as the trans and gauche arrangements in polyethylene that we described in chapter 7. If there is sufficient thermal energy then we can imagine each bond "clicking" between each of these conformations, but on average more bonds will be found in the lowest energy conformation (trans) than in higher energy arrangements (gauche) at any instant of time. If the energy of each of these states is known, then we can use the methods of statistical mechanics to calculate the relative population of the conformational states (e.g., trans vs gauche). This distribution will change with temperature (more bonds will be found in the higher energy gauche states at higher temperatures than at lower temperatures).

Now consider a hypothetical single chain, isolated in space, in equilibrium at a given temperature. There is a certain distribution of trans and gauche states and an average distance between the ends of a chain as it flops around over time. Imagine that we could take some molecular tweezers, grasp the ends of this chain and apply a force. The chain would no longer be at equilibrium, in that a more stretched-out overall chain conformation would now be favored. This, in turn would favor a new distribution of trans and gauche local bond conformations. However, if we imagine the chain sitting in a viscous medium that approximates the effect of other chains, then because of frictional forces it would not be able to follow the applied load immediately, as it would take time for the bonds to rearrange themselves to give the new equilibrium population of states. We say that the chain *relaxes* to this new equilibrium state. Note that we use the word "relaxation" in a broader manner than just "stress relaxation", which simply relates to a specific type of experiment. Relaxation processes that we will consider here are concerned with a whole range of time dependent molecular transitions and rearrangements.

If rotations around each backbone bond in a polymer chain were totally independent of one another, and if in addition we only had to consider transitions between two local states, say trans and gauche, then the relaxation process could be described in terms of a single relaxation time, τ_t^*. This relaxation time characterizes the time scale necessary for the rearrangement to take place. We will give a mathematical definition later on when we consider simple mechanical models.

Of course, in real polymer materials the situation is far more complex; it is only an approximation to consider two allowed bond conformational states; bond rotations are not independent of one another but are affected by both short and long range steric interactions; we don't have isolated chains that are acted on by simple viscous forces and have to consider the effect of entanglements; and so on! Accordingly we can imagine various types of complex coupled motions or relaxation processes and these are usually referred to as "modes" (note the analogy to vibrational modes that we discussed earlier in our chapter on spectroscopy). Each of these modes will have a different characteristic relaxation time or range of times. Even this is a simplification and it has been argued by de Gennes that the actual concept of modes might be invalid. The various relaxation processes that occur in entangled chains may be so complex and coupled that only a smooth distribution or spectrum of relaxation times may actually exist. Nevertheless, the concept of modes and even models that are characterized by single relaxation times are useful when you start out in this part of the subject, particularly if you are struggling with the ideas we have just introduced, so we will persist with them for a time. Just keep in mind that the dynamics of entangled polymer chains is a complex subject and one where we at present only have rough theoretical models.

How fast are relaxations in polymers? As we will see, relaxation is a strong function of temperature. In small or low molecular weight liquids relaxations are very fast ($\sim10^{-10}$ secs). In a polymer melt ($T>T_g$) they are much slower. Maconnachie et al[**] used a neutron scattering method to determine how quickly a

---

[*] Many texts simply use the symbol τ, but we have used that symbol to designate shear stress. Accordingly, we use the subscript t to indicate we are talking about a time constant.
[**] A. Maconnachie, G. Allen and R. W. Richards, *Polymer*, **21**, 1157 (1981).

stretched polystyrene sample (M_w ~144,000) would relax and found that it took about five minutes to approach completion. At temperatures close to the T_g or below the T_g relaxation times are obviously much longer, and this brings us back to a consideration of the four regions of viscoelastic behavior, but now in terms of the types of relaxations that can occur.

In the glassy state conformational changes involving coupled bond rotations are severely inhibited, although they can occur over very long time periods, as we have seen, and local conformational transitions and side chain motions can and do occur. Nevertheless, for small loads, (hence strains) and short times the response can be regarded as essentially elastic.

In the region of the glass transition there is now sufficient thermal energy that various cooperative motions involving longer chain sequences occur. Frictional forces are such that motions are sluggish, however, and retarded or anelastic responses are observed. In dynamic mechanical experiments we are in a region where there is a severe mismatch between the imposed frequency of oscillation and the time scale of the relaxation process. Frictional losses then rise to a maximum, giving the observed peak in the tan δ curve that we discussed earlier. We considered theories or the glass transition in chapter 8, so we will not elaborate on these any further here.

As the temperature is raised the time scale of conformational relaxation processes becomes shorter; in other words the chains can adjust their shape to an imposed deformation in the material as a whole more readily. The time scale for disentanglement of the chains is longer than the time scale for these conformational adjustments, however, so in this region the entanglements act something like cross-links. The chains can stretch out between the entanglement points and we observe a rubbery plateau in the appropriate experiments. Finally, as the temperature is increased still further the time necessary for disentanglement decreases, chain diffusion becomes faster than the measurement time of the experiment and we enter the terminal flow region.

One would expect similar relaxation processes to occur in the amorphous domains of semi-crystalline polymers, but here motion is suppressed and constrained by the crystallites. As we mentioned before, behavior depends on a number of variables and the behavior of individual polymers can be dominated by their own peculiarities, making it much more difficult to make general points. However, relaxation processes observed at temperatures below the T_g are usually associated with the motion of short chain segments or substituent groups, while those observed above the T_g are more complex, often involving coupled processes in amorphous and crystalline domains. The various relaxation processes often show up in tan δ measurements, as illustrated in figure 11.28. This figure compares data obtained from a linear polyethylene sample (LPE) and a low density polyethylene (LDPE) sample. It is conventional to label these transitions α, β, γ, starting from the highest temperature process. The β transition has the characteristics of a glass transition. The γ process involves a local relaxation while the α transition is considerably more complex, involving coupled relaxations in both the crystalline and amorphous domains. In the linear polyethylene sample the β (glass) transition is suppressed, because of the severe constraints on the amorphous regions imposed by the crystalline domains in this

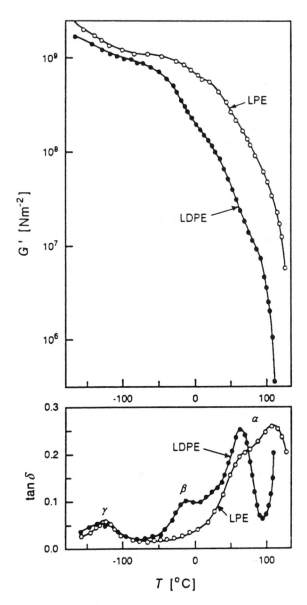

Figure 11.28 *Temperature dependence of the storage shear modulus (top) and the loss tangent (bottom) of linear (LDP) and branched polyethylene (LDPE). Reproduced with permission from H. A. Flocke,* Kolloid–Z. Z. Polym., **180**, *188 (1962).*

highly crystalline polymer. Some confusion can arise as a result of the convention of labeling the highest temperature transition α, and so on down. In amorphous polymers the α-transition identical to the T_g, but in semi-crystalline polymers if a T_g is observed at all it is labeled the β-transition.

E. MECHANICAL AND THEORETICAL MODELS OF VISCOELASTIC BEHAVIOR

In the preceding section we have seen that in general there is a non-instantaneous component of mechanical behavior (e.g., creep) that is non-linear with time (re-examine the deformation vs. time curves shown in figure 11.20 and 11.21. Nevertheless, we can seek to model this *non-linear* time dependence by the assumption of a *linear* relationship between stress and strain. This is the fundamental assumption of *linear viscoelastic models*. There are also non-linear models, but this is an advanced topic that we will not consider.

If the force imposed on a polymer in a creep experiment is small, then the curve of the strain as a function of time $\varepsilon(t)$ is indeed found to be proportional to the load. In other words, if we conduct two creep experiments, one with a constant applied stress of σ_0 and the second with a constant applied stress of $2\sigma_0$, then the strain measured in experiment 2, at any specified instant of time (say 10 minutes after the load is applied), is double that observed in experiment 1.

It is usual to describe the properties in tensile creep in terms of a *tensile creep compliance*, D(t):

$$D(t) = \frac{\varepsilon(t)}{\sigma_0} \tag{11.45}$$

where σ_0 is the constant tensile stress. A shear creep compliance can also be defined as:

$$J(t) = \frac{\gamma(t)}{\tau_0} \tag{11.46}$$

In a similar manner a tensile relaxation modulus and shear relaxation modulus can be defined to describe the results of stress relaxation experiments:

$$E(t) = \frac{\sigma(t)}{\varepsilon_0} \tag{11.47}$$

$$G(t) = \frac{\tau(t)}{\gamma_0} \tag{11.48}$$

It is crucial to realize that although

$$E = \frac{1}{D} \tag{11.49}$$

and:

$$G = \frac{1}{J} \tag{11.50}$$

when we have linear time independent behavior, the same is not true when these quantities become a function of time. For example, consider a polymer that has been subjected to a constant uniaxial stress, σ_0, for a period of, say, 10 hours. After this time the sample has stretched out and the measured strain is $\varepsilon(10)$. Now consider an alternative experiment where the sample is stretched instantaneously to give this value of the strain [i.e., $\varepsilon_0 = \varepsilon(10)$]. After 10 hours the stress required to maintain this strain will be $\sigma(10)$. In general, $\sigma(10)$ will not be the same as σ_0 (it will usually be less), so that:

$$E(10) = \frac{\sigma(10)}{\varepsilon_0} \neq \frac{\sigma_0}{\varepsilon(10)} = \frac{1}{D(10)} \tag{11.51}$$

or, in general:

$$E(t) \neq \frac{1}{D(t)} \tag{11.52}$$

and:

$$G(t) \neq \frac{1}{J(t)} \tag{11.53}$$

The quantities we have defined [E(t), D(t) and G(t), J(t)] provide, in principle, a complete description of the tensile and shear properties of a sample in creep and relaxation experiments. (Modulus and compliance complex functions can also be used to define dynamic properties.) In other words, the observed mechanical behavior of polymers can be represented as plots of E(t), D(t) etc., vs time, temperature, frequency and so on, depending upon the nature of the experiment. Clearly, what we would like to do is find ways of relating the data determined in one type of experiment [e.g., E(t) in stress relaxation] to data obtained in another [e.g., D(t) in creep], but even more critically we would like to be able to interpret this data in terms of fundamental molecular relaxation processes.

The first problem is relatively straightforward as long as the assumption of linear viscoelastic behavior holds. It can be solved using the *Boltzmann Superposition Principle*, originally proposed by Boltzmann in 1874 in an attempt to reduce the complex manifestations of primary creep to some simple scheme. It is not a physical theory, in the sense of having a basis in propositions concerning fundamental molecular behavior, but is extraordinarily useful and has resulted in a coherent phenomenological theory of linear viscoelasticity (i.e., a theory that simply describes the formal structure of the data and does not concern itself with any hypothesis concerning microscopic or molecular behavior). The principle essentially states that the deformation of a sample at any instant is not just a result of the load acting at that instant, but depends upon the entire previous loading history. Furthermore, if a specimen has been subjected to a number of loading steps, then each of these steps makes an independent and additive contribution to the final deformation. It has been shown that this superposition principle is actually a consequence of the assumption of linear viscoelasticity, but we will not go into this any further at this point. Nor will we show how the principle can be used to relate quantities such as compliance and modulus, or to relate static and dynamic properties to one another. The relationships are in the form of integral equations and their derivation necessitates the use of transform methods. This gets beyond the scope of the basic ideas we wish to present in this book. We will revisit the principle of linear additivity of effects later in our discussion, however, and you should have a basic appreciation of the importance of this superposition principle (which can also be applied to the description of other phenomena, such as dielectric properties).

The topics we wish to consider in the remainder of this section concern the relationship of experimentally determined quantities [e.g., E(t), D(t)] to the relaxation behavior of polymers. There are essentially two approaches. The first uses mechanical models, essentially integral equations that attempt to reproduce the data in terms of certain characteristic parameters. This approach uses and combines mathematical representations of various mechanisms of deformation (purely viscous, purely elastic, retarded elastic) and we will discuss these in some detail. This is not a microscopic theory, however, so we will finish this section by describing theories based on the dynamics of polymer chains. This is an advanced topic so our description will be largely qualitative.

Simple Models of Mechanical Behavior

In the next few pages we are going to consider mechanical models that consist of various combinations of springs and dashpots, representing linear elastic behavior and linear viscous (Newtonian) behavior, respectively. These provide picture representations of the differential equations that we are actually using to describe behavior. Many modern texts, particularly those that are more advanced, ignore and even advise their readers to forget springs and dashpots completely and just jump into the equations. They are correct in the sense that these models do not accurately reproduce observed mechanical responses, and are inconsistent in that even if a particular model can be used to approximate say stress relaxation, it cannot be applied to creep. However, this is an introductory text and it has been our experience that students who have not been exposed to aspects of mechanical properties in previous courses can struggle with even simple concepts. Accordingly, our approach will be to first present the simple models to give students a better feel for the combination of elastic, retarded elastic and flow properties that real polymers exhibit. Then we will consider more complex models to show in which direction the mathematics needs to be developed. At this point we then hope that the student has become comfortable with the concept of a spectrum of relaxation (and retardation) times that we will introduce and will be ready to abandon simple models for more advanced treatments, if this is where their inclinations lead them.

Let us start by considering simple uniaxial extension and assume linear behavior (i.e., Hooke's law):

$$\sigma = E\varepsilon \tag{11.54}$$

As mentioned above, we are going to represent this type of behavior using a picture of a spring. Similarly, Newtonian viscous flow

$$\sigma = \eta \frac{d\varepsilon}{dt} \tag{11.55}$$

will be represented by a picture of a dashpot. Both of these are illustrated in figure 11.29, along with some combinations we will get to shortly.

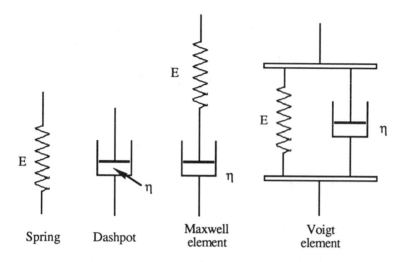

Figure 11.29 *Picture representations of simple elastic, viscous and viscoelastic models.*

The spring is easy to understand and all you have to do is imagine we have an unlimited supply of these things, each with a different value of the modulus, so we can model the behavior of anything from a rubber band (but remember, only at small deformations!) to steel girders. In contrast to the spring, some students have trouble visualizing the nature of a dashpot. Think of it as a piston in a cylinder that can be filled with various perfect liquids (i.e., Newtonian), each of which has a different viscosity. If we wish to consider flow over an extended period of time this will be a very long cylinder, of course, but don't take this model too literally! These are picture representations that will allow us to more easily see how we are combining mathematical equations. It is the same as designing an electric circuit using symbols for resistors and capacitors. This is a nice analogy because those of you who have been exposed to such things will immediately grasp two initial possibilities, combining the spring and dashpot in series and combining them in parallel, as also illustrated in figure 11.29. The former is called the *Maxwell model* (or *element*) while the latter is called the *Voigt model* (or *element*).

We will consider behavior in terms of plots of strain versus time. If we apply a stress to our hypothetically perfect spring, then it would deform instantaneously an amount that depends on its modulus. The strain or deformation would then remain at this value until the stress was removed, upon which it would instantaneously return to its original dimensions, as illustrated in figure 11.30.

In contrast, the dashpot displays perfectly viscous behavior, so the strain would increase in a linear fashion with time, as also illustrated in figure 11.30. The slope of this line is given by:

$$\frac{d\varepsilon}{dt} = \frac{\sigma}{\eta}$$

(11.56)

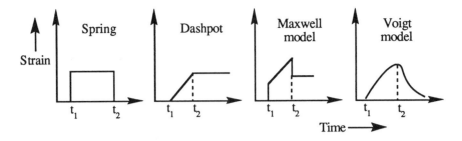

Figure 11.30 *Stress versus time plots for the models shown above in figure 11.29. The point t_1 corresponds to when the stress is first applied, t_2 to when it is removed.*

Upon removal of the stress the piston in the dashpot just stays where it is, it does not return to its original position, so we obtain some permanent deformation that depends upon the time frame of the experiment.

Now consider the Maxwell model, which is a picture representation of a differential equation derived in Maxwell's classical paper "On the dynamical theory of gases"*. In the introductory paragraphs of this work Maxwell discussed the elastic and viscous forces acting in a general "body". He considered both creep and stress relaxation phenomena by noting first that if stress varies with time, then Hooke's law for elastic behavior can be differentiated to give:

$$\frac{d\sigma}{dt} = E \frac{d\varepsilon}{dt} \tag{11.57}$$

Maxwell then assumed viscous forces that obeyed equation 11.56 above. If it is now assumed that the rate of strain is given by a simple sum of these two contributions, which is essentially what the picture of the Maxwell model shown in figure 11.29 represents, it follows that:

$$\frac{d\varepsilon}{dt} = \frac{\sigma}{\eta} + \frac{1}{E} \frac{d\sigma}{dt} \tag{11.58}$$

In a simple stress-relaxation experiment the sample is stretched to a given length and held at this extension, which means:

$$\frac{d\varepsilon}{dt} = 0 \ \text{(stress relaxation)} \tag{11.59}$$

and we can integrate the resulting expression to obtain σ as a function of time:

$$\int_0^t \frac{E}{\eta} dt = -\int_{\sigma_0}^{\sigma} \frac{d\sigma}{\sigma} \tag{11.60}$$

where the initial stress (i.e., at $t = 0$) is equal to σ_0. Then:

---

* J. C. Maxwell, *Phil. Trans. Roy. Soc. London*, **157**, 49-88 (1867), reproduced in *Maxwell on Molecules and Gases*, edited by E. Gerber, S. G. Roush and C. W. F. Everitt.

$$\ln \left(\frac{\sigma}{\sigma_0} \right) = - \left(\frac{E}{\eta} \right) t \qquad (11.61)$$

or:

$$\sigma = \sigma_0 \exp\left[- \left(\frac{E}{\eta} \right) t \right] = \sigma_0 \exp\left[-\frac{t}{\tau_t} \right] \qquad (11.62)$$

where τ_t is equal to η/E and has units of time. It is called the relaxation time (if you understand how exponentials work then you will realize that this is equal to the time for the stress to reach $^1/e$ of its original value).

In a stress-relaxation experiment it is usual to plot the relaxation modulus, $E(t)$ against time on a log/log plot (see figure 11.22). The Maxwell model gives behavior that appears as in figure 11.31. At short time periods ($t \ll \tau_t$) the material is largely elastic, but when viewed over long times ($t \gg \tau_t$) the overall behavior appears more or less purely viscous. For time $t \approx \tau$ the response is a combination, or viscoelastic.

This is a decent, bare-bones description of the gross features of viscoelastic behavior and if we compare figure 11.31 to the data given in figure 11.22, the Maxwell model appears to qualitatively model some of the curves determined at certain temperatures. If a plot of experimental values of $E(t)$ obtained over very long time periods were obtained, however, it would appear as in figure 11.25 (for an amorphous polymer), showing the characteristic four regions of viscoelastic behavior. Clearly the Maxwell model does not represent this range of behavior at all and if you have been thinking about this the problem should be immediately apparent. In our discussion of relaxation we noted that various

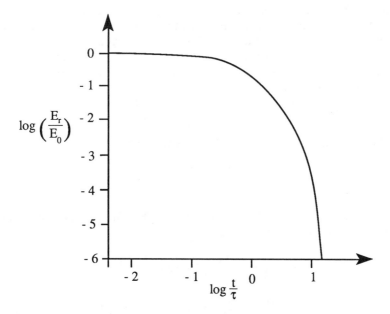

Figure 11.31 *Schematic representation of the behavior of the Maxwell model (note that E_0 is the initial modulus).*

coupled molecular processes are involved and each will be characterized by a different relaxation time or perhaps range of relaxation times. Maxwell's simple model only allows for a single relaxation time, and is therefore incapable of reproducing the rich spectrum of real behavior. Nevertheless, it is the classical initial treatment of the problem of viscoelasticity and in describing relaxation it results in an expression for a relaxation modulus:

$$E(t) = \frac{\sigma(t)}{\varepsilon_0} = \frac{\sigma_0}{\varepsilon_0}\exp\left(-\frac{t}{\tau_t}\right) = E_0\exp\left(-\frac{t}{\tau_t}\right) \tag{11.63}$$

that goes beyond the limitations of the model, as we will see later.

Unfortunately, the Maxwell model does a lousy job in describing the retarded elastic response that is characteristic of creep. If we let the stress have the constant value σ_0 in equation 11.58, then we obtain the result that the creep rate is constant with time:

$$\frac{d\varepsilon}{dt} = \frac{\sigma_0}{\eta} \tag{11.64}$$

which it isn't!

Another way of looking at the Maxwell model is to note that it essentially assumes a uniform distribution of stress (i.e., the stress is common to the spring and dashpot elements). We can make the alternative assumption that the strain is identical in all parts of the system and this gives us the Voigt model, where the spring and dashpot can be pictured as being placed in parallel, as in figure 11.29. It should be intuitively obvious that this model will represent a *retarded* elastic type of behavior. When the stress is applied, the spring wants to deform instantaneously, but is prevented from doing so by its connection to the dashpot. Similarly, when the load is removed, the spring wants to instantaneously return to its original dimensions, but again cannot do so because it is now pushing on the dashpot, which slows it down. This can be expressed mathematically by noting that the strain (or strain rates) in both elements are the same and the total stress applied to the system must be the sum of the stresses in each of the two elements. This gives us:

$$\sigma(t) = E\,\varepsilon(t) + \eta\,\frac{d\varepsilon(t)}{dt} \tag{11.65}$$

In creep experiments the stress σ_0, is constant, so that:

$$\frac{d\varepsilon(t)}{dt} + \frac{\varepsilon(t)}{\tau_t'} = \frac{\sigma_0}{\eta} \tag{11.66}$$

where we have again substituted a time τ_t' for η/E. This is a linear differential equation which has the following solution:

$$\varepsilon(t) = \frac{\sigma_0}{E}\left[\,1-\exp(-t/\tau_t')\,\right] \tag{11.67}$$

or:

$$\frac{\varepsilon(t)}{\sigma_0} = D(t) = D_0\left[\,1-\exp(-t/\tau_t')\,\right] \tag{11.68}$$

which results in plots of strain against time that are of the exponential form shown earlier in figure 11.30. The strain approaches its maximum value (σ_0/E) at a rate that depends on τ_t' (which we now call a retardation time instead of a relaxation time, but is given by the ratio of the same parameters, η/E) and exponentially decays from this value over time once the stress is removed.

Like the Maxwell model, the Voigt model is seriously flawed. First, it is again a single relaxation (or retardation) time type of model, and we know that real materials are characterized by a spectrum of such times. Second, although it produces the retarded elastic response characteristic of creep, it cannot model stress relaxation; under a constant load the Voigt element does not relax (look at the model and think about it)! However, just as we obtained an equation for the relaxation modulus from the Maxwell model (equation 11.63) whose form is of general applicability, we obtain from the Voigt model an equation for creep compliance whose form is also of general applicability (11.67). More on this later.

The simplest flaws of the two models we have so far considered, the fact that the Maxwell model does not account for the retarded elastic response characteristic of creep, while the Voigt model does not account for stress-relaxation, can be easily fixed. Elements can be combined in various ways and one example is the so-called four parameter model, obtained by putting the Maxwell and Voigt models in series, as shown in figure 11.32. The four parameters are the Maxwell spring modulus E_M and dashpot viscosity η_M and the equivalent Voigt parameters E_V and η_V. Plots of strain vs time for a creep exper-

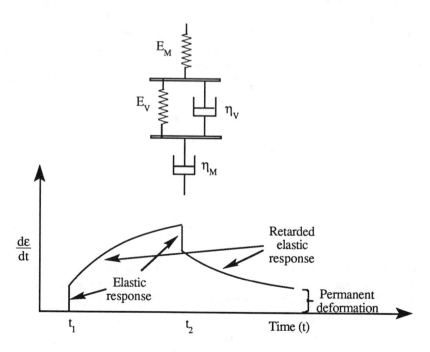

Figure 11.32 Schematic representation of the four parameter model.

iment are also shown in figure 11.32. In this constant load experiment we simply sum the terms for Maxwell and Voigt components to obtain:

$$\varepsilon = \frac{\sigma_0}{E_M} + \frac{\sigma_0 t}{\eta_M} + \frac{\sigma_0}{E_M}\left[1 - \exp(-t/\tau_t)\right] \qquad (11.69)$$

which is a sum of three components, describing elastic, viscous and anelastic behavior, respectively. However, we will not go any further with these simple models. What we have covered gives a flavor of this approach and, hopefully a feel for the range of viscoelastic response.

There are two things we wish to mention in closing this section. First, we have considered simple uniaxial stresses. We could have considered shear instead, but would have simply finished up with equations in τ's and G's instead of σ's and E's. Finally, as we have noted repeatedly, real polymer materials display behavior that can best be described in terms of a spectrum or distribution of relaxation and retardation times and we will now turn our attention to this problem.

Distributions of Relaxation and Retardation Times

One obvious way of introducing a range of relaxation and retardation times into the problem is to essentially construct models consisting of a number of Maxwell and/or Voigt elements connected in series and/or parallel. The Maxwell-Wiechert model, for example, consists of an arbitrary number of Maxwell elements connected in parallel. Figure 11.33 illustrates three elements connected in this manner. Hopefully, you will recall that this picture represents the assumption that the strain is the same in each of the connected Maxwell elements, while the total stress is given by the sum of stresses experienced by each element. Using the first assumption we can immediately write for the three element model shown in figure 11.33:

$$\frac{d\varepsilon}{dt} = \frac{\sigma_1}{\eta_1} + \frac{1}{E_1}\frac{d\sigma_1}{dt}$$

$$= \frac{\sigma_2}{\eta_2} + \frac{1}{E_2}\frac{d\sigma_2}{dt}$$

$$= \frac{\sigma_3}{\eta_3} + \frac{1}{E_3}\frac{d\sigma_3}{dt} \qquad (11.70)$$

(If you're getting a bit confused go back and see how we obtained equation 11.58 for the single Maxwell element then note that all 3 elements must deform identically). For a stress relaxation experiment we put dε/dt equal to zero and obtain (see equation 11.62):

$$\sigma_1 = \sigma_{01}\exp(-t/\tau_{t1}) \qquad (11.71)$$

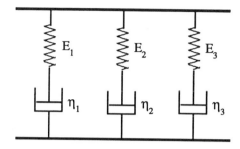

Figure 11.33 *Maxwell-Wiechert model consisting of three Maxwell elements in parallel.*

$$\sigma_2 = \sigma_{02}\exp(-t/\tau_{t2}) \tag{11.72}$$

$$\sigma_3 = \sigma_{03}\exp(-t/\tau_{t3}) \tag{11.73}$$

You will recall that the stress relaxation modulus is defined as:

$$E(t) = \frac{\sigma(t)}{\varepsilon_0} \tag{11.74}$$

and we have the condition that:

$$\sigma(t) = \sigma_1 + \sigma_2 + \sigma_3 \tag{11.75}$$

so that:

$$E(t) = \frac{\sigma_{01}}{\varepsilon_0}\exp(-t/\tau_{t1}) + \frac{\sigma_{02}}{\varepsilon_0}\exp(-t/\tau_{t2}) + \frac{\sigma_{03}}{\varepsilon_0}\exp(-t/\tau_{t3}) \tag{11.76}$$

We can generalize this result to an n-component model to obtain:

$$E(t) = \sum_n E_n \exp(-t/\tau_{tn}) \tag{11.77}$$

where, of course:

$$E_n = \frac{\sigma_{0n}}{\varepsilon_0} \tag{11.78}$$

The result presented in equation 11.77, that the total modulus is the summation of the responses of the individual elements, is essentially a statement of the superposition principle, in the sense that the resulting total dynamics is described as a superposition of a large number of independent modes of relaxation, each represented by a characteristic relaxation time τ_t. Furthermore, Aklonis and Macknight[*] demonstrated that even a very simple model, consisting of just two Maxwell elements in parallel, reproduces qualitatively the major features of stress relaxation in amorphous polymers and we reproduce their result schematically in figure 11.34.

---

[*] J. Aklonis and W. Macknight, *Introduction to Polymer Viscoelasticity*, John Wiley and Sons, New York (1983).

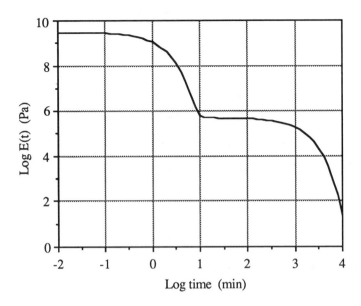

Figure 11.34 *Behavior of a two component Maxwell-Wiechert model in stress relaxation. Calculated using equation 11.77 with values of $E_1 = 3 \times 10^9$ Pa, $E_2 = 5 \times 10^5$ Pa, $\tau_1 = 1$ min and $\tau_2 = 10^3$ min.* After J. Aklonis and W. Macknight, Introduction to Polymer Viscoelasticity, *John Wiley and Sons, New York (1983).*

As you might expect, one can also create a model involving a range of retardation times by combining Voigt elements (this time in series) to obtain for the creep compliance:

$$D(t) = \sum_n D_n \left[1 - \exp(-t / \tau'_{tn}) \right] \qquad (11.79)$$

Here one assumes that the stress in each element is identical, while the total strain is given by summing the contributions from each element (compare this to the assumptions used in the Maxwell-Wiechert model) and the development of equation 11.79 is left as an elementary homework problem.

The next step is to extend these models to the situation where there is, in effect, a continuous distribution of relaxation times, which allows us to replace the summations in equations 11.77 and 11.79 with integrals:

$$E(t) = \int_0^\infty E(\tau_t) \exp(-t / \tau_t) \, d\tau_t \qquad (11.80)$$

$$D(t) = \int_0^\infty D(\tau'_t) \left[1 - \exp(-t / \tau'_t) \right] d\tau'_t \qquad (11.81)$$

The integral form of the equations allows relationships to be established between variables using transform techniques, but the methodology requires the appropriate mathematical background and the results are complicated. There are also mathematical methods for extracting the distribution functions from experimental data, but again this is an advanced topic and we intend to go no further. One final observation is in order, however. You will find in more advanced treatments that it is convenient for various purposes to put the integrals in equations 11.80 and 11.81 onto a logarithmic time scale, so that τ terms are replaced by $\ln \tau$ and other variables are modified accordingly. The significance of this $\ln \tau$ time scale will resurface when we turn our attention to the time-temperature superposition principle in the following sub-section.

To summarize, by using simple models we have developed equations that describe relaxation processes in terms of a distribution of relaxation and retardation times and this at least allows a rationalization of experimental behavior. We have used the classical approach that relies upon mechanical models to illustrate various principles, but you should know that this is anathema to many physicists. Their more rigorous approach is to first assume relaxation occurs at a rate which is linearly proportional to the distance from equilibrium. With this simple assumption and the superposition principle the integral equations given above can be obtained, together with equations describing dynamic behavior that we have not presented here. We are firm in our belief that most students get a better feel for viscoelastic properties if they first study simple models, however. It is still astonishing to us that a simple combination of two Maxwell elements in parallel qualitatively reproduces stress relaxation curves showing the four regions of viscoelastic behavior, as illustrated above in figure 11.34. The fact that real molecular processes are more accurately described by a distribution of relaxation times is then a simple extension of this basic approach.

The Time Temperature Superposition Principle

We now come to the second superposition principle that is important in describing viscoelastic properties. We have discussed previously the observation of a time-temperature equivalence in behavior and to remind yourself of this you should go back and reexamine figures 11.25 and 11.26. Thus, if we take a polymer melt and subject it to a very fast distortion it has the elastic response of a glass (i.e., a high modulus). In the same fashion, if we load a polymer glass and observe the strain over many years we would observe a final deformation and hence a calculated modulus that would be characteristic of a rubber or melt.

This correspondence of time-temperature behavior has considerable significance and can be expressed formally in terms of a superposition principle. To illustrate this let us first consider some creep curves shown schematically in figure 11.35. The quantities plotted are the strain divided by the (constant) stress used in the experiment, multiplied by the temperature (to account for the slight temperature dependence of the elastic moduli, which will be discussed in the section on rubber elasticity), $[\varepsilon(t)/\sigma_0]T$, against log t (t = time). Plots for three different temperatures are shown and as you would no doubt now expect, creep is faster at higher temperatures. For many polymer materials such curves might take months or even years to produce, particularly those describing behavior at low temperatures. The interesting thing is that these curves all appear to have the

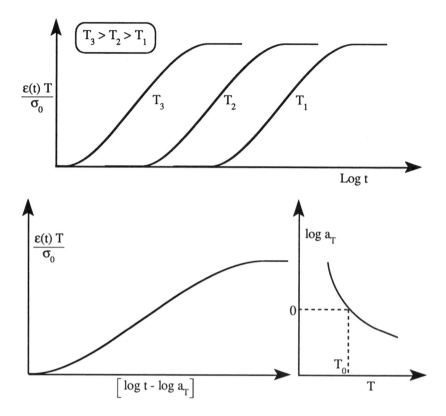

Figure 11.35 Top: creep curves obtained at different temperatures. Bottom: a master creep curve obtained by superimposing the curves obtained at different temperatures and a plot of the shift factor, a_T, as a function of temperature.

same shape and can be superimposed by shifting along the time axis, as also illustrated in figure 11.35. Accordingly, if we could determine an expression for a *shift factor* that connects these curves, we could (in principle) use data obtained at a high temperature, which could perhaps be obtained in a reasonable amount of time, to determine creep properties at lower temperatures, which might ordinarily take years to study experimentally.

Stress relaxation (and also dynamic mechanical) measurements can be superimposed in the same fashion. In previous sections we have presented curves showing the four regions of viscoelastic behavior of amorphous polymers and you may have wondered how these results were obtained, given that for a glassy polymer measurements of stress-relaxation at temperatures well below the Tg could take hundreds of years to complete (i.e., to obtain the data points characteristic of behavior in the terminal flow region). The figures we showed actually represented master curves, assembled by superimposing data obtained at different temperatures (and in ordinary experimental time frames). Examples of individual curves are shown in figure 11.36 and also earlier in this chapter, in figure 11.22). If you examine these curves closely you can see that the curve

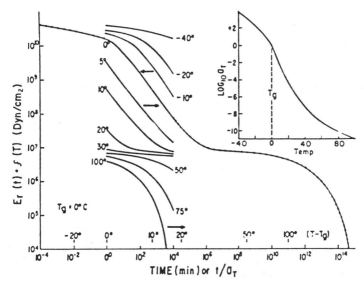

Figure 11.36 Construction of a stress relaxation master curve for a hypothetical polymer with a T_g of 0°C. Reproduced with permission from L. E. Nielson, Mechanical Properties of Polymers and Composites, Vol. 1, *Marcel Dekker, New York, 1974.*

obtained at -20°C can be shift along the horizontal (x) axis to partially overlap and extend the curve obtained at -40°C. Similarly, the -10°C curve can be shifted along the same axis to overlap and extend the range of the first two curves. And so on until a final master curve is constructed. The master curve shown in figure 11.36 actually uses 0°C as a *reference temperature* and so was constructed by shifting curves obtained at lower temperatures to the left (horizontally) and those obtained at higher temperatures to the right. Note also that both figures 11.35 and 11.36 use a log time scale x-axis.

The log of the shift factor required to superimpose curves obtained at one temperature to those obtained at another (log a_T) is plotted vs temperature in both figures 11.35 and 11.36. It was found empirically by Williams, Landel and Ferry that at temperatures above the T_g this shift factor has a temperature dependence of the form:

$$\log a_T = \frac{-C_1(T - T_s)}{C_2 + T - T_s} \tag{11.82}$$

This is called the WLF equation and we discussed it briefly in Chapter 8. C_1 and C_2 are constants for a particular polymer (values of $C_1 = 17.44$ and $C_2 = 51.6$ are often used, but there is some variation from polymer to polymer). In their original paper Williams, Landel and Ferry arbitrarily defined the reference temperature, T_s, to be 243°K for the particular system they were studying (high molecular weight polyisobutylene), but it is now more usual to use T_g as the reference temperature (i.e., put $T_s = T_g$ in equation 11.82). But to emphasize the key point, data obtained from creep, stress relaxation or dynamic experiments obtained at a temperature T can be superimposed upon those obtained at another, say T_s (providing the polymer is amorphous and $T_s > T_g$) by shifting along a log time axis and amount of a_T.

These purely experimental observations bring up a number of interesting questions. What is the molecular basis of this time-temperature superposition principle and are there other ramifications? Also we have been careful to confine our discussion to amorphous polymers and essentially superimpose data obtained in the glass transition range and terminal flow range, where the experimentally determined shift factors can be described by the WLF equation. What about transitions observed at temperatures below T_g in amorphous polymers, or those transitions observed in semi-crystalline polymers? To examine these questions let us first turn our attention to the significance of the log scale and what this superposition principle implies in terms of the temperature dependence of relaxation behavior.

It is convenient to examine stress relaxation in order to develop the arguments and note that an equivalent treatment can be presented for creep and dynamic mechanical data. To keep things simple we will consider the expression for E(t) written in terms of a sum of relaxation processes rather than the integral form necessary to describe a continuous distribution:

$$E(t) = \sum_n E_n \exp(-t / \tau_{tn})$$
(11.83)

Now let us consider a particular mode of relaxation and leave out the n subscript. We will assume that the molecular rearrangements that give rise to this relaxation process have a characteristic time $\tau_{t,0}$ at temperature T_0, but are naturally characterized by a different relaxation time $\tau_{t,1}$ at temperature T_1 (remember, molecules have more kinetic energy at higher temperatures, so that the relaxation time of a particular process will change). Now let us simply *define* the ratio of these two relaxation times to be a_T;

$$a_T = \frac{\tau_{t,1}}{\tau_{t,0}}$$
(11.84)

Accordingly, the exponential term in equation 11.83 can be written:

$$\frac{t}{\tau_{t,1}} = \frac{t}{a_T \tau_{t,0}}$$
(11.85)

or taking logs:

$$\log (t / \tau_{t,1}) = \log (t / \tau_{t,0}) - \log a_T$$
(11.86)

demonstrating that relaxation behavior observed at one temperature can be superimposed upon that at another by shifting an amount a_T along a log time axis. However, real mechanical behavior has to be described in terms of a distribution rather than a single relaxation time. The experimental observation that time dependent mechanical behavior can be shifted and superimposed means that all the relaxation processes involved have the same temperature dependence. Relaxation processes consist of conformational rearrangements involving different length scales, by which we mean they involve coupled motions ranging from just a few chain segments to motions of the chain as a whole, so this correspondence in temperature dependence presumably means that the frictional forces encountered by a chain act in the same or very similar manner on all segments. This is an important point in the development of molecular theories of polymer dynamics, so you should keep it in mind if you intend to study this part of the subject in more depth.

Given that the superposition principle works (but it is not entirely accurate, as curves obtained at different temperatures can actually have slightly different shapes), why does the temperature dependence of the shift factor have the form given by the WLF equation? There is not really a good answer to this, as derivations of the WLF equation usually start from other empirical equations describing viscosity. We will use the Doolittle equation, which we mentioned in our discussions of the nature of the glass transition in chapter 8 and which can be written:

$$\eta = A \exp B \left[\frac{V - V_f}{V_f} \right] \tag{11.87}$$

where V is the total volume of the system and V_f is the "free" volume. We can relate the viscosity to relaxation times and the shift factor by recalling that τ_t is defined as:

$$\tau_T = \frac{\eta}{E} \tag{11.88}$$

and because the modulus changes only slightly with temperature relative to the viscosity we can write for the ratio of τ_t at two temperatures T_1 and T_o, which you will recall is the definition of the shift factor:

$$a_T = \frac{\tau_{t,1}}{\tau_{t,0}} \approx \frac{\eta_1}{\eta_0} \tag{11.89}$$

where η_1 is the viscosity at T_1 and η_0 is the viscosity at T_0 (it is more accurate to multiply the right hand side of equation 11.89 by $T_0 \rho_0 / T_1 \rho_1$, to account for changes in the density (ρ) and modulus with temperature, but we will neglect these terms for the sake of simplicity).

Accordingly we can now relate the shift factor to the free volume of the system using equations 11.89 and 11.87. To obtain the temperature dependence Williams et al. assumed that free volume is a linear function of temperature and could be described by:

$$V_f = V_g \left[0.025 + \Delta\alpha \, (T - T_g) \right] \tag{11.90}$$

where $\Delta\alpha$ is the difference between the coefficients of thermal expansion of the liquid and the glass:

$$\Delta\alpha = \alpha_1 - \alpha_g \tag{11.91}$$

and V_g is the free volume at T_g. Using the Doolittle equation and substituting from equation 11.90 the following equation can be obtained:

$$\log a_T = \frac{(-B/2.303\, f_0)\, (T - T_g)}{f_0 / \Delta\alpha + (T - T_g)} \tag{11.92}$$

where f_0 is the fractional free volume at T_g, V_f/V. This has the form of the WLF equation and the constants C_1 and C_2 can then be identified with corresponding terms in equation 11.92 and are related to free volume.

Because the WLF equation can be related to free volume terms and the T_g, it should be clear that it only applies to amorphous polymers and to temperatures greater than T_g (remember, the viscosity effectively tends to an infinite value as $T \rightarrow T_g$). What about transitions below the T_g and in semi-crystalline polymers?

Relaxation processes below the glass transition temperature in amorphous polymers involve local motions, either in the main chain or a side chain of the polymer. These also obey a time-temperature superposition principle, but the temperature dependence of the shift factor is not governed by a WLF type equation but one that has an Arrhenius form. You will hopefully recall from your physical chemistry classes that the Arrhenius equation describes the relationship between rate constants and temperature and can be written:

$$k = A \exp(-\Delta E_a / RT) \tag{11.93}$$

where E_a is an activation energy and describes the barrier which molecules must cross in order to react upon colliding. Here the activation energy will be related to the height of the energy barrier between conformational states and we have a relationship of the form:

$$\tau_t^{-1} \approx \exp(-\Delta G_a / RT) \tag{11.94}$$

where τ_t^{-1}, the reciprocal of the relaxation rate, is a measure of the frequency of transitions between conformational states separated by a free energy barrier, ΔG_a.

The temperature dependence of transitions in the amorphous regions of semi-crystalline polymers with a low degree of crystallinity are naturally also governed by WLF or Arrhenius type expressions. In addition, an Arrhenius temperature dependence governs coupled crystalline/amorphous relaxations, such as the α-process in polyethylene mentioned in a previous section. In general, however, the observed behavior involves a "vertical" shift factor as well as the "horizontal" shifts we described in terms of log a_T, usually as a result of a change in crystallinity with temperature. These vertical shifts obviously depend strongly on the thermal history of the sample, greatly complicating time-temperature behavior, so we will consider them no further.

The Dynamics of Polymer Chains

In the preceding sub-sections we have considered descriptions of viscoelastic behavior that are essentially phenomenological and can be based on mechanical models. We will round out our discussion by qualitatively discussing some microscopic or molecular theories.

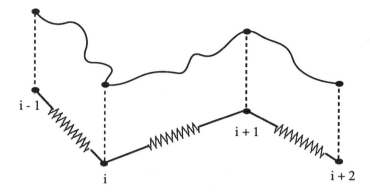

Figure 11.37 *Schematic representation of the model of a chain used in the Rouse-Beuche theory.*

As we have mentioned before, an understanding of the dynamics of polymer chains requires a commitment to work through a lot of mathematics and this is well beyond the scope of this introductory treatment. Those students who intend to pursue this subject will need to consult the books by Beuche, Ferry, de Gennes, and Doi and Edwards. The first two books listed deal with the older Rouse-Beuche theory, while modern concepts are dealt with by de Gennes and Doi-Edwards.

The Rouse-Beuche theory is based on a model where a polymer chain is assumed to behave as a set of beads linked by springs, as illustrated in figure 11.37. The springs vibrate with frequencies that depend upon the stiffness or force constant of the spring, but these vibrations are considered to be modified by frictional forces between the chain and the surrounding medium. There are various types of vibrations of a chain of this type, called the normal modes of the chain, each characterized by a certain frequency (see Chapter 6 on spectroscopy; note that the nature of vibrational spectroscopic normal modes is different because their motions are not damped, but the idea is the same). These can be calculated using straightforward classical mechanics. There is a characteristic relaxation time associated with each of these damped normal modes and each of these contributes to viscoelastic properties. In terms of this model it was found that the flow behavior of polymer melts is dominated by the vibrational mode with the longest relaxation time, corresponding to a coordinated movement of the molecule as a whole.

The Rouse theory predicts that the viscosity should be directly proportional to the molecular weight, which we have seen is only true up to the entanglement limit, beyond which the viscosity becomes proportional to the molecular weight to the power 3.4. This is because in the Rouse model each chain is considered to be moving independently, so entanglements are neglected. The model can be modified by allowing certain modes to be more strongly hindered or damped, but there are other problems and disagreements with experiment. We noted previously that de Gennes questions whether the concept of discrete modes is appropriate at all, as non-linear effects would result in a mixing of such modes to

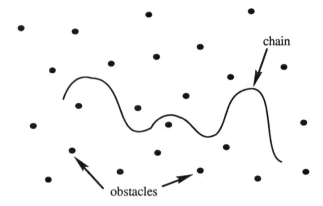

Figure 11.38 *The de Gennes model of reptation; the chain moves among the fixed obstacles with a snake-like motion, but cannot cross any of them.*

give a spectrum of relaxation times. You should not go away with the feeling that this model is useless, however, it served for many years as a good tool for analyzing experimental data and provides the foundation on which modern theories have been built. It helped to firmly establish the idea that chain motion is responsible for creep, stress relaxation and other viscoelastic properties.

One of the principal problems of the Rouse-Beuche approach is that it essentially allows chains to "pass through one another" to get to new positions and neglects the effects of entanglements. de Gennes proposed a different approach based on the concept of *reptation*. This is illustrated in figure 11.38 where a chain is "trapped" by other chains that are considered to act as a set of obstacles. The chain is not allowed to cross these obstacles, but can slither through them with motions similar to a snake moving through the grass, hence the name reptation.

Using *scaling arguments* de Gennes found that the diffusion coefficient D of a chain in the melt should depend inversely on the square of the molecular weight.

$$D \propto \frac{1}{M^2} \tag{11.95}$$

Which it does. He also obtained a relationship between melt viscosity and molecular weight.

$$\eta_0 \propto M^3 \tag{11.96}$$

Which is not quite right ($\eta_0 \approx M^{3.4}$). At this time it is not exactly clear (at least to us) whether this is just a minor problem that can be fixed by tinkering with the model, or if there is a more fundamental difficulty. Regardless, the concept of reptation has played a fundamental role in our understanding of polymer dynamics and the purpose of these paragraphs is to give you a feel for the ideas behind the theory.

F. NON-LINEAR MECHANICAL AND RHEOLOGICAL BEHAVIOR

We started our discussion of mechanical and rheological properties with the simplest linear models, Hooke's law for the description of elastic deformation and Newton's law for viscous flow. We also considered linear viscoelastic models, where the non-linear response that is observed as a function of time was accounted for by the assumption of a linear relationship between stress and strain. Mechanical models were then constructed by combining linear elastic and viscous elements in various ways to reproduce non-linear time dependent behavior.

All of these linear models essentially depend upon the assumption of small deformations or small strain rates. In actually using or processing polymers we often reach the limits of these assumptions and earlier in this chapter we briefly described two types of non-linear effects, the stress/strain properties of rubber networks and shear thinning in polymer melts at high strain rates (see figures 11.14 and 11.17). In this section we will return to this subject, but we will only consider rubber elasticity in anything more than a qualitative fashion. The treatment of flow by microscopic models or empirical constitutive equations gets beyond the scope of an introductory treatment. Nevertheless, there are some consequences of non-linear behavior that you should appreciate and we will describe these in a qualitative fashion.

Rubber Elasticity

The best place to start this discussion is through a reconsideration of some simple thermodynamics. Consider a strip of rubber and the change in the free energy of this rubber that occurs upon stretching. We will use the Helmholtz free energy, F (which means we are holding the volume V and temperature T, constant). Starting from:

$$F = E - TS \qquad (11.97)$$

where E is the internal energy and S the entropy (and if you have forgotten what free energies are go back and reread the introduction to chapter 8), we obtain for changing L, the length of the strip in the direction we are stretching:

$$\left(\frac{\partial F}{\partial L}\right)_{V,T} = \left(\frac{\partial E}{\partial L}\right)_{V,T} - T\left(\frac{\partial S}{\partial L}\right)_{V,T} \qquad (11.98)$$

The force (f) required to give this extension is related to the change in free energy by:

$$f = \left(\frac{\partial F}{\partial L}\right)_{V,T} \qquad (11.99)$$

This relationship between force and free energy can be obtained more formally through considering the first and second laws of thermodynamics, and if your knowledge of this subject is weak you should go back to your old physical chemistry notes and work this out. For our purposes the important part of equations 11.98 and 11.99 is that it demonstrates that when a force is applied to a material there is a change in both the internal energy and entropy. Earlier in this chapter we considered changes in internal energy upon stretching an idealized crystalline lattice and in materials such as metals it is these types of changes that dominate the mechanical response (i.e., we neglect $\partial S/\partial L$). In elastic networks, however, it is the other way round and the entropic changes dominate behavior. This is because the response to (non-destructive) loads largely depends upon the changes in conformations of the chains, rather than a stretching of bonds. If we neglect any energetic contribution, thus assuming that the rubber is "ideal"[*] , we obtain:

$$f = -T \left(\frac{\partial S}{\partial L} \right)_{V,T}$$

(11.100)

(note that $f \propto T$, we will come back to this shortly). Accordingly, if we could obtain from statistical mechanics a relationship between S and L, we would have a theory of rubber elasticity.

It is useful to start by considering a single chain and imagine that we could grab its ends and stretch it along (say) the z direction in a Cartesian coordinate system. We considered the distribution of conformations available to such a chain in Chapter 7 and noted that to a first approximation this could be described as a Gaussian function. Accordingly, the distribution of end-to-end distances P(R) is given by:

$$P(R) = P(x,y,z) = \left[\frac{\beta^2}{\pi} \right]^{3/2} \exp(-\beta^2 R^2)$$

$$= \left[\frac{\beta^2}{\pi} \right]^{3/2} \exp(-\beta^2 \{ x^2 + y^2 + z^2 \})$$

(11.101)

where:

$$\beta^2 = \frac{3}{2Nl^2} = \frac{3}{2<R_0^2>}$$

(11.102)

and $<R_0^2>$ is the mean-square of all possible end-to-end distances of the chain, which you recall is equal to Nl^2, where N is the number of segments in a chain and l is the length of each segment.

This probability distribution is related to the number of conformations available to a chain if one end is located at the origin of a Cartesian system while the other is fixed at the point (x,y,z). You will hopefully recall that Boltzmann equation for the entropy:

$$S = k \ln \Omega$$

(11.103)

---

[*] If you pursue the thermodynamics and statistical mechanics of rubber in more detail you will find an analogy to a perfect or ideal gas. In each treatment interactions between molecules are neglected.

and we can equate the number of configurations available to the system, Ω, to P(R). For uniaxial extension along the z-axis we have:

$$\Omega = P(R) = P(x,y,z) = P(0,0,z) \qquad (11.104)$$

and hence:

$$s = \text{constant} - k\beta^2 z^2 \qquad (11.105)$$

(We have used lower case s here to indicate we are describing a single chain. A capital S will be used later to describe the entropy of a collection of such chains.)

If we neglect changes in internal energy we can use equation 11.100 to obtain the following force/displacement relationship for the Gaussian chain (i.e., we substitute z for L and differentiate:

$$f = 2kT\beta^2 z \qquad (11.106)$$

This says that if the chain's ends are a distance z apart, they are acted on by a force given by equation 11.106. This force is zero when z = zero and the force is linear with distance between the end of the chain. What we have just obtained, of course, is Hooke's law! Except that a few things have to be kept in mind when considering this correspondence. First, this relationship should be good for a wide range of deformations, much larger than just a couple of percent. It will only fail at extensions where the assumption that the distribution of distances between the chain ends can be described by a Gaussian function breaks down (maybe extensions of 300% or more, depending on the chain length). Also, note that the modulus is equal to $2kT\beta^2$ and so varies directly with T. The higher the temperature, the higher the modulus. Hence if we heat a stretched chain, we would expect it to contract (if we raise the modulus the extension is smaller for a given load). This is actually observed for a real rubber (consisting of a collection of chains)[*]. Finally, you should keep in mind that left to itself R will not be a constant value, but fluctuates around an average. Accordingly, if we could clamp the ends of the chain a set distance apart, the force of retraction would fluctuate. Fortunately, we don't have to worry about this when we average over *billions* and *billions* of chains, which is what we need to do when considering the behavior of actual rubber samples. This we will do next.

For practical use as a material we have to cross-link the chains to form a network, so that they do not flow (slip past one another under a load). This introduces all sorts of complications into a theoretical treatment. Some of the difficulties are experimental, in that it is very difficult to precisely control cross-linking. There will be dangling ends that are not part of the network and there will be different chain lengths between cross-link points. If that were not enough, the chains will tangle up and these entanglement points might also act somewhat like cross-link points (except that the chains will still be able to slip past one another to some degree). Some of these possibilities are illustrated in figure 11.39. It is possible, however, to make model network systems by care-

---

[*] But keep in mind that the rubber will also expand on heating, so the chain must be stretched a sufficient amount that thermal expansion effects are overcome.

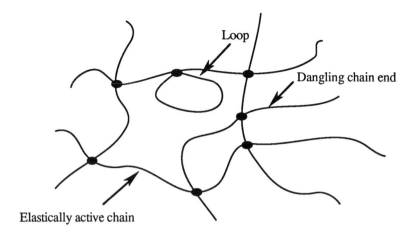

Figure 11.39 *Schematic of network structure and defects.*

fully synthesizing monodisperse (or, more accurately, nearly monodisperse chains) and linking them by their ends in solution (to minimize entanglements), but nothing in this world is perfect (except a sunny spring morning squandered on a lush, green, golf course). Notwithstanding this, in order to construct a useful model we will assume that we can make a perfect network, see what its properties should be and then compare these predictions to the behavior of real elastomers. This brings us to the second set of difficulties, which are theoretical in nature. We will need to account for the cross-link points or junctions in the model and to do so we will make the simplest assumption, that they are fixed in the sample and change their positions in proportion to sample deformation.

We will consider a cubic block of lightly cross-linked rubber that undergoes strain along a set of axes defined parallel to a Cartesian system, so that the extension ratios in the x, y, z system are equal to λ_1, λ_2 and λ_3, as illustrated in figure 11.40 [note: the extension ratios are equal to l/l_0, while the strain is equal to $(l - l_0)/l_0$].

If there is no change in volume then:

$$\lambda_1 \lambda_2 \lambda_3 = 1 \qquad (11.107)$$

Now we focus on an individual chain within this block and compare its entropy in the stretched (s) and unstretched (s_0) states. Instead of using the end-to-end distance R, we need to consider deformations parallel to the set of Cartesian axes. The definition we are going to use is also illustrated in figure 11.40. One end of the chain is at the origin of our Cartesian system (0, 0, 0), while the other end happens to be at a point x_0, y_0, z_0 in the unstrained state, so that the end-to-end distance R_0 is given by:

$$R_0^2 = x_0^2 + y_0^2 + z_0^2 \qquad (11.108)$$

Accordingly, the entropy is given by:

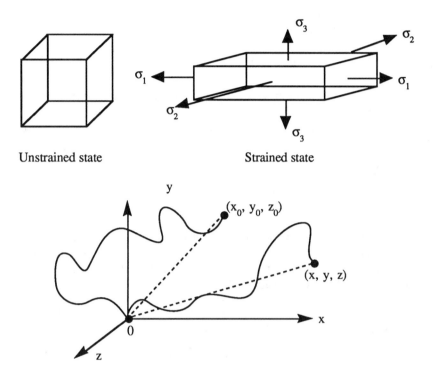

Figure 11.40 *Schematic illustration of homogeneous strain (top) and affine deformation of a chain (bottom).*

$$s_0 = c - k\beta^2 R_0^2 = c - k\beta^2 (x_0^2 + y_0^2 + z_0^2) \tag{11.109}$$

where c is a constant (see equation 11.105).

We now assume that this chain deforms in exact proportion to the parent cube (this is called the *affine* assumption), so that if the end of the chain at x_0, y_0, z_0 goes to x, y, z, then from the definition of extension ratios:

$$x = \lambda_1 x_0 ; \quad y = \lambda_2 y_0 ; \quad z = \lambda_3 z_0 \tag{11.110}$$

The entropy of the deformed chain is given by:

$$s = c - k\beta^2 (x^2 + y^2 + z^2) = c - k\beta^2 (x_0^2 \lambda_1^2 + y_0^2 \lambda_2^2 + z_0^2 \lambda_3^2) \tag{11.111}$$

So that the entropy change on deformation is:

$$\Delta s = -k\beta^2 \{ (\lambda_1^2 - 1) x_0^2 + (\lambda_2^2 - 1) y_0^2 + (\lambda_3^2 - 1) z_0^2 \} \tag{11.112}$$

Now, in our perfect network we are going to assume that the chain lengths, defined by the molecular weight between cross-link points, are all identical (this means that the parameter β is a constant, see equation 11.96), and we can obtain our total entropy of deformation by summing over all chains.

$$\Delta S = \sum \Delta s = -k\beta^2 \{ (\lambda_1^2 - 1) \sum x_0^2 + (\lambda_2^2 - 1) \sum y_0^2 + (\lambda_3^2 - 1) \sum z_0^2 \} \quad (11.113)$$

The $\sum x_0^2$, $\sum y_0^2$, $\sum z_0^2$ terms are just the sums of the squares of the positions of the chain ends, *relative to the positions of each of their other ends.* If in the unstrained state there is a perfectly random distribution of end-to-end vectors, then:

$$\sum x_0^2 + \sum y_0^2 + \sum z_0^2 = \sum R_0^2 \quad (11.114)$$

and

$$\sum x_0^2 = \sum y_0^2 = \sum z_0^2 = \frac{1}{3} \sum R_0^2 \quad (11.115)$$

(Remember there is a random distribution of the ends in all directions.)

By definition, the average of the square of the end-to-end distance $< R_0^2 >$ must be equal to the sum of all the squares of the end-to-end distances divided by the number of chains, N, so that:

$$\sum R_0^2 = N < R_0^2 > \quad (11.116)$$

and we obtain:

$$\Delta S = -\frac{1}{3} Nk\beta^2 < R_0^2 > (\lambda_1^2 + \lambda_2^2 + \lambda_3^2 - 3) \quad (11.117)$$

However:

$$\beta^2 = \frac{3}{2 < R_0^2 >} \quad (11.118)$$

hence:

$$\Delta S = -\frac{1}{2} Nk(\lambda_1^2 + \lambda_2^2 + \lambda_3^2 - 3) \quad (11.119)$$

For simple extension in the x direction (assuming constant volume):

$$\lambda_2 = \lambda_3 \text{ and } \lambda_1 = \frac{1}{\lambda_2 \lambda_3} \quad (11.120)$$

therefore:

$$\Delta S = -\frac{1}{2} Nk \left(\lambda_1^2 + \frac{2}{\lambda_1} - 3 \right) \quad (11.121)$$

Recalling that if we neglect internal energy:

$$f = -T \frac{dS}{dl} \quad (11.122)$$

or if we use λ as our variable:

$$f = -T \frac{dS}{d\lambda_1} \quad (11.123)$$

so that:

$$f = NkT \left(\lambda_1 - \frac{1}{\lambda_1^2} \right) \qquad (11.124)$$

We can write this as:

$$f = E \left(\lambda_1 - \frac{1}{\lambda_1^2} \right) \qquad (11.125)$$

where the modulus E is equal to NkT. This is equivalent to the first term in the Mooney-Rivlin equation, discussed earlier in this chapter.

There are a number of things about this simple result that need to be considered, relating to the simplified nature of the derivation (which follows that given by Treloar) and the assumptions that have gone into the model. Before considering these however, it is useful to first see how well equation 11.125 matches experimental results. A comparison is shown in figure 11.41. It does not look very good, but it is not as bad as it first seems! The major deviation is at high extensions and there are two reasons for this. First, elastomers like natural rubber that have a very regular chain structure crystallize at high extensions and that is one reason for the deviation. The second is that above a certain extension (~ 500% in figure 11.41) the Gaussian approximation breaks down. This is not therefore a conceptual failure of the theory, but more of a limitation imposed by one of the assumptions we have used to obtain a simple solution. Theories that account for finite chain lengths show the upturn in the curve at high elongations, as illustrated in figure 11.42, so we can live with this deviation. The fact that the experimental curve still falls below the theoretical curve in the intermediate range of strain is more difficult to explain, however, and requires that more complicated treatments be applied. We won't do that here, but you should be aware of the following points:

1) To obtain the total deformation we simply summed contributions from all the chains (equation 11.113). Flory has described a more rigorous statistical mechanical treatment that results in an extra term in equation 11.124. This term plays no role in simple extension, however (but is important in swelling).

2) The affine assumption was originally used by Flory and in this assumption the cross-link points move in proportion to the deformation of the sample. An alternative model, originally proposed by James and Guth, allows the junction points to fluctuate significantly about mean positions. This so-called "phantom network" only changes the pre-factor in equation 11.124, so both approaches give an equation of the form:

$$f = FkT \left(\lambda_1 - \frac{1}{\lambda_1^2} \right) \qquad (11.126)$$

for simple extension.

3) A more recent constrained junction model described by Flory goes a long way to fixing up the theory by allowing for fluctuations of the junctions, the effect of entanglements etc.

In conclusion, we can say that the main features of the statistical theory are certainly correct and the crucial point you should grasp is the entropic nature of rubber elasticity. The simpler theories do a reasonable job of reproducing the

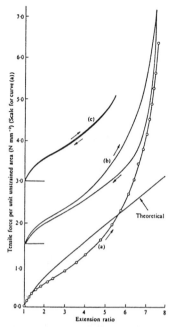

Figure 11.41 *A comparison of the stress-strain curve for natural rubber with experimental predictions (a). Curves (b) and (c) show reversibility at "low" extensions and hysteresis effects at higher elongations. Reproduced with permission from L. R. G. Treloar,* The Physics of Rubber Elasticity, Third Ed., *Clarendon Press, Oxford, 1975.*

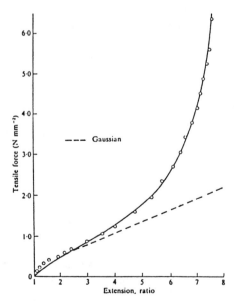

Figure 11.42 *The results of fitting a non-Gaussian force/extension relationship to experimental data. Reproduced with permission from L. R. G. Treloar,* The Physics of Rubber Elasticity, Third Ed., *Clarendon Press, Oxford, 1975.*

stress-strain data up to strains of 300% or more, but more complicated treatments are necessary to more precisely reproduce the nuances of the data.

Melt Flow at High Strain Rates

Non-linear behavior manifests itself in a number of ways in polymer melts, but we will discuss only two. The first of these is *shear thinning*. In our preliminary discussions of polymer melt rheology we noted that only at low strain rates (γ) do we find that the viscosity, η, is constant (i.e., independent of γ). At higher strain rates η decreases with increasing γ. A qualitative understanding of this phenomenon follows in a straightforward manner from a consideration of the effect of entanglements. As the strain rate increases chain segments become increasingly aligned along the shear direction and this effectively decreases the number of entanglements relative to the unstrained state.

The second manifestation of non-linear behavior concerns flow through small diameter tubes or capillaries. This is of considerable practical importance because of its relevance to processing by extrusion. Essentially, we have the situation where a polymer is forced into a narrow tube, as illustrated schematically in figure 11.43, and when it emerges at the other end it swells, so

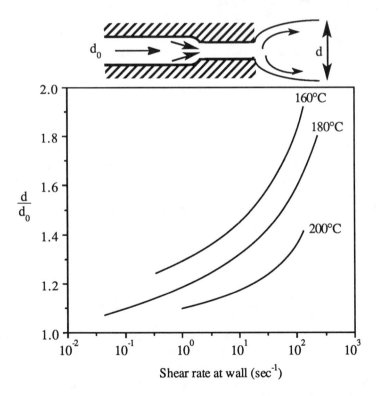

Figure 11.43 *Die swelling observed for a melt of polystyrene at various shear rates and temperatures. Drawn schematically from the data of Burke and Weiss,* Characterization of Materials in Research, *Syracuse University Press, 1975.*

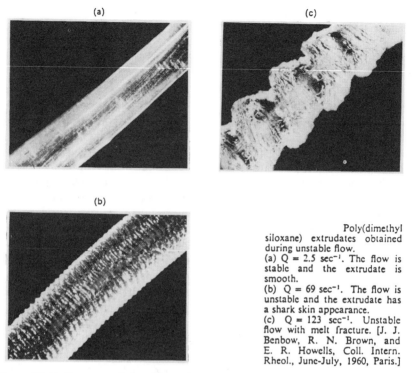

Poly(dimethyl siloxane) extrudates obtained during unstable flow.
(a) $Q = 2.5$ sec^{-1}. The flow is stable and the extrudate is smooth.
(b) $Q = 69$ sec^{-1}. The flow is unstable and the extrudate has a shark skin appearance.
(c) $Q = 123$ sec^{-1}. Unstable flow with melt fracture. [J. J. Benbow, R. N. Brown, and E. R. Howells, Coll. Intern. Rheol., June-July, 1960, Paris.]

Figure 11.44 Stages on the road to melt fracture during the extrusion of poly(dimethyl siloxane). Reproduced with permission from J. J. Benbow, R. N. Browne and E. R. Howells, Coll. Intern. Rheol., Paris, *June-July 1960*.

that the diameter of the extrudate, d, is greater than that of the capillary, d_0. Die swelling is a complex phenomenon and depends upon a number of factors, including strain rate, the length of the capillary relative to its diameter, temperature, and so on. Clearly, it is governed by elastic forces acting to deform the chains as they are pushed into the die and as they are sheared in the narrow channel. Normal stresses are developed in a direction perpendicular to the flow (i.e., axially in the capillary). These stresses are relieved and deformation is recovered when the polymer exits the die, so the extrudate swells.

There is a limit to the rate at which polymers can be extruded that is governed by the onset of *melt fracture*. Above a certain critical shear stress polymers no longer extrude smoothly. The problem starts with barely visible surface defects, but at higher shear stresses the extruded polymer can have a matte (loss of surface gloss) or so-called "sharkskin" appearance. The instability increases with shear rate and the distortion develops to a banded structure and eventually leads to a loss of cohesion, as illustrated in figure 11.44. There is again a qualitatively simple explanation. Above a certain strain rate the polymer chains no longer have time to relax and the material behaves like an elastic almost glassy solid, distorting and breaking. This excursion into fracture brings us to the final topic in this chapter, strength and failure.

G. ULTIMATE PROPERTIES

So far we have largely considered the mechanical and rheological properties of polymers under non-destructive loads. To complete our discussion we need to consider the conditions under which a polymer can fail, which to most people new to the subject usually means breaks apart, but must really mean the conditions under which a material can no longer fulfill its intended purpose, perhaps because it has irreversibly changed its dimensions beyond some critical amount (through yielding or creep, for example).

Failure can occur under various circumstances, some examples being under a tensile load, under impact, as a result of continuously applied oscillatory stress (fatigue), and so on. The way a material fails will also depend upon the rate of loading. For example, although we do not think of a liquid as being able to fail in the same way as a solid, we have seen that if a polymer melt is extruded too quickly it does not come out of the die as a continuous stream, but as globs of stuff in discontinuous lumps. We call this melt fracture. Similarly, if we take a piece of polyethylene and pull on it slowly, it will neck and yield, deforming considerably after some critical stress has been reached, as the stress-strain curve presented earlier as figure 11.12 would indicate. If we pull on this piece of polyethylene, quickly, however, it will fail with little or no yielding in what can be thought of as a brittle fashion.[*]

It is useful to think of failure in terms of so-called brittle and ductile behavior, although this is an arbitrary division and in the failure of a real material you often get bits of both. Nevertheless, if you think of the failure of a piece of glass, then this is clearly of the brittle type. A sharp crack propagates through the material giving a fracture surface where there is little evidence of yielding (i.e., bits of material being deformed and pulled out). Ductile behavior, on the other hand, often (but not always) involves the formation of a neck and large scale deformations. Materials that fail in this fashion are generally regarded as tough, because yielding absorbs a lot of energy.

Polymers nearly always yield to some degree during failure and this process can proceed by two mechanisms, shear yielding or crazing. These are not mutually exclusive processes, however, and they can show up together or one can precede the other. The mechanism that dominates in a particular situation will depend upon factors such as polymer morphology, temperature, rate of loading, and the rest of the usual suspects. Clearly, for most practical applications we wish to use polymers in situations where they do not "yield" (although we always have to be aware of creep and stress relaxation). This is not how most engineers define strength, however, which is usually in terms of the stress at fracture, where the sample finally breaks into two (or more) pieces. This serves to complicate the life of the design engineer who has been dragged kicking and screaming into working with plastics, but that is not our problem! We are just going to consider how strong polymers are compared to how strong they could be and what, if anything, we can do about this. This will lead us to a

---

[*] You can do this experiment yourself by taking the bits of plastic used to connect six packs of canned pop (soft drinks) together and comparing the results of pulling slowly to what happens when you pull quickly.

brief discussion of the processes of yielding and what this means in terms of the toughness of polymer materials.

The Theoretical Strength of Polymeric Materials

We are used to thinking of materials as having an inherent strength that is a consequence of their chemical character and most people believe that metals, for example are just naturally stronger than polymers. The problem with this view is that the materials we usually come in contact with are seriously flawed and none are as strong as they should be. As Gordon has pointed out in his wonderful books on structures and strong materials*, our understanding of the basis of cohesion and fracture is fairly new, dating from the early work of Griffiths ($\approx$1920), which did not really take hold until the 1950's, when jet airliners started falling out of the sky at an alarming rate. Up until then engineers dealt with things called "allowable stresses", which were empirically based on enormous numbers of tensile tests. Griffiths asked the essential question, "How strong should a material be?" and performed some simple calculations on glass by determining the force necessary to break the bonds joining two adjacent layers of atoms. Glass was chosen because it not only breaks in the required clean brittle fashion, without significant yielding that would complicate the analysis, but it is also a relatively straightforward matter to measure its surface tension. Griffiths wanted to do this because he had the brilliant idea of relating the energy of the fracture surfaces to the strain energy in the two layers of bonds where fracture is assumed to occur. This allowed Griffiths to calculate that the strength of glass should be between 0.7×10^4 and about 2×10^4 MPa (or, in "real" units, between 1×10^6 and 3×10^6 psi). He found that glass fibers that were about 1 mm thick broke at an average stress of 170 MPa ($\approx$ 25,000 psi). In other words, glass often has about 1% of the strength it should have!

When Griffiths drew glass fibers down to smaller diameters, however, something peculiar happened, their measured stress-to-break increased. A plot of tensile strength vs. diameter, taken from Gordon's book (cited above) is reproduced in figure 11.45, and it can be seen that really thin fibers have a much higher stress-to-break. The curve can be extrapolated to give values close to the theoretical value.

We now understand that the weakness of glass is predominantly due to flaws, which were greatly reduced in the thinner fibers. We won't go into this because we are concerned with polymers, which are much more interesting materials! However, the availability of polydiacetylene single crystals, whose stress/strain properties we discussed earlier in the content of the validity of Hooke's law, means that more recent Griffiths-type experiments can also be performed on polymer fibers of different thickness and the results extrapolated to give their theoretical strength, as shown in figure 11.46. The theoretical strength was determined to be 3 GPa, which is somewhat lower than the theoretical strengths calculated for polyethylene and some other polymers ($\approx$10 to

---

* J. E. Gordon, *The Science of Structures and Materials*, Scientific American Library (1988); *The New Science of Strong Materials, Second Edition*, Penguin Books (1976).

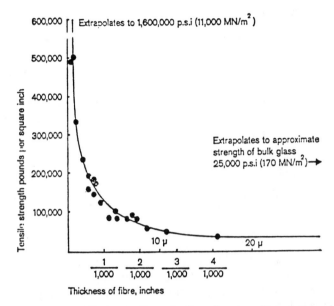

Figure 11.45 *Plot of tensile strength vs. diameter for glass fibers. Reproduced with permission from J. E. Gordon,* The New Science of Strong Materials, Second Edition, *Penguin Books (1976).*

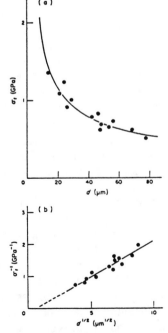

Figure 11.46 *Top: dependence of fracture stress, σ_f, upon fiber diameter, d, for poly-diacetylene single crystal fibers. Bottom: plot of $1/\sigma_f$ versus $d^{0.5}$. Reproduced with permission from C. Galiotis and R. J. Young,* Polymer, *24, 1023 (1983).*

Table 11.1 Theoretical Tensile Strength and Young's Modulus of Linear Polymers.

Polymer	Predicted modulus (GPa)	Theoretical tensile strength (GPa)
Polyethylene	182-340	19-36
Nylon 6,6	196	-
Poly(ethylene terephthalate)	122	-
Polytetrafluoroethylene	160	15-16
Cellulose (parallel to the chain)	56.5	12-19

40 GPa) listed in table 11.1. This may be because many polydiacetylenes have very long side chains, but this is not important here. The crucial point is that even this polymer which is about as perfectly aligned as we can make it, is weaker than it should be. In glass fibers the flaws are usually surface cracks while for the polydiacetylene the flaws are largely within the crystal.

If we now turn our attention to more common polymers such as polyethylene, we are no longer looking at a more or less homogeneous structure, but one where there are chain folded crystals, amorphous regions, and so on. These materials have measured strengths that are far lower than their theoretical strength. If we are to make a strong material it is obvious that we must line up the chains in the direction of stress as perfectly as possible, so that the strong covalent bonds rather than weak intermolecular forces "take the strain". Also, it is desirable to make the chain length as long as possible. Theoretically you would want chains to extend from one end of the sample to the other, but at very high molecular weights there are usually enough intermolecular contacts and entanglements to ensure a high cohesive strength. The problem is that these entanglements are also present in the melt and, as we noted above, the melt viscosity of a polymer is proportional to the molecular weight to the power of 3.4, so it is difficult to produce highly oriented fibers from very high molecular weight polymers. Nevertheless, in the past few years a number of ingenious methods have been developed (e.g., gel spinning and ultrafast extrusion) and polyethylene fibers with a strength of about 4 GPa have been produced. This is still considerably less than the theoretical strength, but is pretty good, as can be seen from a comparison of values of specific tensile strength and modulus plotted in figure 11.47 for a range of polymers and other materials.

Although we have clearly come a long way in developing polymer fibers with superior strength and stiffness, this is not the end of the story when it comes to desirable mechanical properties. Toughness, or the ability to withstand an impact, can be more important in certain applications. Clearly elastomers, which have a low strength and modulus, are very tough, as you can pound them

---

* From the data listed by R. J. Young in *Comprehensive Polymer Science*, Vol. 2, C. Booth and C. Price, Editors, Pergamon Press, Oxford, 1989.

with a hammer until the cows come home and they just bounce back after each blow. Tough materials are those which will absorb a large amount of energy during fracture. In polymers this occurs through various yielding processes and we now turn our attention to these.

Yielding in Polymers

As we mentioned in the introduction to this section, there are two mechanisms of yielding in polymers and you should keep in mind that one does not necessarily occur to the exclusion of the other. We will start by considering *shear yielding*. In figure 11.6, at the beginning of this chapter, we showed micrographs of polystyrene and polymethylmethacrylate which showed shear bands preferentially oriented at 45°, the direction of maximum shear stress. These bands are due to an orientation of the polymer chains along this direction and various molecular theories have been advanced to explain this phenomenon.

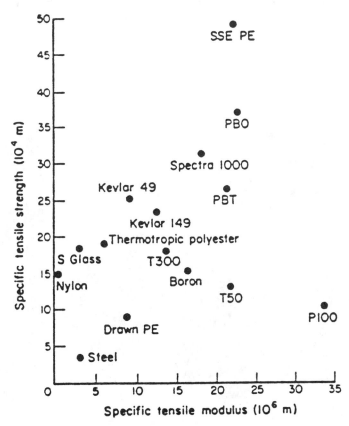

Figure 11.47 *Variation of specific tensile strength with specific tensile modulus for high performance fibers. T300, T50 and P100 are carbon fibers; spectra 1000 is a gel-spun polyethylene; and SSE PE is a solid state extruded polyethylene. Reproduced with permission from S. J. Krause, et al.,* Polymer, **29**, *1354 (1988).*

Some of these involve the creation of free volume under a tensile load (but these cannot explain yielding in compression), while others involve various types of molecular motion. As far as we can tell, none of these theories are entirely satisfactory.

In contrast to these glassy polymers, which do not yield significantly before fracture, flexible semi-crystalline polymers (i.e., those whose amorphous regions are rubbery) can show considerable amounts of yield or "cold-drawing" before failure. We showed a typical stress-strain diagram for such a polymer earlier in figure 11.12 and illustrated the characteristic appearance of "necking". For small deformations there is a Hookean elastic type behavior, but beyond a certain stress, known as the yield point, irreversible deformation occurs. This is related to a complex process of slip between lamellar planes of the crystal and unfolding of the chains to give a fiber type morphology, as illustrated in figure 11.48.

In many brittle thermoplastics at stresses just below fracture, a "whitening" of the sample occurs. This is due to the phenomenon of *crazing*, where small (microscopic) crack-like entities form in a direction perpendicular to the applied stress. These tiny cracks scatter light, giving an opaque or "white" appearance to

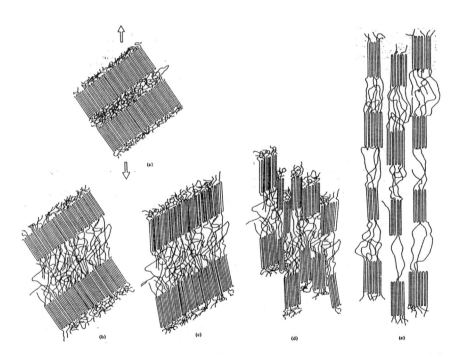

Figure 11.48 A model of the stretching mechanism for semi-crystalline polymers showing progressive (a through e) slip between lamellar planes. Reproduced with permission from J. Schultz, Polymer Material Science, *Prentice-Hall, New Jersey, 1974.*

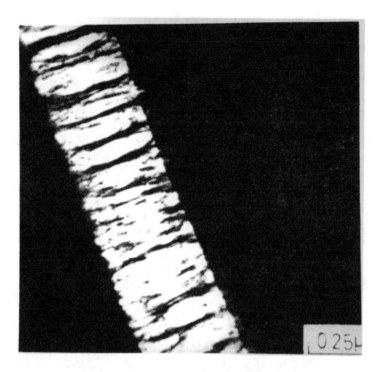

Figure 11.49 *Craze band in polystyrene. Reproduced with permission from P. Beahan et al.*, Proc. Roy. Soc. London, *A343, 525 (1975).*

the material, and if examined under the microscope tiny fibrils can be seen spanning the cracks and helping to "hold them together," as shown in figure 11.49. The formation of these tiny cracks and fibrils obviously absorbs some energy, but at some critical stress macroscopic cracks are formed which propagate through the material and it breaks.

This suggests two possible mechanisms for toughening polymers that fail in this fashion. The first method promotes the formation of crazes by the addition of a small amount of rubber, either by blending or copolymerization, to an otherwise rigid polymer. If the rubber particles are present as small phase separated domains they can act as stress concentrators, because there is a big mismatch between the modulus of the rubber and the rigid matrix, promoting the formation of crazes throughout the body of the sample rather than near a crack tip, where they are usually found in homogeneous rigid polymers. This is illustrated in figure 11.50. This type of rubber reinforcement is now being used to produce impact resistant plastics.

The second mechanism involves crack stopping in materials such as glass fiber reinforced rigid polymers. If the strength of adhesion between the polymer matrix and the fiber is less than that of either bulk material, then as a crack approaches a fiber the matrix can be pulled away, blunting the crack and absorbing energy. This accounts for the toughness of glass fiber boat hulls, which are often smacked into docks by inebriated weekend boaters.

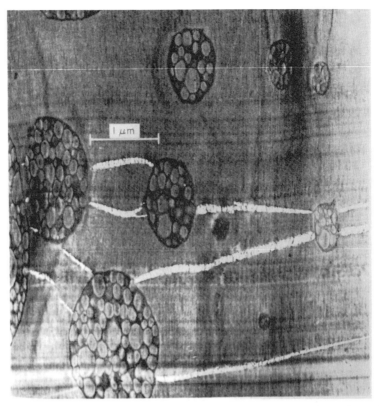

Figure 11.50 Crazing in rubber toughened polystyrene. Reproduced with permission from R. P. Kambour and D. R. Russell, Polymer, *12, 237 (1971).*

The processes of yielding we have just described must obviously depend upon specific relaxation processes within the polymer. Accordingly, they will depend upon the ratio of the strain rate to these relaxation rates. That is why polyethylene can be drawn slowly at room temperatures to give a fiber like morphology. The rate of loading is less than the relaxation times of the molecular rearrangements. If the sample is stretched quickly, however, these rearrangements or relaxations do not have time to occur and the sample will fail in a brittle fashion. Similarly, as we lower the temperature relaxation slows down and again the material becomes more brittle. What we would really like to make for many engineering applications is a polymer with a high strength and modulus, yet with good toughness or impact strength. This is a difficult task, but a first step is to understand the nature of the problem and the factors involved and that is hopefully what you have gleaned from this final section in this final chapter of our book.

H. STUDY QUESTIONS

1. (a) What is the difference between creep and stress relaxation and why are these subjects so important to a fundamental understanding of polymeric materials?

(b) Sketch a diagram of the 4-parameter model used to describe the creep of polymeric materials.

(c) Assume that the polymer is uncrosslinked and is subjected to a stress at time t_1, held for a period of time t_h, and the stress removed after time t_2. Sketch a creep curve predicted from the 4-parameter model and label all relevant parts of the curve.

2. (a) What is the principle of time-temperature superposition and how can it be used to predict the creep of a particular polymeric material after 25 years?

(b) Obtain the following equation (equation 11.92 in the text) for the shift factor:

$$\log a_T = \frac{(-B / 2.303\, f_0)\, (T - T_g)}{f_0 / \Delta\alpha + (T - T_g)}$$

from the Doolittle equation.

3. (a) Defining the shear stress and shear strain in a dynamic-mechanical experiment as:

$$\gamma^* = \gamma_0 \exp i\omega t$$

$$\tau^* = \tau_0 \exp i(\omega t + \delta)$$

Obtain expressions for the complex shear modulus ($G^* = G' + i\,G'$) and compliance ($J^* = ?$) and show how the real and imaginary parts of these equations (G', G'') correspond to expressions given in the text (equations 11.42 and 11.43 for the modulus; components of the compliance are not given).

(b) Although $G(t) \neq 1/J(t)$, show that $G^* = 1/J^*$ (at the same temperature and frequency).

(c) Consider the Maxwell model (equation 11.58) subjected to a sinusoidal tensile stress:

$$\sigma(t) = \sigma_0 e^{i\omega t}$$

Differentiate and substitute this equation into equation 11.58 and obtain the following expression for the complex tensile compliance D^*, defined as the difference in strain at times t_2 and t_1 divided by the difference in stress at these two times:

$$D^* = \frac{\varepsilon(t_2) - \varepsilon(t_1)}{\sigma(t_2) - \sigma(t_1)} = D - \frac{i\,D}{\tau\,\omega}$$

(d) Show that the complex tensile modulus at the same temperature and frequency is given by:

$$E^* = \frac{E\,\tau^2\,\omega^2}{1 + \omega^2\,\tau^2} + \frac{i\,\tau\,\omega\,E}{1 + \omega^2\,\tau^2} = E' + i\,E''$$

Also show that:

$$\tan \delta = \frac{1}{\omega \tau}$$

5. In a similar fashion obtain expressions for D', D", E' and E" for a Voigt element, again assuming the application of a sinusoidal stress (and a strain that is out-of-phase with the stress by an angle δ; see text).

6. A certain polymer ($T_g = 10°C$) has a melt viscosity of 1.5×10^5 poises at 25°C. Use the WLF equation to *estimate* its viscosity at 60°C. What assumptions did you make?

7. A strip of lightly vulcanized natural rubber at room temperature is heated in an oven and observed to *expand*. The same piece of rubber is stretched at room temperature to 100% of its original longitudinal dimensions by clamping one end and attaching a small weight to the other. It is then placed in the same oven where it is observed to *contract*. How do you explain these phenomena?

8. A rubber is subjected to a shear stress such that $\lambda_x = \lambda$, $\lambda_y = 1$ and $\lambda_z = 1/\lambda$. Obtain an expression relating the shear stress, τ, and the shear strain, γ, assuming $\gamma = \lambda - 1/\lambda$. Comment on the difference between this stress/strain relationship and the one for simple elongation.

9. You have just purchased a cheap car that has flexible vinyl seat coverings (i.e., plasticized PVC). After a few months you notice some yellow gunk has started to build up on the inside of the windshield. Because you are an intelligent student and don't smoke (having enthusiastically taken the advice of your elders and the Surgeon General—fat chance!) you guess that the plasticizer is evaporating out of the PVC and condensing on your windshield. Given that the T_g of pure PVC is about 80°C, what changes in the stress/strain behavior of your seat coverings would you anticipate over time? What do you think will happen to your cheap seats and why do you wish you could afford real Corinthian leather?

10. Discuss and explain how you would expect the viscosity of a polystyrene melt to vary with:
    (a) Molecular weight
    (b) Strain rate

11. A metal ball is dropped from a height of 2 meters onto a sample cut from a plastic sheet. The plastic sheet punctures in a ductile fashion leaving a hole roughly the size of the ball. The same metal ball is now dropped onto another sample of the same plastic sheet from a height of 4 meters. This time the plastic sample shatters in a brittle manner. How would you explain this phenomenon?

I. SUGGESTIONS FOR FURTHER READING

(1)   P. J. Flory, *Principles of Polymer Chemistry*,
       Cornell University Press, Ithaca, New York, 1953.

(2)   L. R. G. Treloar, *Physics of Rubber Elasticity,
       Third Edition*, Clarendon Press, Oxford, 1975.

(3)   J. J. Aklonis and W. J. MacKnight, *Introduction to Polymer
       Viscoelasticity, Second Edition*, Wiley Interscience, New York, 1983.

(4)   J. D. Ferry, *Viscoelastic Properties of Polymers, Third Edition*,
       John Wiley & Sons, New York, 1980.

(5)   R. J. Young, "Strength and Toughness", in *Comprehensive Polymer
       Science, Vol. 2, Polymer Properties*, C. Booth and C. Price, Editors,
       Pergamon Press, 1989.

*If anybody is interested, the movie rights to this book are still available.